Moment Formation In Solids

NATO ASI Series

Advanced Science Institutes Series

A series presenting the results of activities sponsored by the NATO Science Committee, which aims at the dissemination of advanced scientific and technological knowledge, with a view to strengthening links between scientific communities.

The series is published by an international board of publishers in conjunction with the NATO Scientific Affairs Division

A Life Sciences **B Physics**	Plenum Publishing Corporation New York and London
C Mathematical and Physical Sciences	D. Reidel Publishing Company Dordrecht, Boston, and Lancaster
D Behavioral and Social Sciences **E Engineering and Materials Sciences**	Martinus Nijhoff Publishers The Hague, Boston, and Lancaster
F Computer and Systems Sciences **G Ecological Sciences**	Springer-Verlag Berlin, Heidelberg, New York, and Tokyo

Recent Volumes in this Series

Volume 111—Monopole '83
edited by James L. Stone

Volume 112—Condensed Matter Research Using Neutrons: Today and Tomorrow
edited by Stephen W. Lovesey and Reinhard Scherm

Volume 113—The Electronic Structure of Complex Systems
edited by P. Phariseau and W. M. Temmerman

Volume 114—Energy Transfer Processes in Condensed Matter
edited by Baldassare Di Bartolo

Volume 115—Progress in Gauge Field Theory
edited by G. Lehman, G. t'Hooft, A. Jaffe, P. K. Mitter, I. Singer and R. Storra

Volume 116—Nonequilibrium Cooperative Phenomena in Physics and Related Fields
edited by Manuel G. Velarde

Volume 117—Moment Formation in Solids
edited by W. J. L. Buyers

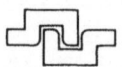

Series B: Physics

Moment Formation In Solids

Edited by

W. J. L. Buyers
Atomic Energy of Canada, Ltd.
Chalk River, Ontario, Canada

Springer Science+Business Media, LLC

Proceedings of a NATO Advanced Study Institute on Moment Formation in Solids, held August 21–September 2, 1983, on Vancouver Island, Canada

Library of Congress Cataloging in Publication Data

NATO Advanced Study Institute on Moment Formation in Solids (1983: Victoria, B.C. and Vancouver Island, B.C.)
Moment formation in solids.

(NATO ASI series. Series B, Physics; v. 117)
"Proceedings of a NATO Advanced Study Institute on Moment Formation in Solids, held August 21—September 2, 1983, on Vancouver Island, Canada"—Verso t.p.
Includes bibliographies and index.
1. Solids—Magnetic properties—Congresses. 2. Valence (Theoretical chemistry)—Congresses. 3. Kondo effect—Congresses. 4. Metals—Magnetic properties—Congresses. 5. Superconductivity—Congresses. I. Buyers, W. J. L. II. North Atlantic Treaty Organization. Scientific Affairs Division. III. Title. IV. Series.
QC176.8.M3N38 1983 530.4′1 84-17952
ISBN 978-1-4757-1540-8 ISBN 978-1-4757-1538-5 (eBook)
DOI 10.1007/978-1-4757-1538-5

Originally published by Plenum Press, New York in 1984

PREFACE

The problem of moment formation in metallic systems lies at the interface of localized and itinerant magnetism. The phenomena observed correspond to destruction rather than to formation of spin-correlations. They give rise to the progression from localized ground states through Kondo and mixed-valence behaviour to itinerant magnetic or non-magnetic systems. Somewhere in the progression superconductivity can occur in the presence of f-moments.

To bring together the disparate ideas and methods, previously the subject of intense debate only at a number of topical conferences, a two-week Advanced Study Institute was held August 21 - September 02, 1983, at Lester B. Pearson College of the Pacific, Vancouver Island. In the stimulating environment provided by the pines and fiord at the site of this remote United World College on the Canadian West Coast, scientists and students from around the globe gathered to hear lectures by experts. The Study Institute involved seventy-six people and followed in the tradition of previous "Banff" summer schools organized by the Canadian Association of Physicists. It was made possible by grants from the North Atlantic Treaty Organization, the Natural Sciences & Engineering Research Council of Canada and Atomic Energy of Canada. The grants permitted many young scientists from Nato and non-Nato countries to learn the fundamentals as well as the latest results in the field.

It is hoped that this book will provide a timely summary for research scientists and graduate students of the state-of-the-art for Kondo and mixed-valence systems. One of the highlights of the Institute was certainly the flowering of the Kondo resonance, which emerged from several different theoretical approaches. Another was the idea that not only does this provide the underpinning for Fermi-liquid theory, but the heavy-fermions thus produced can form a superconducting ground state.

The success of the Institute was in no small measure due to the help of the members of the International Advisory Committee, W. Kohn, L.M. Falicov, P.W. Anderson, J.W. Wilkins, P. Wachter, M.B. Maple, D. Wohlleben, C.M. Varma, G.H. Lander and T. Penney.

It is a pleasure to acknowledge E.C. Svensson who managed the financial affairs of the Study Institute as well as T.M. Holden and B. Bergersen for their role in planning and local arrangements. M. Carey provided experienced secretarial assistance before and after the Institute and was largely responsible for the smooth handling at Pearson College of the needs of participants.

W.J.L. Buyers
Chalk River, 1983

CONTENTS

EFFECTS OF ENVIRONMENT ON THE ELECTRONIC CONFIGURATION IN INTERMEDIATE VALENCE RARE EARTH SYSTEMS
T. Penney

LOCALIZED MOMENTS IN METALS: CRYSTAL FIELD EXCITATIONS AND THEIR COUPLING TO ELECTRONS AND PHONONS
P. Fulde

FUNDAMENTALS OF THE INTERMEDIATE VALENCE PROBLEM
C.M. Varma

PHENOMENA INVOLVING MAGNETIC MOMENT SUPPRESSION

M. Brian Maple*

Department of Physics and
Institute for Pure and Applied Physical Sciences
University of California, San Diego
La Jolla, CA 92093 USA

INTRODUCTION

The following is an abbreviated account of a series of three lectures that were given by the author at the NATO/CAP Advanced Study Institute on Moment Formation in Solids. Various phenomena that are associated with magnetic moment suppression in rare earth and actinide systems were reviewed in these lectures, the first two of which were primarily (but not exclusively) devoted to concentrated systems such as α-Ce, "gold" SmS, and UIr_3B_2, and the last of which focused on dilute systems such as $(\underline{La}Ce)Al_2$, $(\underline{La}Sm)Sn_3$ and $\underline{Th}U$.

Magnetic moments in rare earth (RE) and actinide (A) systems are derived from partially-filled 4f and 5f electron shells of the RE and A ions, and their degree of suppression (or development, depending on one's viewpoint) spans a continuum between well-defined (Curie-like magnetic susceptibility) and nonexistent (Pauli-like magnetic susceptibility). Any degree of magnetic moment suppression may be found in a given rare earth or actinide substance under ambient conditions, or it may often be induced by the application of an external pressure or by alloying with another element.

---

*Research supported by the U.S. Department of Energy under Contract No. AT03-76-ER-70227.

The mechanism for moment suppression involves the interaction of the localized f electrons and the itinerant conduction electrons. Within the context of the Friedel-Anderson model,[1,2] moment suppression occurs via increased hybridization of f electron and conduction electron states, or by proximity of the f state to the Fermi level, while in terms of the s-f exchange or "Kondo" model,[3] it is associated with the gradual formation of a many body singlet in which the f magnetic moments are screened by conduction electron spins. These two models have been shown to bear a relationship to one another. In certain regimes the Anderson Hamiltonian can be "transformed" into an exchange Hamiltonian for both nondegenerate (Schrieffer-Wolf transformation[4]) and degenerate (Coqblin-Schrieffer transformation[5]) orbital cases. Moreover, the renormalization group solution of the nondegenerate orbital Anderson model has been shown to "map" onto the renormalization group solution of the spin 1/2 "Kondo" model.[6] Recent theoretical developments on this problem are described throughout this volume.

Emphasized in the first two lectures on concentrated rare earth and actinide systems were substances that contain an ordered sublattice of RE ions in which the valence (and, in turn, the occupation number of the 4f electron shell) is intermediate between two integral values. The terminology applied to such systems is intermediate valence (IV), although homogeneous mixed valence, valence fluctuations and interconfiguration fluctuations (ICF) are frequently used. Topics that were discussed included the apparent connection between nonmagnetic behavior and intermediate valence, the remarkable physical properties of IV systems, the nature of the IV ground state, and the "narrow gap" semiconducting behavior of certain IV compounds such as "gold" SmS, SmB_6 and TmSe.

Both the normal and superconducting state properties of dilute rare earth and actinide systems were considered in the last lecture. It was pointed out that the anomalous Kondo-like behavior of many dilute rare earth metallic systems is strikingly similar to that of concentrated rare earth systems, a fact which has not, in the author's view, been adequately appreciated. Moreover, many of the recent "Kondo" or IV theoretical models treat an assemblage of noninteracting RE "impurities." Superconductivity provides an additional and very sensitive test of theoretical descriptions of RE and A impurity ions in metals, and is especially relevant in view of the recent observation of "heavy Fermion" superconductivity in rare earth and actinide compounds such as $CeCu_2Si_2$, UBe_{13} and U_6Fe.

The approach taken in these lectures was somewhat historical and the material covered, which by necessity could not be comprehensive, reflected to a certain extent the author's own interests and involvement in the evolution of this subject. One of the main purposes of the lectures was to identify important issues and provide a general background for lectures of a more technical nature that followed during the summer school.

CONCENTRATED RARE EARTH AND ACTINIDE SYSTEMS

One Electron 4f Shell-Cerium

Cerium is one of the most interesting elements in the periodic table, both in terms of its behavior in elemental form and as a constituent of alloys and compounds. The author first encountered the demagnetization of Ce ions through experiments on the effect of pressure P on the superconducting transition temperature T_c of the superconductors La_3In[7] and La[8] containing Ce impurities. A brief description of the pressure experiments on the LaCe system and their interpretation follows.

Shown in Fig. 1 are T_c vs P data on the LaCe system for Ce impurity concentrations of 0, 0.7, 1.3, 2 and 16 at.% that were obtained by the author, J. Wittig and K. S. Kim[8] in 1969. The depression of T_c of the LaCe alloys, relative to La, first increases with pressure, passes through a maximum near 15 kbar, and then decreases with pressure up to ~ 140 kbar, the high pressure limit of the experiment. Particularly striking are the data for the 2 at.% Ce sample; here, the destructive effect of the Ce impurities on superconductivity becomes so strong that regions of superconductivity are separated by a "normal gap" on the pressure axis between ~6 and ~15 kbar. In the inset of Fig. 1 are isobars of T_c vs Ce concentration, the shape of which changes from linear at pressures below ~15 kbar to exponential-like at pressures above ~15 kbar.

The remarkable behavior of T_c as a function of pressure in the LaCe system was first interpreted in terms of a pressure-induced demagnetization of the Ce impurities within the context of the Anderson model.[8] At ambient pressure, the LaCe system exhibits a Kondo effect with a low Kondo temperature $T_K \sim 0.1$ K, and the crystalline electric field splits the Ce J = 5/2 Hunds' rule ground state into an excited state quartet and a ground state doublet

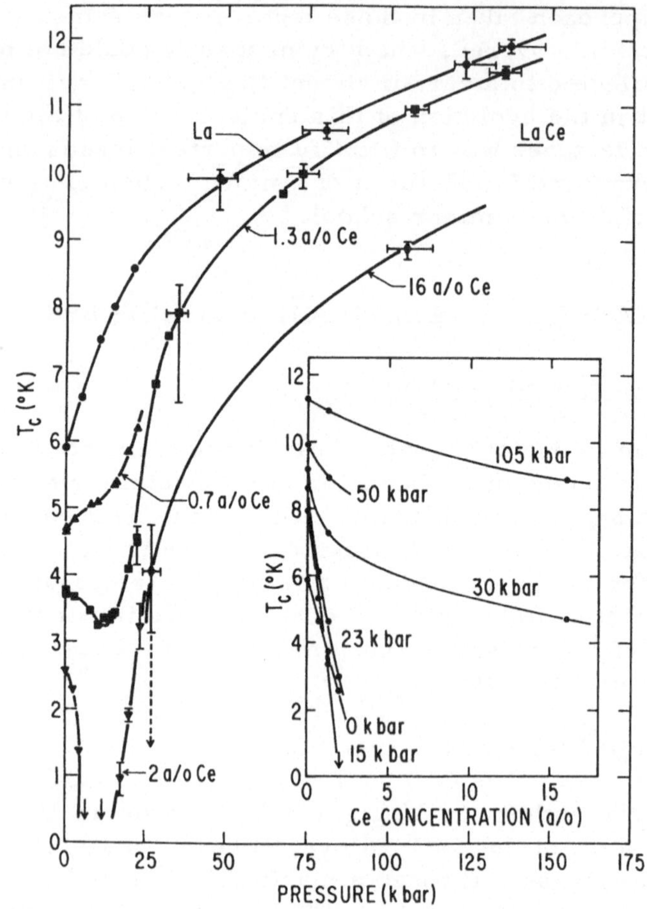

Fig. 1. Pressure dependence of the superconducting transition temperature T_c of as-cast predominantly fcc La and LaCe alloys to very high pressure. The vertical bars represent the transition widths and the horizontal bars the pressure inhomogeneity in the high pressure cell. Isobars of T_c vs Ce concentration are shown in the inset (from Ref. 8).

that are separated by ~ 100 K.[9] Applied pressure was assumed to have the effect of decreasing the energy E_f between the spin-up Ce 4f resonant state and the Fermi level E_F, thereby increasing the magnitude of the 4f electron-conduction electron spin exchange interaction; i.e., $\mathcal{J} \sim -\langle V_{kf}^2 \rangle / E_f$ for $\Delta \ll E_f$. Here, $\langle V_{kf}^2 \rangle$ is the square of the matrix element mixing the localized Ce 4f and conduction electron states, $\Delta = \pi \langle V_{kf}^2 \rangle N(E_F)$ is the Ce 4f electron resonant level width, and $N(E_F)$ is the density of conduction electron states at E_F.

The increase of $|\mathcal{J}|$ with pressure necessarily results in an increase of the rate of depression of T_c with impurity concentration n, $(-dT_c/dn)_{n=0}$, with pressure. This follows from both the Abrikosov-Gor'kov (AG) theory,[10] which is valid to order $\mathcal{J}^2$, and extensions of the AG theory by Müller-Hartmann and Zittartz[11] and others[12, 13] which incorporate the Kondo effect. In models of the latter type, T_K increases exponentially with $|\mathcal{J}|$ [i.e., $T_K \sim T_F \exp(-1/N(E_F)|\mathcal{J}|)$, where T_F is the Fermi temperature], while $(-dT_c/dn)_{n=0}$ increases with increasing T_K in the range $T_K \ll T_{c0}$ to $T_K \sim T_{c0}$. At higher pressures, the Ce ions are expected to demagnetize as the Ce 4f level comes into close proximity with the Fermi level, for which the depression of T_c should pass through a maximum and thereafter decrease with pressure.

The pressure-induced demagnetization of the Ce ions within the context of the Anderson model is schematically represented in Fig. 2. Diagrams A and B correspond to the magnetic and non-magnetic states of Ce, respectively. Demagnetization from A to B by means of increasing E with constant $\langle V_{kf}^2 \rangle$ (or Δ) is indicated by the arrow in the $(E_F - E)/U$ vs $\pi\Delta/U$ phase diagram that separates the magnetic and nonmagnetic Hartree-Fock solutions of the Anderson model (note that E is the original position of the 4f level with respect to E_F before the levels have been broadened by Δ, after which the final position is E_f). The Anderson model

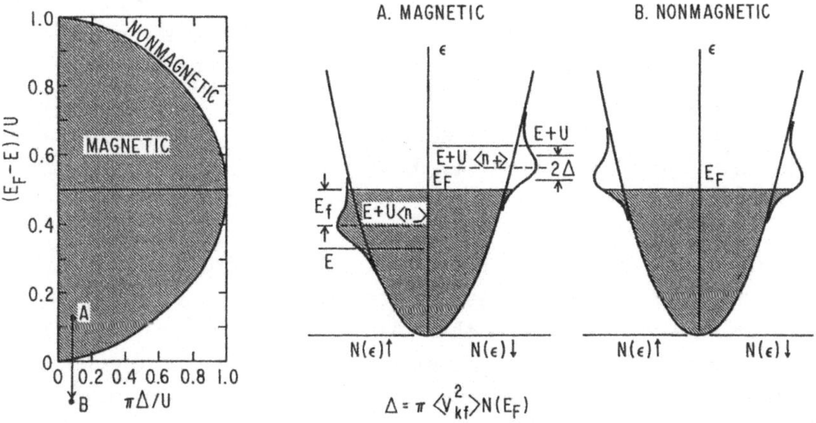

Fig. 2. Schematic diagram of the magnetic-nonmagnetic transition of Cc impurities in the Friedel-Anderson model. The transition is assumed to be driven by a decrease of the energy separating the Ce 4f level and the Fermi level with pressure (from Ref. 15).

parameters were estimated to be $E_f \sim 0.1$ eV, $\Delta \sim 0.01$ eV and $U \sim 5$-10 eV.[14]

An alternative explanation for the decrease of the depression of T_c above ~15 kbar is that the Ce magnetic moments remain well defined and that T_K continues to increase with pressure from $T_K \sim T_{c0}$ to $T_K \gg T_{c0}$. The MHZ theory predicts that $(-dT_c/dn)_{n=0}$ will exhibit a maximum near $T_K \sim T_{c0}$ and then decrease monotonically with further increase of T_K. However, this leads to values of T_K that are of the order of 10^6 K or more at pressures above ~100 kbar, which clearly seems unphysical.[15]

The pressure-induced demagnetization of Ce ions inferred from the T_c vs P measurements on the LaCe system appears to be the Ce impurity analogue of the well-known γ-α electronic phase transition that occurs under pressure in elemental Ce. The γ-α phase transition was first reported by Lawson and Tang[16] who performed x-ray diffraction measurements on Ce metal as a function of pressure at room temperature. They found that the volume collapsed by ~17% at ~7 kbar without change of the fcc crystal structure, and suggested that the transition involved the promotion of the Ce 4f electron into a 5d state. In the T-P phase diagram of elemental Ce,[17] the fcc γ and α phases are separated by a line of discontinuous transitions that terminates in a critical point. Other regions of the T-P phase diagram correspond to a bcc δ phase, a dhcp β phase, and a high pressure α' phase which apparently has the α-U structure,[18] although there seems to be some controversy about the crystal structure of α'-Ce.[17] Magnetic susceptibility measurements as a function of temperature at 10 kbar on α-Ce reveal a weakly temperature dependent Pauli-like susceptibility.[19] Superconductivity has been reported for α'-Ce[20] with $T_c \sim 1.9$ K right above the α-α' phase boundary and for α-Ce[21] with $T_c \sim 50$ mK just below the α-α' phase boundary.

Gschneidner and Smoluchowski[22] have estimated the valences of γ- and α-Ce from lattice parameter data and various physical properties and have arrived at the values 3.06 for γ-Ce at room temperature and atmospheric pressure, and 3.67 for α-Ce at 116 K and atmospheric pressure. However, both the promotional model and the valence of Ce have been called into question recently. A variety of measurements and theoretical considerations have indicated that α-Ce contains substantial f electron character. Thus, a Mott transition from a localized to a delocalized 4f state may be a more appropriate description than simple promotion from a 4f to a

5d state.[23] The issue of Ce valence was recently raised in L_{III} absorption measurements on a large class of Ce compounds by Bauchspiess et al.[24] who concluded that the Ce valences were considerably smaller than estimated from lattice parameter measurements. In fact, on the basis of L_{III} absorption measurements, the valence of certain Ce compounds that were formerly believed to be tetravalent were found to have a valence near 3.3. Even CeO_2 was found to have a valence of 3.3!

These findings led J. E. Allen and the author and our collaborators[25,26] to investigate the compounds $CeCo_2$, $CeRu_2$ and $CeIr_2$ by means of photoemission spectroscopy (PES) and Bremsstrahlung isochromat spectroscopy (BIS). The results of recent resonant PES and BIS experiments by Allen et al.[26] on CeAl, $CeIr_2$ and $CeRu_2$ are shown in Fig. 3. Here it can be seen that in γ-like materials such as CeAl, there are two prominent peaks in the combined resonant PES and BIS spectrum, one about 2 eV below E_F and another 4 eV above E_F, representing the spectral weights for $4f^1$ and $4f^2$. However, for the $CeIr_2$ and $CeRu_2$ spectra, a peak of unresolved sharpness is observed just above E_F and seems to grow larger, the smaller the $4f^1$ and $4f^2$ spectral weights. It is tempting to attribute the sharp feature near E_F to the so-called Kondo resonance which should appear near the Fermi level according to recent theories based on the Anderson model and discussed throughout this volume. The PES and BIS measurements yield $E_f \sim 2$ eV, $\Delta \sim 0.5$ eV and $U \sim 6$ eV for the Anderson model parameters. These values of E_f and Δ are much larger than those estimated from thermodynamic properties, as discussed earlier. This discrepancy is not very well understood at the present.

Many Electron 4f Shell-Samarium

Two Sm compounds, SmS and SmB_6, appear to be the many 4f electron analogues of Ce metal and certain Ce compounds. In particular, the behavior of SmS under pressure is remarkably similar to that of Ce metal under pressure in several respects.

At zero pressure, the compound SmS is a semiconductor with the cubic NaCl crystal structure.[27] The Sm ions are divalent, and electrical conduction occurs via thermal activation of localized electrons from the Sm 4f electron shells into the conduction band with a small activation energy ~ 0.2 eV. The Sm 4f electron shell contains six electrons and the compound exhibits ionic Van Vleck paramagnetism with a nonmagnetic $J = 0$ ground state.[27] Shown in

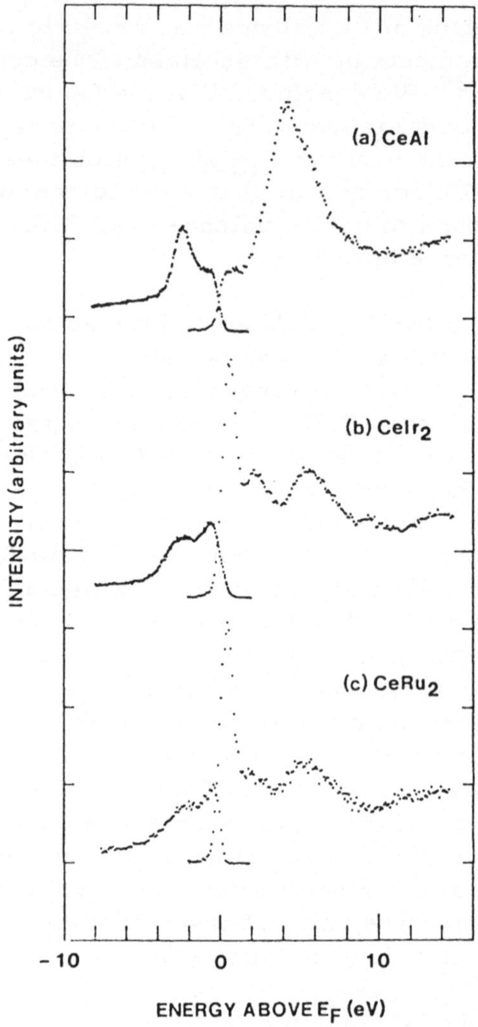

Fig. 3. Combined resonant PES and BIS spectra at room temperature for (a) CeAl, (b) $CeIr_2$ and (c) $CeRu_2$ (from Ref. 26).

Fig. 4(b) are magnetic susceptibility χ vs temperature data[28] at zero pressure for SmS[27] and several other Sm and Eu compounds in which the RE ion has the configuration $4f^6$.

Under pressure, SmS undergoes a discontinuous transition from a black semiconducting phase to a gold and, apparently, metallic phase as first reported by Jayaraman et al.[29] The transition is accompanied by a large decrease in volume ($\Delta V/V \sim -8\%$) without a

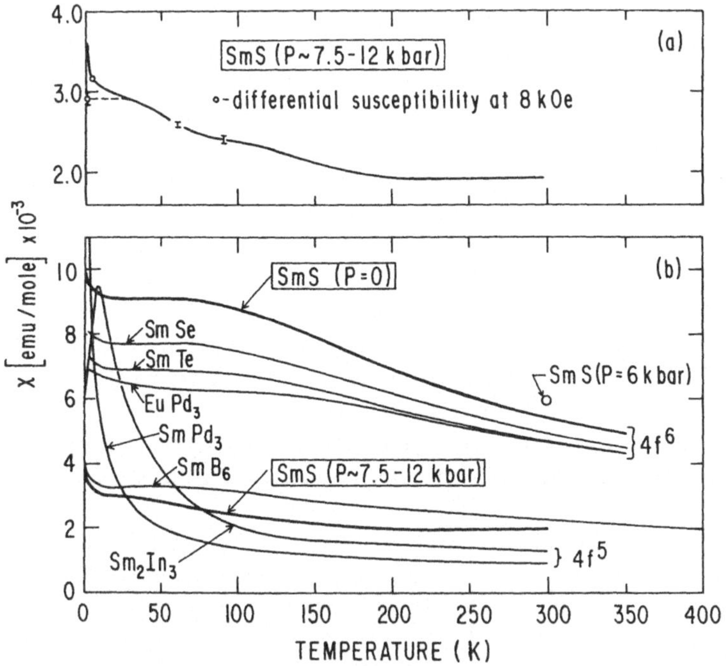

Fig. 4. (a) Temperature dependence of the magnetic susceptibility χ of SmS in the collapsed high pressure phase. The pressure was 12 kbar at 300 K and 7.5 kbar at liquid helium temperature. (b) Comparison of $\chi(T)$ for ions in the $4f^6$ and $4f^5$ configuration with "collapsed" SmS and SmB_6. (From Ref. 28).

change in crystal structure. In the collapsed phase, the lattice parameter[29] of SmS indicates that the Sm valence is ~ 2.7 or, equivalently, the average occupation number of the Sm 4f electron shell is ~ 5.3.[28] Thus, two 4f electron shell configurations, $4f^6$ and $4f^5$, are involved in the ground state of SmS in its collapsed gold phase. Additional evidence for a nonintegral Sm valence in the collapsed phase has been provided by Mössbauer isomer shift measurements by Coey et al.[30] on SmS in the collapsed high pressure phase and the "chemically collapsed" pseudobinary compound $Sm_{0.77}Y_{0.23}S$, and X-ray photoemission spectroscopy (XPS) studies by Campagna et al.,[31] Freeouf et al.[32] and Pollak et al.[33] on various chemically collapsed pseudobinary compounds found by alloying SmS with a third element.

The magnetic susceptibility of SmS in the gold phase as a function of temperature was measured by the author and D. K. Wohlleben in 1971.[28] The results are shown in Fig. 4(a) where it can be seen that χ exhibits weak, but definite temperature dependence below ~ 200 K, and then saturates to a constant value below ~ 40 K with no sign of magnetic order. The constant value of χ at low temperature contrasts sharply with what would be expected if the transition proceeded directly to the trivalent $4f^5$ configuration. In this case the ground state would have to be at least a doublet, which results in a low temperature divergence or magnetic order, as shown in Fig. 4(b) for two normal $4f^5$ compounds ($SmPd_3$ and Sm_2In_3). Note that at high temperature, the susceptibility of SmS is intermediate between normal $4f^6$ and $4f^5$ behavior. Also shown in Fig. 4(b) is the magnetic susceptibility of SmB_6 which has a weak temperature dependence,[34] similar to that of gold SmS.

It was noted by Maple and Wohlleben[28] that the nonmagnetic behavior of α-Ce, gold SmS, and SmB_6 correlated with their having intermediate valences which, from lattice parameter measurements, were estimated to be 3.7, 2.7 and 2.7, respectively. We interpreted the correlation between the "nonmagnetic" susceptibility at low temperatures and intermediate valence in terms of an ionic model proposed by Hirst,[35] and suggested that the 4f shell undergoes temporal valence or interconfiguration fluctuations (ICF) between configurations $4f^n$ and $4f^{n-1}$, accompanied by the emission and absorption of a conduction electron. The lifetime $\tau_o \sim h/k_B T_o$ associated with the fluctuations between the states $4f^n$ and $4f^{n-1}$ can be inferred from the temperature T_o below which the magnetic susceptibility saturates to a constant value as $T \rightarrow 0$ and is of the order of 10^{-13}-10^{-12} s for most RE compounds. This notion is consistent with "slow" Mössbauer measurements as well as "fast" XPS measurements. Mössbauer measurements, which have a characteristic measuring time $\sim 10^{-9}$-10^{-7} s, cannot resolve the isomer shift for the two configurations $4f^n$ and $4f^{n-1}$, but rather yield an isomer shift that is intermediate between that expected for the two configurations. On the other hand, XPS measurements, which have a measuring time $\sim 10^{-17}$ s, are able to resolve f-hole excitation spectra which correspond to the simultaneous presence of both configurations $4f^n$ and $4f^{n-1}$.

These interconfiguration fluctuations are believed to occur when the energies E_n and E_{n-1} of the configurations $4f^n$ and $4f^{n-1}5d^1$ are approximately degenerate in energy within the 4f resonant level width $\Delta = \pi\langle V_{kf}^2\rangle N(E_F)$. Thus at temperatures well below T_o, the physical properties of lanthanide ICF systems should resemble those of a metal with a narrow band of width Δ that overlaps the Fermi level. In such cases one expects to observe large linear specific heat coefficients γ and the characteristic saturation of the magnetic susceptibility in the limit $T \to 0$. Observed values of γ are often as high as several hundred mJ/mole K^2 (and can be as high as 1.6 J/mole K^2 in the case of $CeAl_3$![36]).

In contrast, in systems where the excitation energy between the two configurations, $E_{exc} = E_n - E_{n-1}$, is much larger than Δ, interconfiguration fluctuations are effectively "blocked" by energy considerations, and the simple ionic limit ($T_o \to \infty$) which typifies the behavior of most lanthanide ions in a metallic environment is recovered. Here, the lanthanide ions carry well-defined magnetic moments which are consistent with Hund's rules, giving rise to Curie-Weiss behavior of the magnetic susceptibility as a function of temperature with crystal-field and Van Vleck modifications, magnetic order, and so on.

It is interesting to note that the criterion for the formation of a local moment in the ICF model is roughly given by the relation

$$\Delta / E_{exc} \leqslant 1 \tag{1}$$

which can be compared to the Hartree-Fock criterion which is given by

$$\Delta / U \leqslant 1 \; . \tag{2}$$

Thus, in the ICF model, one can retain the strong correlations between electrons within the 4f shell, which have a characteristic energy $U \sim$ 5-10 eV, and still describe nonmagnetic situations even though Δ is only of the order of 10^{-2} eV, as long as $E_{exc} < \Delta$.

A phenomenological expression for the magnetic susceptibility which qualitatively described the behavior of many compounds for $T > T_o$ was suggested by Maple and Wohlleben[37] in 1973

$$\chi(T) = \frac{\epsilon(n)[\mu_{eff}(n)]^2 + [\epsilon(n) - 1][\mu_{eff}(n-1)]^2}{3k_B(T + T_o)} \tag{3}$$

where $\epsilon(n)$ is the fraction of time the 4f shell has the configuration

$4f^n$ and T_0 represents the randomizing influence of the interconfiguration fluctuations. These ideas have been extended by Sales and Wohlleben[38] and Sales[39] and rather remarkable quantitative agreement between the calculated and measured $\chi(T)$ curves of several RE compounds has been obtained.

Specific heat measurements on SmS in its gold phase reveal an electronic contribution with a very large electronic specific heat coefficient $\gamma \sim 145$ mJ/mole K^2, a lattice contribution, and a broad feature at higher temperature.[40] The broad feature suggests a decrease in entropy with decreasing temperature as the system approaches its highly correlated ground state which is comprised of a quantum mechanical admixture of the configurations $4f^6$ and $4f^5 5d^1$. However, the ground state of gold SmS appears to be insulating as evidenced by the electrical resistivity which increases with decreasing temperature in a manner that is reminiscent of a semiconductor with a very small energy gap.[40] Recently, several groups have measured the electrical resistivity of gold SmS under pressure and found that the temperature dependence of the resistivity changes from semiconductor-to metallic-like in the neighborhood of 20 kbar.[41-43]

At atmospheric pressure, the physical properties of the compound SmB_6 closely resemble those of gold SmS.[34] As mentioned previously, SmB_6 has a Sm valence of ~ 2.7 and a weakly temperature dependent magnetic susceptibility that approaches a constant value as $T \to 0$.[34] Like gold SmS, the electrical resistivity increases with decreasing temperature in a thermally activated manner and then, below 3 K, saturates to a value that can be as large as 10^4 times the room temperature value.[34, 44, 45] NMR,[46] electron tunneling,[47-51] far infrared absorption,[47, 48] and low temperature specific heat measurements[47] are all consistent with the existence of a small energy gap of several meV. Recently, the electrical resistance of SmB_6 was measured under pressure[52] and some of the data are shown in Fig. 5. Similar to gold SmS under pressure, the electrical resistance changes from semiconductor-to metallic-like in the vicinity of 60 kbar.

The similarity of the physical properties of gold SmS and SmB_6 suggests that a common mechanism may be responsible for the insulating ground state, the small energy gap and its closing with pressure. A number of theories have been advanced to account for the small energy gap in SmB_6 such as the d-f hybridization gap model proposed by Mott[53] and the disordered Wigner lattice model of Kasuya et al.[54]

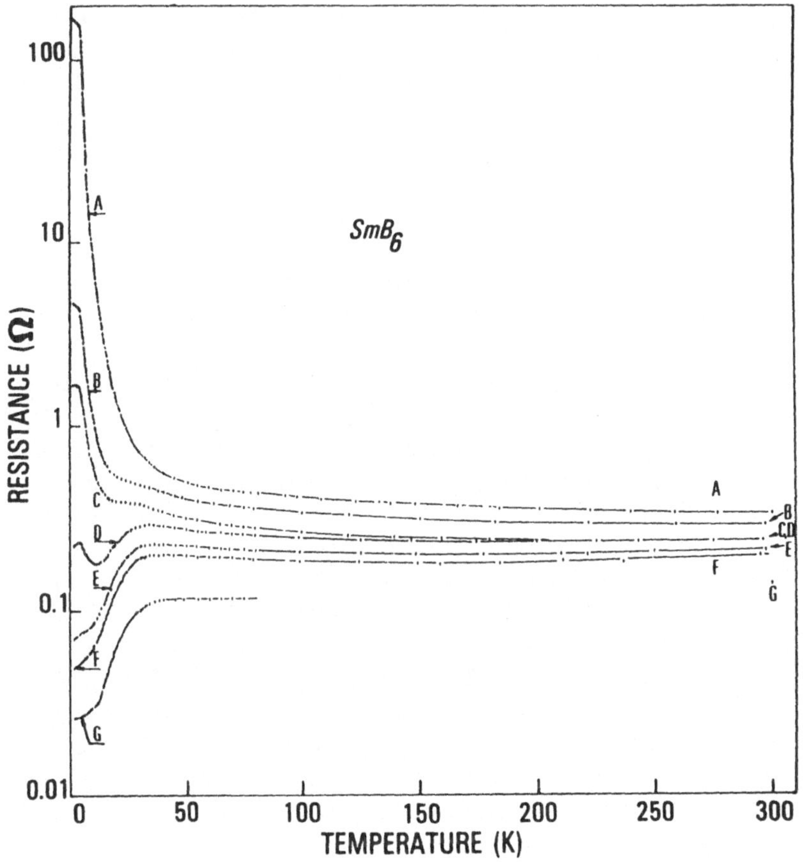

Fig. 5. Electrical resistance vs temperature for SmB_6 at various pressures (A - 18 kbar, B - 33 kbar, C - 47 kbar, D - 62 kbar, E - 75 kbar, F - 89 kbar, and G - 145 kbar). (From Ref. 52).

In contrast to the semiconducting behavior of the electrical resistivity of gold SmS and SmB_6, the electrical resistivity of most intermediate valence rare earth compounds exhibits metallic, although often complex behavior. There are, however, some common features. First, for $T \gg T_0$, $\Delta\rho(T) \equiv \rho(T) - \rho(0)$ is generally of the order of several hundred $\mu\Omega$-cm which implies strong charge and/or spin disorder scattering. Second, when $T \sim T_0$, pronounced features in ρ are often observed such as maxima or minima that are indicative of Kondo-like behavior and/or crystalline electric field effects. Finally, for $T \ll T_0$, there is a rapid decrease in ρ with decreasing T as the system approaches its highly correlated ground state. For some systems, the electrical resistivity takes the form $\rho = AT^2$ for $T \ll T_o$, suggesting a description in terms of Fermi liquid theory. In fact, a number of Fermi liquid models have been applied to describe the physical properties of intermediate valence systems.[23,56,57]

Intermediate Valence Ternary Compounds

In recent years, the interaction between superconductivity and magnetic order has been extensively investigated in ternary rare earth compounds.[58] However, ternary rare earth and actinide compounds are also a vast reservoir of potential new intermediate valence materials. During the past year, we have been investigating a series of compounds with the formula MT_3B_2 where M is Ce or U and T is Co, Ru, Rh or Ir.[59] The first indication that the CeT_3B_2 compounds might be candidates for IV behavior was provided by their reduced unit cell volumes, relative to the values expected for trivalent Ce. This is illustrated in Fig. 6 where plots of unit cell volume vs RE for the RET_3B_2 compounds with T = Ru, Rh and Ir display characteristic depressions at RE = Ce. Deviations of plots of lattice parameter or unit cell volume vs RE from integral valent behavior is one of the simplest and most common methods for identifying IV rare earth compounds. We investigated the U compounds because of the similarity between the physical properties of many actinide and IV rare earth compounds.

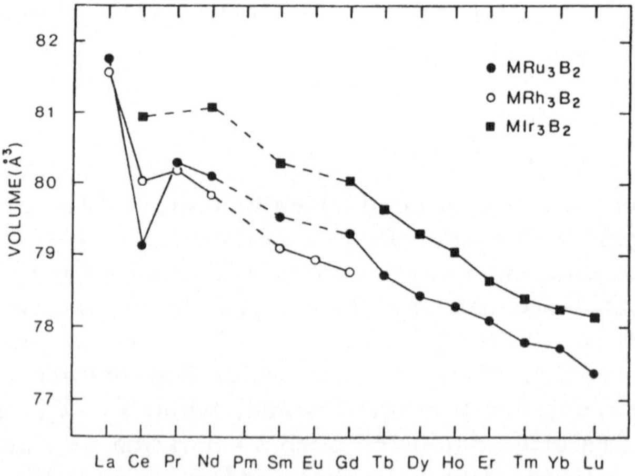

Fig. 6. Unit cell volume of RET_3B_2 compounds with T = Ru, Rh and Ir vs RE (from Refs. 59 and 60).

Electrical resistivity ρ vs temperature data are shown in Fig. 7 for CeT_3B_2 and UT_3B_2 compounds with T = Ir, Ru and Co. The ρ(T) curve of the $CeIr_3B_2$ compound exhibits characteristic IV behavior with a maximum at $T_{max} \sim 180$ K. Noteworthy is the enormous magnitude of ρ of $CeIr_3B_2$ compared to the other CeT_3B_2 and UT_3B_2 compounds. The ρ(T) curves of $CeRu_3B_2$ and $CeCo_3B_2$ are more α-Ce like, and $CeRu_3B_2$ actually becomes superconducting at 0.68 K. The ρ vs T data of UIr_3B_2 and UCo_3B_2 show strong negative curvature and do indeed resemble those of IV rare earth compounds.

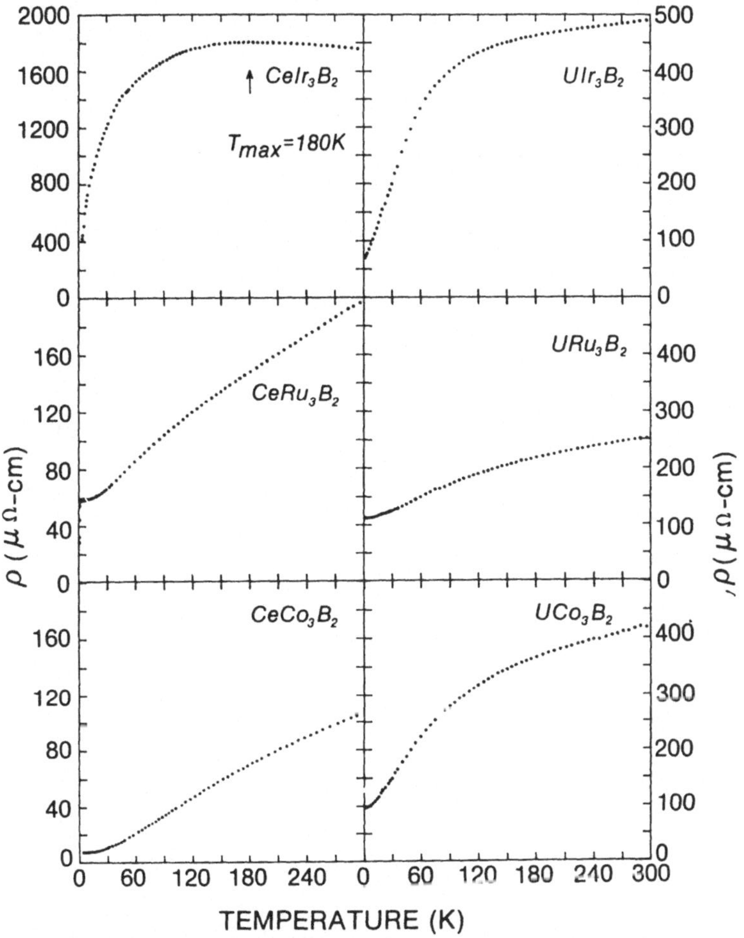

Fig. 7. Electrical resistivity ρ vs temperature for CeT_3B_2 and UT_3B_2 compounds with T = Ir, Ru and Co (from Ref. 60).

Electrical resistivity ρ vs temperature data for the compound $CeRh_3B_2$ are shown in Fig. 8. The rapid decrease in ρ is associated with the decrease in spin disorder scattering when the compound orders ferromagnetically. Our magnetization measurements, which yield a Curie temperature of 113 K and a saturation moment of $0.37\ \mu_B$, are in satisfactory agreement with those of Dhar et al.[61] who previously reported that $CeRh_3B_2$ is an itinerant ferromagnet. This is a rather unusual situation since Ce has an intermediate valence in this compound.

Low temperature specific heat measurements have also been made on CeT_3B_2 and UT_3B_2 compounds with T = Co, Ru, Rh and Ir. The highest values of the electronic specific heat coefficients for the CeT_3B_2 and UT_3B_2 compounds are $\gamma = 44$ mJ/mole K^2 for $CeIr_3B_2$ and $\gamma = 64$ mJ/mole K^2 for UIr_3B_2. Magnetic susceptibility measurements indicate that the $CeIr_3B_2$ compound is nonmagnetic.

DILUTE RARE EARTH AND ACTINIDE SYSTEMS

The properties of systems that consist of small concentrations of RE or A impurity ions dissolved in metallic matrices are of interest for several reasons. First, in the limit of very small

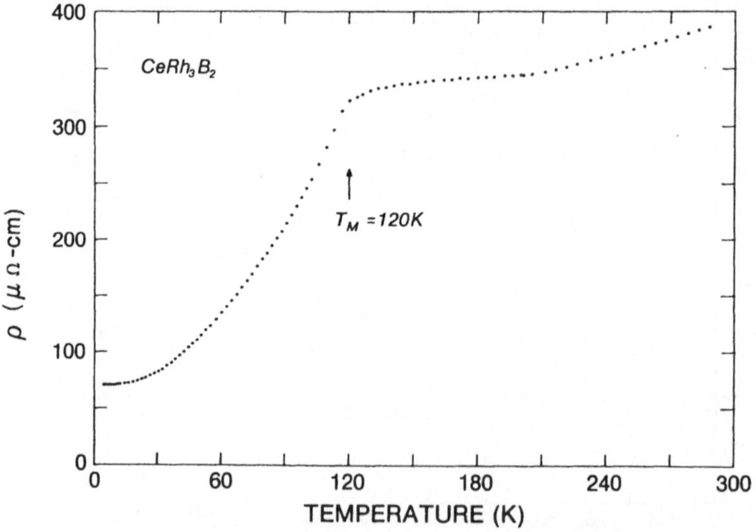

Fig. 8. Electrical resistivity ρ vs temperature for $CeRh_3B_2$ (from Ref. 60).

concentrations of impurity ions, the influence of interimpurity interactions can be minimized. This makes it possible to examine models that are based on an array of noninteracting RE or A ions. Second, if the metallic matrix is superconducting, the effect of RE and A impurities on superconductivity can be assessed. By correlating the magnetic state of the impurity with its effect on superconductivity, it is possible to establish a type of superconducting spectroscopy for determining the magnetic state of an impurity in a (superconducting) metal. The delicate nature of superconductivity makes it a sensitive means for testing theoretical models. A brief description of this superconducting spectroscopy follows.

Superconducting Spectroscopy

There are three regimes of superconducting behavior that have been established by experiments on various superconducting matrix-impurity systems. These regimes can be distinguished by the detailed dependence of (1) the reduced superconducting transition temperature T_c/T_{c0} on impurity concentration n (T_c refers to the alloy, T_{c0} to the matrix), and (2) the reduced specific heat jump $\Delta C/\Delta C_0$ on T_c/T_{c0} (ΔC refers to the alloy, ΔC_0 to the matrix), both of which reflect the magnetic state of the RE or A impurity ion.[9] The three types of behavior are depicted in Fig. 9 and exemplary systems are indicated. The reader is referred to references 9 and 15 and references cited therein for a more detailed discussion of the exemplary systems.

The first two categories (IA and IB) refer to long-lived local moments; i.e., $T_0 \ll T_{c0}$ where T_0 is the characteristic magnetic moment fluctuation temperature. In this regime, the f electrons and conduction electrons interact via the exchange interaction $\mathcal{H} = -2\mathcal{J}\underset{\sim}{S}\cdot\underset{\sim}{s}$ where $\underset{\sim}{S}$ is the spin of the RE or A impurity ion and $\underset{\sim}{s}$ is the conduction electron spin. For simplicity, we confine ourselves to S-state ions like Gd^{3+} or Eu^{2+} where $S = 7/2$ and $L = 0$, or Ce systems where the Ce ground state is a Γ_7 doublet, separated from the Γ_8 excited state by $\sim 10^2$ K, with an effective spin $S_{eff} = 1/2$.

Case IA corresponds to $T_0 \ll T_{c0}$, and $\mathcal{J} > 0$. In this limit, the hybridization between the localized f and conduction electron states is weak. The superconducting properties can be described well by the temperature independent pair breaking theory of Abrikosov and Gor'kov.[10] The exemplary systems for this limit are (<u>La</u>Gd)Al_2 and <u>Th</u>Gd.

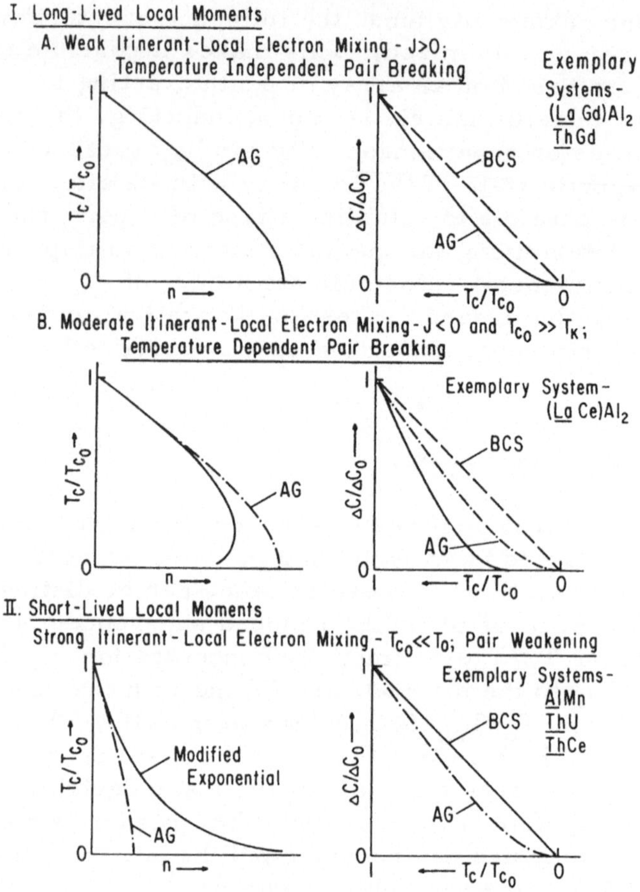

Fig. 9. The systematics of superconductivity in the presence of local moments. The solid lines represent the behavior of the curves of T_c/T_{c0} vs n and $\Delta C/\Delta C_o$ vs T_c/T_{c0} for each of the three regimes of solute magnetic behavior (from Ref. 9).

Case IB corresponds to $T_0 \ll T_{c0}$, but with $\mathcal{J} < 0$. Here the hybridization between the localized f and conduction electron states is moderately strong. Temperature dependent pair breaking theories such as those due to Müller-Hartmann and Zittartz[11] and Ludwig and Zuckermann[12] describe the striking superconducting properties in this regime, particularly the reentrant superconductive behavior in which a sample within a certain range of impurity concentration first becomes superconducting at an upper critical temperature T_{c1} and then loses its superconductivity at a lower critical temperature T_{c2}.

These remarkable properties appear to originate from the competition between singlet spin pairing of conduction electrons in the superconducting state ($\underset{\sim}{k}\uparrow$, $-\underset{\sim}{k}\downarrow$) and the formation of a Kondo many body singlet in which the conduction electron spins tend to compensate the impurity spin, the degree of compensation increasing as the temperature is lowered through T_K. For $S = 1/2$, the impurity spin should be completely screened at $T = 0$. Since the binding energy of the superconducting state is $E_B^{SC} \sim k_B T_{c0}$, whereas the binding energy of the Kondo quasibound state is $E_B^K \sim k_B T_K$, the strongest modifications of superconductivity are expected where $T_K \sim T_{c0}$, as observed experimentally. The exemplary system in this limit is ($\underline{La}$Ce)Al_2 for which $T_K \sim 0.1$ K. Recently, it was shown that the low temperature specific heat of the ($\underline{La}$Ce)Al_2 system[62] can be described by the Bethe-Ansatz solution of the Kondo model for spin 1/2.[63] Evident in the magnetic susceptibility of ($\underline{La}$Ce) Al_2 alloys are the ~ 100 K splitting by the crystalline electric field of the $J = 5/2$ Hund's rule multiplet into a Γ_7 doublet ground state and Γ_8 quartet excited state, and a reduction of the effective magnetic moment of Ce by ~0.7 as $T \to 0$.[9]

Regime II corresponds to short-lived local moments for which $T_0 \gg T_{c0}$. In this limit, the localized impurity f state can be treated as a broad nonmagnetic resonance of width Δ whose centroid lies a distance E_f from the Fermi level. Modification of the superconducting properties occurs via pair weakening, rather than pair breaking, interactions. The depression of T_c with impurity concentration n can be described by a modified exponential relation first proposed by Kaiser[64]

$$T_c/T_{c0} = \exp[-An/(1 - Dn)] \tag{4}$$

where A and D depend on the Anderson model parameters E_f, Δ and U, and the reduction of ΔC with T_c follows the BCS law of corresponding states; i.e., $\Delta C/\Delta C_0 = T_c/T_{c0}$. The modified exponential relation for T_c/T_{c0} vs n is due to (1) the reduction of $N(E_F)$, a one body effect first considered by Zuckermann[65] in 1965, and (2) a reduction of the attractive electron-phonon interaction by Coulomb repulsion between electrons that scatter into the localized resonant state, a two body effect first investigated by Takanaka and Takano[66] in 1966, and Ratto and Blandin[67] in 1967.

Exemplary systems include $\underline{Al}$Mn, $\underline{Th}$Ce and $\underline{Th}$U (3d, 4f and 5f resonances).[9] Analysis of the detailed form of T_c/T_{c0} vs n for

ThCe in terms of the modified exponential relation [Eq. (4)] gives $\Delta \simeq 0.01$ eV and $U \simeq 0.1$ eV, which are presumably renormalized local Fermi liquid parameters for this system. The values of T_0 are generally extracted from normal state properties such as an impurity contribution to the electrical resistivity that varies as $\Delta\rho = \rho_0[1-(T/T_0)^2]$, the enhancement of γ, a peak in the thermoelectric power, etc. For AlMn, ThCe and ThU, T_0 has been estimated to be ~ 500 K, ~ 1000 K and ~ 100 K, respectively.

Magnetic-nonmagnetic transitions of impurity ions between case IB (T_0, $T_K \ll T_{c0}$) and case II ($T_0 \gg T_{c0}$) can be induced by the application of pressure or by alloying the matrix with another element. The LaCe system under pressure discussed earlier is an example of a pressure induced magnetic-nonmagnetic transition of Ce impurities. An analogous magnetic-nonmagnetic transition of Ce impurities at ambient pressure was subsequently observed by alloying the La with Th. Whereas the transition was reflected in a maximum in $(-dT_c/dn)_{n=0}$ vs P in LaCe under pressure, it was manifested in a maximum in $(-dT_c/dn)_{n=0}$ vs Th concentration in the (La, Th)Ce system. In addition, low temperature specific heat measurements on various (La, Th)Ce alloys revealed that the depression of the reduced specific heat jump with reduced critical temperature, $[d(\Delta C/\Delta C_0)/d(T_c/T_{c0})]_{T_c = T_{c0}}$, also exhibits a maximum as a function of Th composition.

HEAVY FERMION SUPERCONDUCTIVITY

From the previous discussion, impurities with partially-filled f electron shells lead to a depression of T_c of the matrix into which they have been introduced, even when they are nonmagnetic. Thus, it seems surprising that superconductivity would be discovered in heavy Fermion systems such as $CeCu_2Si_2$[68] and UBe_{13}[69] with effective masses of ~ 200 times the mass of the free electron. Apparently, a necessary condition for the occurrence of superconductivity is the presence of an ordered sublattice of RE or A ions.

The class of heavy Fermion superconductors, that may also include U_6Fe,[70] are characterized by low values of T_c, very high values of the upper critical field H_{c2} and the slope $(-dH_{c2}/dT)_{T_c}$, and extremely large values of γ. The physical properties of the compound UBe_{13} are particularly striking and are described in this volume in a paper by H. R. Ott, H. Rudigier, Z. Fisk and J. L. Smith.

Recently, the temperature dependence of H_{c2} of UBe_{13} was measured in applied magnetic fields between 0 and 60 kOe and found to be extremely unusual.[71] The H_{c2} vs T curve has an enormous initial slope $(-dH_{c2}/dT)_{T_c} \sim 790$ kOe/K, where $T_c = 0.854$ K, bends over and then becomes linear in T with a slope $(-dH_{c2}/dT) =$ 91 kOe/K between ~0.65 K and 0.37 K, at which $H_{c2} = 60$ kOe. A standard analysis of the H_{c2} vs T curve near T_c yields a value of ~ 470 kOe for the orbital critical field at T = 0, $H^*_{c2}(0)$, and indicates that the temperature dependence of H_{c2} in the vicinity of T_c is determined by the paramagnetic limiting field $H_p(T)$. However, the unusual shape and anomalously large initial slope of the H_{c2} vs T curve suggests the intriguing possibility of an unconventional type of superconductivity such as p-wave superconductivity. The resolution of these intriguing questions will have to await further theoretical and experimental developments.

REFERENCES

1. J. Friedel, Nuovo Cimento Suppl. 12, 1861 (1958).
2. P. W. Anderson, Phys. Rev. 124, 41 (1961).
3. See various articles, in: "Magnetism; A Treatise on Modern Theory and Materials," H. Suhl, ed., Academic Press, New York (1973), Vol. V.
4. J. R. Schrieffer and P. A. Wolf, Phys. Rev. 149, 491 (1966).
5. B. Coqblin and J. R. Schrieffer, Phys. Rev. 185, 847 (1969).
6. H. R. Krishna-murthy, K. G. Wilson and J. W. Wilkins, Phys. Rev. Lett. 35, 1101 (1975); H. R. Krishna-murthy, J. W. Wilkins and K. G. Wilson, Phys. Rev. B21, 1044 (1980).
7. M. B. Maple and K. S. Kim, Phys. Rev. Lett. 23, 118 (1969).
8. M. B. Maple, J. Wittig and K. S. Kim, Phys. Rev. Lett. 23, 1375 (1969).
9. M. B. Maple, Appl. Phys. 9, 179 (1976), and references cited therein.
10. A. A. Abrikosov and L. P. Gor'kov, Sov. Phys. JETP 12, 1243 (1961).
11. E. Müller-Hartmann and J. Zittartz, Z. Physik 234, 58 (1970); Phys. Rev. Lett. 26, 428 (1971).
12. A. Ludwig and M. J. Zuckermann, J. Phys. F1, 516 (1971).
13. P. Schlottmann, Solid State Commun. 16, 1297 (1975); J. Low Temp. Phys. 20, 123 (1975).
14. B. Coqblin, M. B. Maple and G. Toulouse, Int. J. Magn. 1, 333 (1971).

15. M. B. Maple, in "Magnetism: A Treatise on Modern Theory and Materials," H. Suhl, ed., Academic Press, New York (1973), Vol. V, pp. 289-325.
16. A. W. Lawson and T.-Y. Tang, Phys. Rev. 76, 301 (1949).
17. D. C. Koskenmaki and K. A. Gschneidner, Jr., in "Handbook on the Physics and Chemistry of Rare Earths," K. A. Gschneidner, Jr. and L. Eyring, eds., North-Holland, Amsterdam (1978), pp. 337-377.
18. F. H. Ellinger and W. H. Zachariasen, Phys. Rev. Lett. 32, 773 (1974).
19. M. R. MacPherson, G. E. Everett, D. Wohlleben and M. B. Maple, Phys. Rev. Lett. 26, 20 (1971).
20. J. Wittig, Phys. Rev. Lett. 21, 1250 (1968).
21. C. Probst and J. Wittig, in "Low Temperature Physics-LT14," M. Krusius and M. Vuorio, eds., North-Holland, Amsterdam (1975), Vol. 5, pp. 453-456.
22. K. A. Gschneidner, Jr. and R. Smoluchowski, J. Less-Common Metals 5, 374 (1963).
23. J. M. Lawrence, P. S. Riseborough and R. D. Parks, Rep. Prog. Phys. 44, 1 (1981) and references cited therein.
24. K. R. Bauschspiess, W. Boksch, E. Holland-Moritz, H. Launois, R. Pott and D. Wohlleben, in "Valence Fluctuations in Solids," L. M. Falicov, W. Hanke and M. B. Maple, eds., North-Holland, Amsterdam (1981), pp. 417-421.
25. J. W. Allen, S.-J. Oh, I. Lindau, M. B. Maple, J. F. Suassuna and S. B. Hagström, Phys. Rev. B26, 445 (1982).
26. J. W. Allen, S.-J. Oh, M. B. Maple and M. S. Torikachvili, Phys. Rev. B28, 5347 (1983).
27. E. Bucher, V. Narayanamurti and E. Bucher, J. Appl. Phys. 42, 1741 (1971).
28. M. B. Maple and D. Wohlleben, Phys. Rev. Lett. 27, 511 (1971).
29. A. Jayaraman, V. Narayanamurti, E. Bucher and R. G. Maines, Phys. Rev. Lett. 25, 1430 (1970).
30. J. M. D. Coey, S. K. Ghatak, M. Avignon and F. Holtzberg, Phys. Rev. B14, 3744 (1976)
31. M. Campagna, E. Bucher, G. K. Wertheim and L. D. Longinotti, Phys. Rev. Lett. 33, 165 (1974).
32. J. L. Freeouf, D. E. Eastman, W. D. Grobman, F. Holtzberg and J. B. Torrance, Phys. Rev. Lett. 33, 161 (1974).
33. R. A. Pollak, F. Holtzberg, J. L. Freeouf and D. E. Eastman, Phys. Rev. Lett. 33, 820 (1974).
34. A. Menth, E. Buehler and T. H. Geballe, Phys. Rev. Lett. 22, 295 (1969).

35. L. L. Hirst, Phys. Kondens. Mat. 11, 255 (1970).
36. K. Andres, J. E. Graebner and H. R. Ott, Phys. Rev. Lett. 35, 1779 (1975).
37. M. B. Maple and D. Wohlleben, AIP Conf. Proc. 18, 447 (1974).
38. B. C. Sales and D. K. Wohlleben, Phys. Rev. Lett. 35, 1240 (1975).
39. B. C. Sales, J. Low Temp. Phys. 28, 107 (1977).
40. S. D. Bader, N. E. Phillips and D. B. McWhan, Phys. Rev. B7, 4686 (1973).
41. F. Lapierre, M. Ribault, F. Holtzberg and J. Flouquet, Solid State Commun. 40, 347 (1981).
42. M. Konczykowski, J. Morillo and J. P. Senateur, Solid State Commun. 40, 517 (1981).
43. F. Holtzberg and J. Wittig, Solid State Commun. 40, 315 (1981).
44. J. C. Nickerson, R. M. White, K. N. Lee, R. Bachmann, T. H. Geballe and G. W. Hull, Phys. Rev. B3, 2030 (1971).
45. J. W. Allen, B. Batlogg and P. Wachter, Phys. Rev. B20, 4807 (1979).
46. O. Peña, M. Lysak, D. E. McLaughlin and Z. Fisk, Solid State Commun. 40, 539 (1981).
47. S. von Molnar, T. Theis, A. Benoit, A. Briggs, J. Flouquet, J. Ravex and Z. Fisk, in "Valence Instabilities," P. Wachter and H. Boppart, eds., North-Holland, Amsterdam (1982), pp. 389-395.
48. B. Batlogg, P. H. Schmidt and J. M. Rowell, in "Valence Fluctuations in Solids," L. M. Falicov, W. Hanke and M. B. Maple, eds., North-Holland, Amsterdam (1981), pp. 267-269.
49. M. Takigawa, H. Yasuoka, Y. Kitoaka, T. Tanaka, H. Nozaki and Y. Ishizawa, J. Phys. Soc. Japan 50, 2525 (1981).
50. I. Frankowski and P. Wachter, Solid State Commun. 41, 577 (1982).
51. G. Güntherodt, W. A. Thompson, F. Holtzberg and Z. Fisk, in "Valence Instabilities," P. Wachter and H. Boppart, eds., North-Holland, Amsterdam (1982), pp. 313-317.
52. J. Beille, M. B. Maple, J. Wittig, Z. Fisk and L. E. DeLong, Phys. Rev. B28, 7397 (1983).
53. N. F. Mott, Philos. Mag. 30, 403 (1973); J. Phys. (Paris) Colloq. 41, C5-51 (1980).
54. T. Kasuya, K. Takegahara, T. Fumita, T. Tanaka and E. Bannai, J. Phys. (Paris) Colloq. 40, C5-308 (1979).

55. M. B. Maple, L. E. DeLong and B. C. Sales, in the "Handbook on the Physics and Chemistry of the Rare Earths," K. A. Gschneidner, Jr. and L. Eyring, eds., North-Holland, Amsterdam (1978), pp. 797-846.
56. M. T. Beal-Monod and J. M. Lawrence, Phys. Rev. B21, 5400 (1980).
57. D. M. Newns and A. C. Hewson, J. Phys. F10, 2429 (1980).
58. See, for example, M. B. Maple, in "Advances in Superconductivity," B. Deaver and J. Ruvalds, eds., Plenum, New York (1983), pp. 279-346, and references cited therein.
59. H. C. Ku, Ph.D. thesis, University of California, San Diego, 1980, unpublished.
60. K. N. Yang, M. S. Torikachvili, M. B. Maple and H. C. Ku, Bull. Am. Phys. Soc. 28, 341 (1983); and to be published.
61. S.K. Dhar, S. K. Malik and R. Vijayaraghavan, J. Phys. C14, L321 (1981).
62. S. D. Bader, N. E. Phillips, M. B. Maple and C. A. Luengo, Solid State Commun. 16, 1263 (1975).
63. V. T. Rajan, J. H. Lowenstein and N. Andrei, Phys. Rev. Lett. 49, 497 (1982).
64. A. B. Kaiser, J. Phys. C3, 409 (1970).
65. M. J. Zuckermann, Phys. Rev. A140, 899 (1965).
66. K. Takanaka and F. Takano, Progr. Theor. Phys. 36, 1080 (1966).
67. C. F. Ratto and A. Blandin, Phys. Rev. 156, 513 (1967).
68. F. Steglich, J. Aarts, C. D. Bredl, W. Lieke, D. Meschede, W. Franz and H. Schäfer, Phys. Rev. Lett. 43, 1892 (1979).
69. H. R. Ott, H. Rudigier, Z. Fisk and J. L. Smith, Phys. Rev. Lett. 50, 1595 (1983).
70. L. E. DeLong, J. G. Huber, K. N. Yang and M. B. Maple, Phys. Rev. Lett. 51, 312 (1983).
71. M. B. Maple, J. W. Chen, S. E. Lambert, Z. Fisk, J. L. Smith and H. R. Ott, to be published.

PHOTOEMISSION AND RELATED TECHNIQUES

Yves Baer

Institut de Physique
Université de Neuchâtel
CH-2000 Neuchâtel, Switzerland

1. INTRODUCTION

Many different types of electron spectroscopies are nowadays widely used to probe the electronic states of solids. Quite often the results are directly compared to theoretical calculations or introduced in models supposed to explain other experimental results. The validity of these approaches is seldom discussed since the spectroscopic results are implicitly assumed to reflect simply a rigid single-particle density of states (DOS). In fact this way to proceed is in most cases an approximation which may become very poor or even break down when the correlation of the states involved in the considered process increases. These problems are very closely related to the formation of moments in solids, which is only possible in highly correlated systems where the exchange energy can become strong enough to favour a particular ground state.

The aim of this paper is to investigate and to illustrate how correlation and moments can be studied by electron spectroscopies. A very simple and intuitive language will be used to describe the meaning of the information extracted from the different spectra. Fig. 1 shows the processes involved in the most common electron spectroscopies. It must be emphasized that the simple but misleading rigid band scheme has only been used in order to illustrate the mechanisms of the different transitions. The techniques represented in the upper part of Fig. 1 can be expected to account for the occupied states, those of the lower part for the unoccupied states. More correctly, one should distinguish between processes decreasing or increasing the outermost level population. The strength of the perturbation induced by these transitions controls the difference between initial and final states. For this reason, spectroscopies based on the excitation of core levels will account

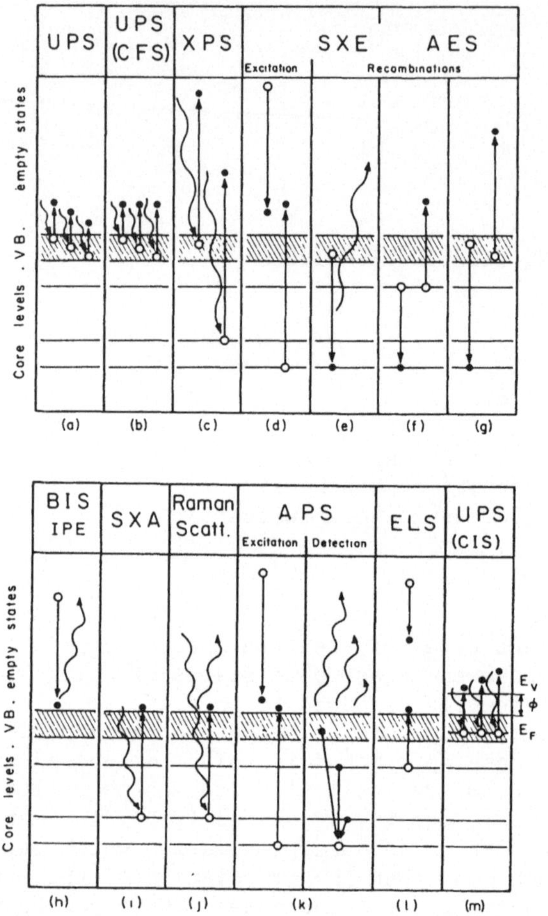

Fig. I Transitions involved in the electron spectroscopies. (a) UPS: UV-photoemission spectroscopy. (b) UPS(CFS): UV-photoemission at constant final state. (c) XPS: X-ray photoemission spectroscopy. (d): core-hole excitation by high-energy electrons. (e) SXE: soft X-ray emission. (f),(g) AES: Auger electron spectroscopy. (h) BIS: bremsstrahlung isochromat spectroscopy or IPE: inverse photoemission. (i) SXA: soft-ray absorption. (j): Raman scattering. (k) APS: appearance potential spectroscopy. (l) ELS: energy loss spectroscopy. (m) UPS (CIS): UV- photoemission spectroscopy at constant initial state.

for the important reaction of the system to a strong potential change, whereas, photoemission (XPS, UPS) and BIS of extended band states can be considered to correspond to a weak perturbation.

2. THE KOOPMANS' APPROXIMATION

The concept of single-particle density of states is commonly used without particular justification to describe the electronic states in solids and transitions between these states. In fact, the single-particle eigenvalues ε_i obtained for example from a Hartree-Fock (HF) ground state calculation are never directly observed in electron spectroscopies, at least not in the usual sense. For the sake of simplicity we shall consider only the photoemission process. If the different single-particle states i of a solid are populated by n_i electrons in the ground state, the adiabatic binding energy of a level is defined as the lowest energy required to liberate from the solid one electron occupying this particular level. It can be expressed as the difference between the two HF ground state total energies of the considered populations :

$$E_i^B = E_{HF}^t\ (\ldots(n_i-1)\ldots) - E_{HF}^t\ (\ldots n_i \ldots) \qquad (1)$$

If we consider a system containing a large number $N = \sum n_i$ of interacting electrons, the emission of one electron induces a very small change of the self-consistent field experienced by the (N-1) electrons occupying the "passive" orbitals not directly involved in the emission process. In first approximation one can maintain these (N-1) orbitals frozen in their initial state to calculate the final state energy :

$$E_i^B \lesssim E_{fr}^t\ (\ldots(n_i-1)\ldots) - E_{HF}^t\ (\ldots n_i \ldots) = -\varepsilon_i \qquad (2)$$

It is easy to show in the HF formalism that this energy difference corresponds precisely to the energy eignevalue $-\varepsilon_i$ of the electron which has been eliminated from the system. This is the Koopmans' approximation which is certainly well satisfied for broad bands since the electrons are spread throughout the solid. Attempts have been made to extend this approximation beyond the HF formalism to correlated systems [1,2]. The frozen final state energy alwaysoverestimates somewhat the HF energy and the difference is commonly interpreted as a relaxation energy ΔE_{rel}^i :

$$E_i^B = -\ \varepsilon_i - \Delta E_{rel}^i \qquad (3)$$

The density-functional formalism provides an elegant language to describe the physical meaning of the Koopmans' approximation. Since the partial derivatives of the total energy with respect to occupation are rigorously given by orbital eigenenergies, we can write [3,4] :

$$E^t(\ldots(n_i-1)\ldots) - E^t(\ldots n_i \ldots) = = -\frac{\partial E^t}{\partial n_i} + \frac{1}{2}\frac{\partial^2 E^t}{\partial n_i^2} - \ldots = -\varepsilon_i + \underbrace{\frac{1}{2}\frac{\partial \varepsilon_i}{\partial n_i} - \ldots}_{\Delta E^i_{rel}} \quad (4)$$

The limitation of the Taylor expansion to the first term corresponds to a linearization of the function $E^t(\ldots n_i \ldots)$ at the point corresponding to the ground state population. Physically this linear dependence means that the interaction of the emitted electron with the (N-1) "passive" electrons is simulated by a fixed potential or, in other words, that the self-consistency has been switched off in the ground state. This is precisely the consequence resulting from the freezing of the (N-1) orbitals in their initial states, which is used to derive the Koopmans' approximation in the HF formalism. The calculation of binding energies or transition energies can be improved by using the transition-state technique which is merely a mathematical device without physical meaning. In metals, the situation is complicated by the final state screening which can be accounted for in the density-functional formalism [4].

3. THE PHOTOIONIZATION PROCESS IN THE SUDDEN APPROXIMATION

Photoionization at sufficiently high energy is a fast process which transforms suddenly the system by elimination of one electron. It is important to find a general formalism describing this process in order to understand the resulting final states. At first let us consider a large system of electrons ($N \gg 1$) in its ground state $\psi_o(N)$ at $T = 0$. Some unspecified external action is supposed to modify suddenly at $t=0$ the Hamiltonian from H_o to H_1. Since the wave function can not be discontinuous, $\psi_o(N)$ will still describe the system at $t>0$ but it is no longer an eigenstate of the new Hamiltonian H_1. It has to be expressed as a linear combination of the eigenstates $\psi_m(N)$ of H_1 :

$$\psi_o(N) = \sum_m b_m \psi_m(N) \; ; \; b_m = \langle\psi_m|\psi_o\rangle \, . \quad (5)$$

$|b_m|^2$ represents the probability of the stationary final state $\psi_m(N)$ at $t>0$.

An infinitely fast change of H has no physical meaning and it is necessary to describe the real evolution of H by a characteristic time t_o, as shown in Fig. 2. Instead of a sudden change (dashed line) one can approximate the real evolution of H (full line) by an intermediate Hamiltonian H_2 (dotted line) and express simply the continuity conditions at $t=0$ and $t=t_o$. The result [5] is that if the condition

$$t_o \ll h/|E_1^m - E_o| \quad (6)$$

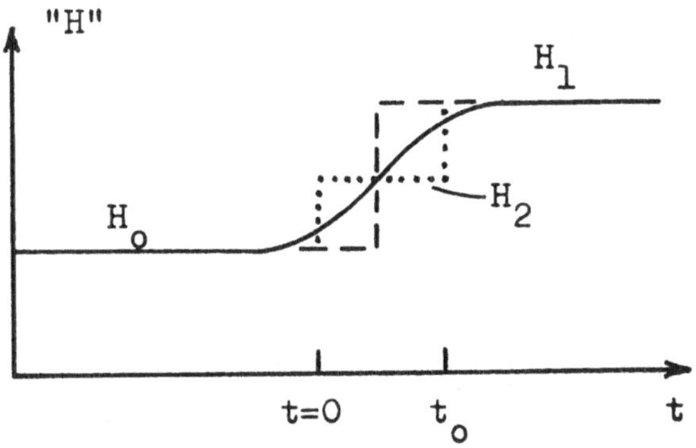

Fig. 2 Different approximations of the real evolution of H (full line).

is fulfilled, the sudden approximation (eq.5) remains valid. If the condition (6) is combined with the uncertainty relation $\Delta E \cdot t_o \sim h$, one obtains $\Delta E > |E_1^m - E_o|$. This means simply that, starting from an initial energy E_o, intermediate stationary states of H_2 can not exist in the energy range of interest. The development (5) can be usually limited to final states with energies E_1^m remaining in a range above E_F not exceeding largely ΔE_{rel}.

The application of the sudden approximation to the photoionization is straightforward. The relevant term in the single-electron Hamiltonian resulting from the presence of a radiation field of vector potential $\vec{A}$ is $H^j = -e/m_j c \cdot \vec{A}(\vec{r}_j) \cdot \vec{P}_j$. It is responsible for transitions between initial and final states $\psi^i(N)$ and $\psi^f(N)$, described by the matrix element :

$$<\psi^f(N)|H^j|\psi^i(N)> = <\psi^f(N-1)|\psi^i(N-1)> <\psi_j^f|H^j|\psi_j^i> . \quad (7)$$

$M_{fi} = <\psi_j^f|H^j|\psi_j^i>$ is the usual dipole matrix element between the single-particle orbitals of the initial bound state ψ_j^i and the free-electron-like final state ψ_j^f. The sudden approximation is then applied to the (N-1) "passive" electrons which feel the sudden emission of the electron as a change of their effective Hamiltonian [6]. The initial state $\psi^i(N-1)$ in the HF formalism corresponds precisely to the frozen wave function $\psi_o^{fr}(N-1)$ introduced in the Koopmans' approximation. The continuity condition (5) has now the following form :

$$\psi_o^{fr}(N-1) = \sum_m b_m \psi_m(N-1) \; ; \; b_m = <\psi_o^{fr}(N-1)|\psi_m(N-1)> . \quad (8)$$

The weight of the different final eigenstates is given by $|b_m|^2$, with the sum rule $\Sigma\, |b_m|^2 = 1$.

The mean energy is easily calculated :

$$\bar{E}_f(N-1) = \sum_m |b_m|^2 E_m(N-1) = \langle\psi_o^{fr}(N-1)|H|\psi_o^{fr}(N-1)\rangle = E_o^{fr}(N-1) \quad (9)$$

This is precisely the final state energy of the Koopmans' approximation, but now we have the exact relation (sometimes referred to as Koopmans' theorem) :

$$\bar{E}_f(N-1) - E_o^{HF} = -\varepsilon \quad (10)$$

It must be stressed that (10) is not equivalent to the Koopmans' approximation (2)! A complete spectrum accounting for the ionization of a particular level must display a large number of final states. Their center of gravity yields the HF eigenvalue $-\varepsilon$ of the ionized level. The condition (6) allowing us to use the sudden approximation seems to be satisfied in practical situations [7], since the emitted or impinging electrons have energies substantially larger than those of the relevant outer level excitations. The screening mechanisms concomitant to the induced local population change has been neglected but it is not expected to modify drastically the nature of the formalism discussed previously [8].

In many respects the behaviour of the outermost states is largely determined by the strength of their correlation. Experiments yield hardly a direct access to this quantity for which it is even difficult to find a precise and unique definition. Qualitatively the simple theoretical framework presented here will allow us to show that the electron spectroscopies are powerful tools for detecting the strength of the correlation and also for investigating the existence of moments in a very short time scale. We will discuss exclusively these aspects with the help of a few typical examples illustrating the evolution of the spectra from extended band states to localized atom-like states. Ambiguous situations arising on the border line between these extreme cases (actinides) will be discussed. Resonance phenomena and the symptoms of mixed valence in electron spectroscopies will be briefly described. Finally, on the controversial problem of Ce, an attempt will be made to present the experimental facts which can be extracted from recent spectra.

4. NARROW-BAND METALS

The Koopmans' approximation is accurate in nearly-free-electron metals [9] where the Bloch states can be assumed to have a weak correlation and to be spread within the solid. The excitation of core levels is experienced by the conduction electrons as the sudden creation of a finite range potential. It has been shown that the ground state in the presence of this potential is largely orthogonal to the initial ground state [10] and is therefore not accessible to the experiment. The observed response in photoemission accounts for a high number of low-energy electron-hole pair excitations around E_F giving rise to asymmetric line shapes from which the adiabatic binding energy can still be obtained [11,12]. The edges of X-ray absorption and emission spectra are more dramatically affected by similar many-body effects [13] which will not be discussed here.

In narrow-band metals, the local character of the states becomes important so that the relaxation energy associated with the decrease or increase of the population can no longer be neglected. It must be stressed that in any metal at finite temperature, rather local population fluctuations at E_F involving infinitesimal energy changes of the system [3] must exist. They allow the itinerancy of the electrons. Photoemission and BIS in an energy range of the order of kT around E_F induce suddenly similar population fluctuations with the only difference that they are not in equilibrium with the phonon gas. This is not a serious practical problem since the electron-phonon coupling energy is small compared to the instrumental resolution. The states at the Fermi energy of metals are not affected by the relaxation and E_F is a fixed point in all samples in electrical equilibrium with the instrument. The practical consequence, illustrated schematically in Fig. 3, is that the bottom of the quasi-particle spectrum (dashed line) is shifted toward E_F when compared to a ground state eigenvalue spectrum (full line). The net effect appears as a narrowing of the excitation spectrum which has still to keep its center of gravity at the frozen position (eq.10). For this reason the narrowing can be anticipated to be enhanced by a weight transfer from the band-like excitations to some higher excited states sketched as a satellite in Fig. 3. The discussion of these excitations is delayed to the next section and for the moment only band-like excitations will be considered. Numerous experimental studies have been devoted to narrow bands in transition metals. They provide impressive demonstrations [14-19] of the power of photoemission and inverse photoemission to analyze occupied and empty band states according to their quantum numbers : a) $\vec{k}$-vector by controlling in vacuum the propagation direction of the electrons relative to the crystal orientation, b) the spin component S_z in magneti-

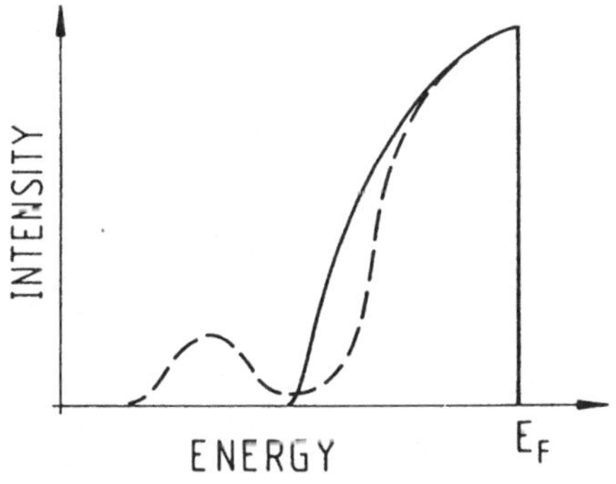

Fig. 3 — Hypothetical narrow DOS obtained from calculated ground state eigenvalues. --- schematic shape of the corresponding photoemission spectrum.

cally ordered samples by controlling the spin polarization of the electron beams, c) the polarity p of the states with respect to a mirror plane by using the symmetry selection rules for different radiation polarizations, d) finally the atomic-like symmetry (ℓ) of band states by identifying typical cross-section dependences as a function of the photon energy.

The direct and detailed observation of the band exchange splitting provides an unambiguous demonstration of the validity of the band magnetism approach in 3d metals. Photoemission processes are very fast [8] so that they provide a nearly instantaneous measure of the electronic properties fluctuating at lower frequency. The microscopic region sampled by photoemission of a band state is given by the inelastic mean free path of the excited electron, i.e. about 20 Å. These conditions have been exploited to follow the evolution of the exchange band splitting in Ni when the temperature is raised above the ordering temperature [20]. The results obtained by angle-resolved photoemission show clearly that a short-range magnetic order persists but their interpretation is impeded by the increasing incoherence of the exchange part of the potential which gradually destroys the validity of a $\vec{k}$-vector analysis. In antiferro- to para-magnetic phase transitions, the structures accounting for hybridization gaps resulting from the back-folding of the paramagnetic Brillouin zone in the ordered state have also been found to persist above T_N [21,22].

XPS of core levels can be a very useful probe of the local moment on the different atoms of solids. In order to avoid complications with electrostatic and spin-orbit interactions, one chooses usually s shells which can be only exchange-split by the presence of a local spin. This effect has been mostly exploited in ionic compounds of the 3d elements [23,24] and in the lanthanides where the spin dependence is obvious, as shown in Fig. 4 [25]. The discrepancy

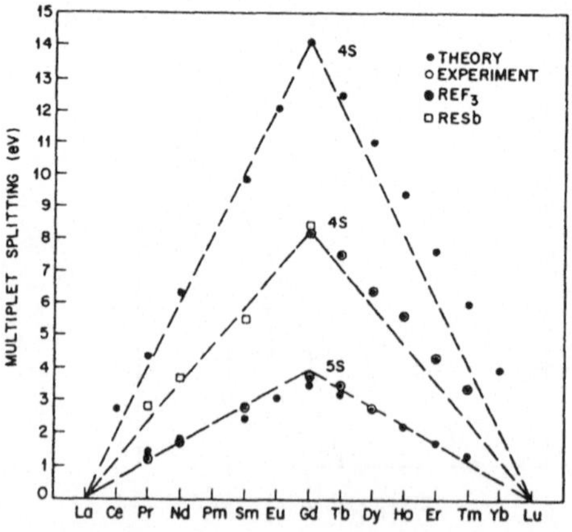

Fig. 4 4s and 5s exchange splitting in the rare earths.

1) Direct photoemission process

(A)

$3p^6$

$3p^6(3d4s)^{10} + \hbar\omega \rightarrow$

$3p^6(3d4s)^9 V_F + e$

(B)

$3d^8$

$3p^6$

$3p^6(3d4s)^{10} + \hbar\omega \rightarrow$

$3p^6 3d^8 4s^2 + e$

2) Photoabsorption process

(A)

$3p^5$

$3p^6(3d4s)^{10} + \hbar\omega \rightarrow$

$3p^5(3d4s)^{11}$

(B)

$3d^{10}$

$3p^5$

$3p^6(3d4s)^{10} + \hbar\omega \rightarrow$

$3p^5 3d^{10} 4s^1$

3) Super-Coster-Kronig Auger decay

(A)

$3p^6$

$3p^5(3d4s)^{11} \rightarrow$

$3p^6(3d4s)^9 V_F + e$

(B)

$3d^8$

$3p^6$

$3p^5 3d^{10} 4s^1 \rightarrow$

$3p^6 3d^8 4s^2 + e$

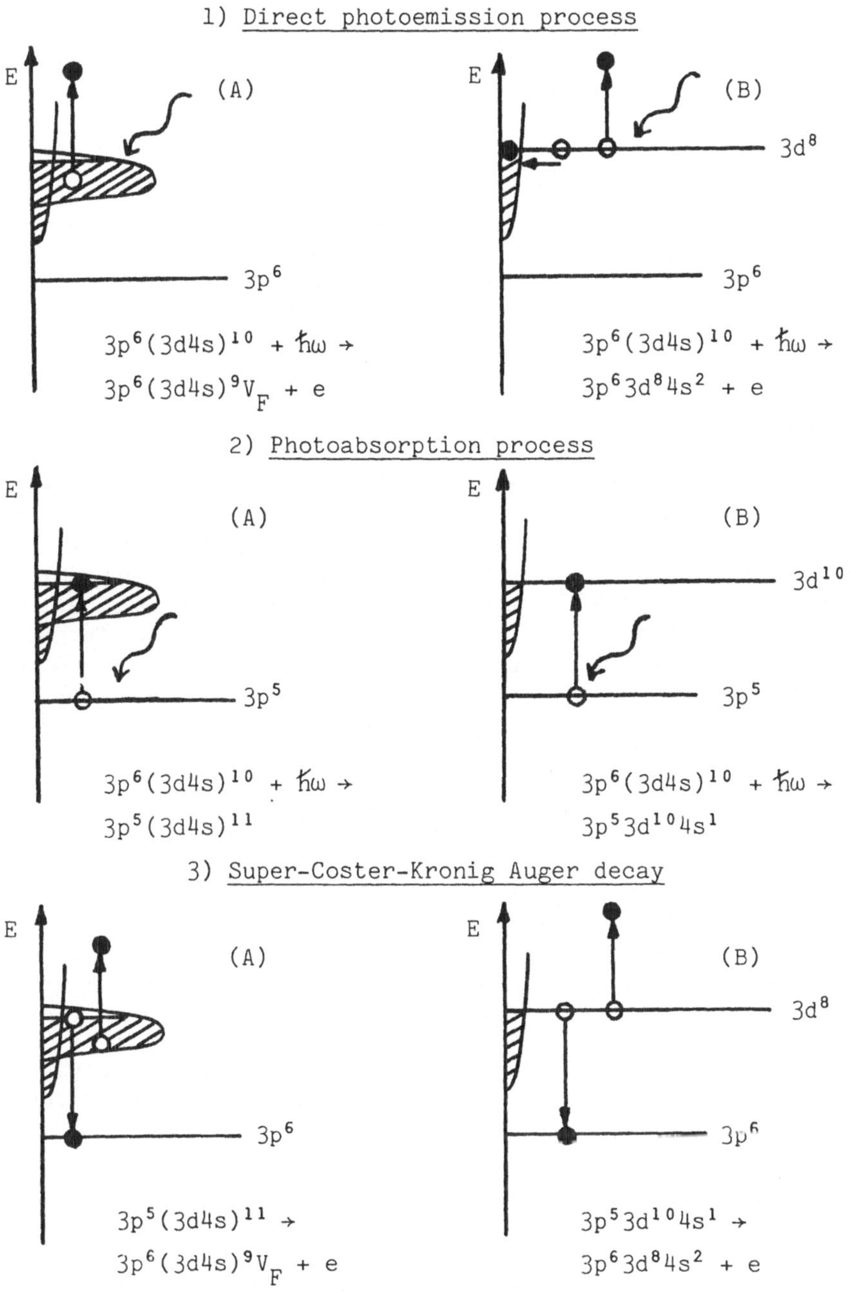

Fig. 5 Resonant photoemission in a narrow-band metal (Ni).

between experiment and theory for the 4s splitting is attributed to the high correlation between 4s and 4f electrons. The persistence above T_c of the 3s exchange splitting in 3d metals has been probably observed [26]. Better resolved spectra could yield a decisive answer to the controversial problem of local moments in magnetically disordered systems.

5. PRECURSORY MANIFESTATION OF THE LOCALIZATION

It has been pointed out in the previous section that the sizable relaxation energy involved in the emission of highly correlated band states has to be "dissipated" by some excitation sketched as a satellite in Fig. 3. We shall consider the case of Ni which has given rise to numerous calculations [27-31]. Let us assume that the ℓ-projected band states correspond to the configuration $3d^{9.4}4s^{0.6}$. The normal band emission process is represented in Fig. 5 (1A) where the created band hole is assumed to be screened by scattering of electrons at E_F (V). Alternatively, one has to consider that the emission of a 3d electron produces a rather well-localized hole corresponding to a $3d^{8.4}$ configuration. The attractive atomic potential may become strong enough to "break" locally the sd hybridization in order to form an atomic $3d^8$ state usually discussed in terms of two correlated holes spending most of their time on the same site. Another, but to some extent equivalent description of the different final states has been given very early in terms of hole-screening achieved by the filling of s or d states [32]. A naive attempt is made in Fig. 5 (1B) to depict the formation of the valence band satellite. The energy of this satellite is found to remain surprisingly constant (6±0.5 ev) from Ni to Cr [16] but it is weak or barely discernible in conventional photoemission spectra. Resonant photoemission is commonly used to make conspicuous these atomic-like final states. When the photon energy is swept through the threshold excitation energy E_0 of a core level, a strong absorption takes place in the narrow photon energy range corresponding to the empty d-band width, (or to the sharp energy of a localized level). The two possible final states produced by photoabsorption are shown in Fig. 5 (2A,B). The existence of the excitation (2B) is confirmed by the observation of its radiative decay occuring exactly at the absorption energy. The non-radiative decay (Super-Coster-Kronig Auger process) represented in Fig. 5 (3A,B) leads exactly to the same final state as those resulting from the direct photoemission process (1A,B).

Let us consider at first only one type of process (A or B). The same initial and final states are linked by two different channels, the first one being continuous, the second one quasi-discrete as a function of $\hbar\omega$. This is a familiar situation giving rise to interference phenomena responsible for a characteristic intensity distribution called Fano profile [33] :

$$I \propto \frac{\left(q + \frac{\hbar\omega - E_0}{\Gamma}\right)^2}{1 + \left(\frac{\hbar\omega - E_0}{\Gamma}\right)^2}$$

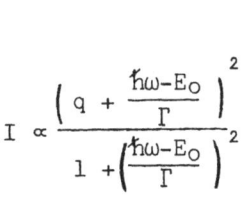

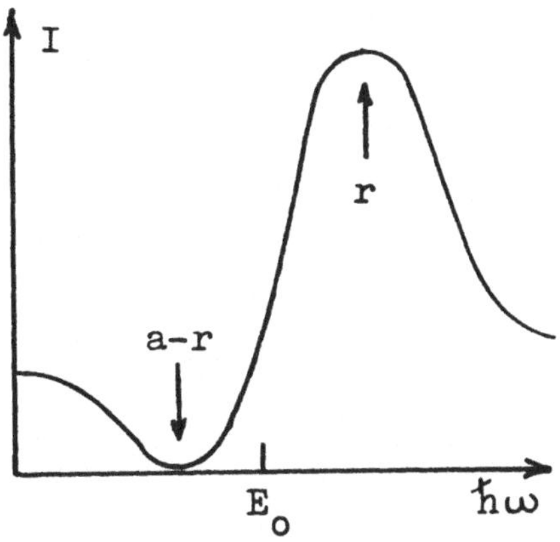

The parameters Γ and q are expressed by the different matrix elements for the considered transitions [33] and it appears that the amplitude of a Fano-profile is enhanced by the localization of the outer level involved in the process. For the rare earths and actinides, recipes have been established for the extraction of the f contribution by subtraction of photoemission spectra recorded at resonance (r) and anti-resonance (a-r). This procedure is only qualitative and fails to account correctly for the fact that to some extent the intensity behaviour of the different outer states must be described by individual Fano-profiles. A very useful visualization of such effects in terms of a classical damped oscillator has been proposed in the time-dependent local density approximation [34].

A theoretical description of these resonances in the situation illustrated by Fig. 5, where both band and localized final states can be formed is a very difficult task [35-38]. In order to understand at least qualitatively the emergence on resonance of satellites barely discernible off resonance, one can simply invoke the enhanced localized character of these excitations. Fig. 6 shows typical photoemission spectra of Ni [39] at different photon energies crossing the 3p threshold. From $\hbar\omega > 65.9$ eV, the d satellite (c) moves to higher energies in an $M_{2,3}$ VV Auger peak. The weak structure marked by an arrow at 14 eV in the 67.7 eV spectrum can be possibly attributed to the decay of an intermediate $3d^9$ state into a three holes $3d^7$ state. The right panel of Fig. 6 illustrates the complexity of the interference phenomena occurring at the 3p threshold: the intensity of the different band peaks and of the satellite have different photon energy dependences. The quasi-atomic origin of the 6 eV Ni satellite has been a controversial problem for many years. It has been now quite often tested and is widely accepted. Lately, the atomic model has been used to calculate the spin polarization of the electron accounting for the satellite [35] and the result has been clearly confirmed by an experiment [40].

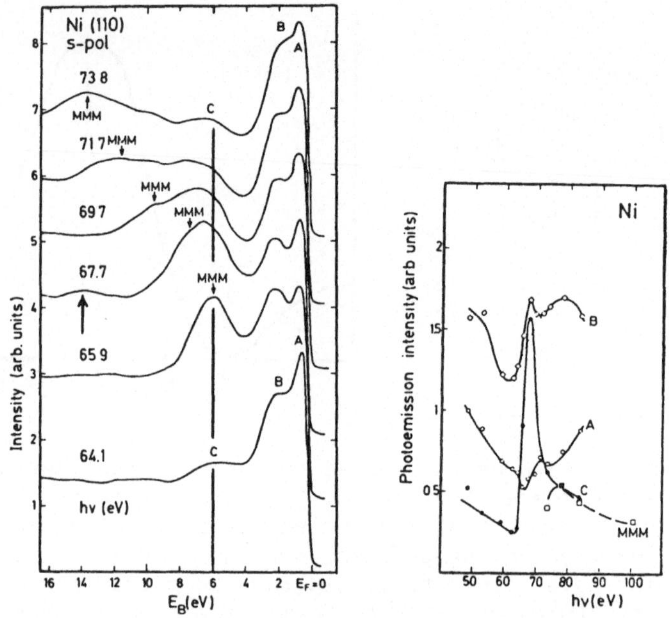

Fig. 6 left panel : resonant photoemission spectra of Ni. right panel: intensity distribution in different regions of the photoemission spectra [39].

The satellites observed in photoemission spectra of correlated band states are precursory manifestations of the localization. They have been anticipated to exist in hole excitation spectra as a consequence of strengthening of the atomic potential induced by the population decrease of the correlated electrons. Similar manifestations are very unlikely in BIS where the induced population increase can not be expected to favour the formation of localized excitations. No evidence of satellites of this nature can be found in BIS spectra of Co and Fe [41].

6. LOCALIZED OPEN SHELLS

Relaxation effects in the ionization of core levels in atoms are important. For example, the relaxation energy of the 1s level in Cr is ~60 eV. On the binding energy scale, however, the Koopmans' approximation is still well satisfied since $\Delta^{1s}_{rel}/E^{B}_{1s} \sim 1\%$. This ratio does not change very much for less bound levels since both terms decrease simultaneously. This situation is completely different if the usual relationship between orbital size and binding energy is broken. The best known examples are the lanthanides where the maximum of the 4f radial charge lies closer to the nucleus than the maxima of the 5s,5p and 6s radial charges. However, for $\ell=3$, the centrifugal contribution to the radial potential is so strong that the 4f shell is very weakly bound and can be only gradually filled as Z increases. The consequence of this unusual situation is : $\Delta^{4f}_{rel}/E^{B}_{4f} > 1$, this means a complete breakdown of the Koopmans' approximation. The energy of the system is now a discrete function of the population of this shell and one can only ask for the total energy required for a given population change. In metals, the screening concomitant to such localized charge variations is dramatically important as shown by renormalized-atom calculations [42].

XPS and BIS spectra of lanthanides show sharp peaks which account for the occupation increase or decrease of the 4f levels and are superimposed on the continuous spectrum of the extended states. These excitation energies must be interpreted as difference between total energies, as defined by eq. (1). In order to relate correctly the continuous and discrete spectra of extended and localized states, it is important to identify the total energy of the ground state configuration f^n with the position of the Fermi energy. As shown in Fig. 7, the energies Δ- and Δ+ measured by XPS and BIS can only be interpreted as minimum many-electron energies necessary to modify by ± 1 the count of the localized f states. The decomposition of XPS (A) and BIS (C) processes into two steps should help to visualize the meaning of the energies Δ_+ and Δ_- when they are compared with those observed in direct processes from or to E_F (B). It must be stressed that this interpretation is based on the fact that in metals, population fluctuations at E_F involve infinitesimal energy changes. One should also notice, (and keep in mind when comparing with the core hole excitations illustrated in Fig. 10) that in Fig.7

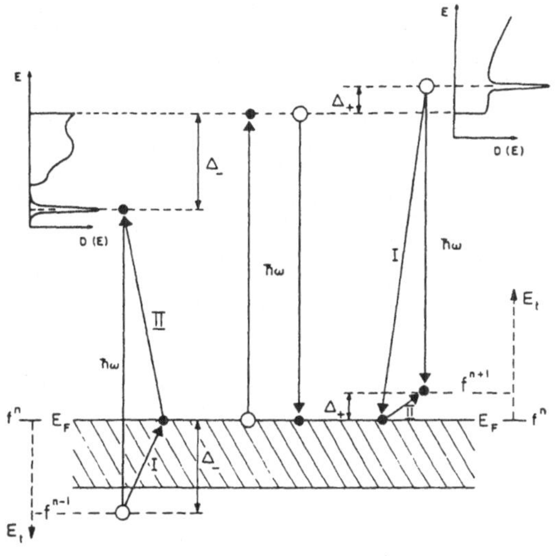

Fig. 7 XPS and BIS excitations in a metal containing a localized and open f shell.

the total energy axes have opposite directions for a decrease (A) or increase (C) of the level population. Finally, the combination of step I in XPS and step II in BIS can be considered as the transition $f^n + f^n \rightarrow f^{n-1} + f^{n+1}$ between two distant atoms. The corresponding energy $\Delta_- + \Delta_+$ is the effective Coulomb correlation energy U_{eff}. In Fig. 8 [43] the combined XPS-BIS spectra of Gd with half-filled 4f shell yield the simplest illustration of the excitation spectrum of a localized outer level. The final states are only split by spin-orbit interaction. For 4f populations different from 0,7 and 14, the electrostatic interaction gives rise to complicated multiplets as demonstrated by the Tb spectrum of Fig. 8. Fortunately the initial ground state $|f^N\rangle$ can be expressed in an unique manner as a linear combination of $|f^{N-1}\rangle|f^1\rangle$ wave functions corresponding to the symmetry-allowed set of atomic quantum numbers labelled symbolically Ω^{N-1} and φ^1 :

$$|\Omega\rangle = \sum_{\Omega^{N-1},\varphi^1} Q(\Omega^N,\Omega^{N-1},\varphi^1)\ |\Omega^{N-1}\rangle\ |\varphi^1\rangle .$$

The photoemission final states in the sudden approximation are just given by the same expression from which the single orbital $|\varphi^1\rangle$ has been eliminated [44]. The weights of the different parent terms $|\Omega^{N-1}\rangle$ are then simply given by the squares of the fractional parentage coefficients Q. Symmetrically, the different terms of the shell can be described by the unoccupied wave functions and a similar expression can be developed for the BIS process [45]. This formalism is a powerful tool for analyzing the initial population and

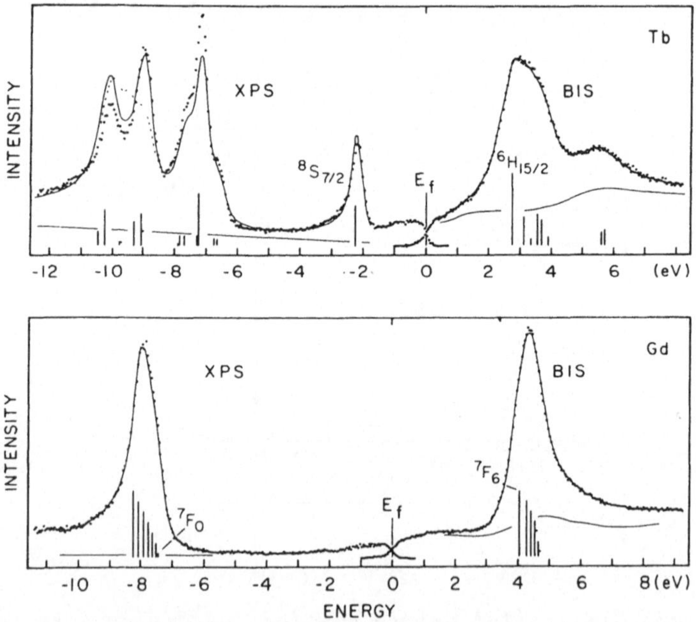

Fig. 8 Combined XPS-BIS spectra of Tb and Gd metals [43].

quantum numbers. For example, the lengths of the vertical bars in the Tb spectrum of Fig. 8 are proportional to the different Q^2, their relative positions are taken from UV measurement and the full line corresponds to the broadening of the bars by a realistic line shape [43]. A systematic study of the whole lanthanide series shows that the large value of U_{eff} is responsible for the localized nature of the 4f states.

In this connection it is interesting to consider briefly the 5f electrons in the actinides. It is now well established [46] that at the beginning of this series the 5f states are less localized than the 4f states in the rare earths and that they form narrow fd-hybridized bands in the pure metals. The comparison of the XPS-BIS spectra of the two corresponding elements Nd and U shown in Fig. 9 [46] provide a glaring demonstration of the efficiency of these two techniques in the determination of the localized or itinerant character of outer states. As predicted theoretically, photoemission spectra [47] confirm that from Am metal the correlation energy becomes strong enough to prevent the formation of a 5f band. Already in the lighter element U,the ionic character of the bond, as for example in UO_2 [48], or the particular stoichiometry and structure of intermetallic compounds (UPd_3 [49]) can favour the tetravalence of U. The resulting increase of the atomic potential is sufficient to localize the two remaining 5f electrons which show the same behaviour as 4f localized states.

The core-level excitation of atoms containing an outer localized f^n shell yields rather complex spectra which can be well accoun-

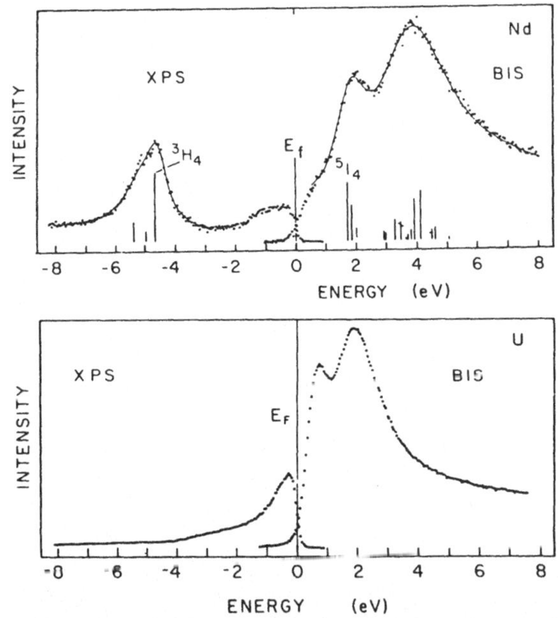

Fig. 9 Combined XPS-BIS spectra of Nd and U metals [46].

ted for in the sudden approximation. The total energy of a Wigner-Seitz cell around a particular atom as a function of its f-population is a discrete function lying on a parabola, as shown in Fig. 10 (a). The creation of a deep core hole in this atom simulates the atomic charge of the next element ((Z+1) approximation) so that in the presence of this hole the total energies lie on a different parabola for which the lowest state is now f^{n+1},(b). Disregarding for the sake of simplicity the reaction of the extended valence electrons to the core level photoionization, the probability of observing in XPS a particular f^m final population is given by (eq. 8) :

$$|b_{f^n \to f^m}|^2 = |\langle f^n | f^m \rangle|^2 = \delta_{nm} + \Delta_{nm}$$

The very small parameter Δ_{nm}($|\Delta_{nm}| \ll 1$) is negative for n=m and positive for n≠m. It expresses the weak loss of orthogonality of the f wave functions due to the different potentials in initial and final states and a possible f-symmetry admixture to the valence states. This rough consideration provides qualitatively an important result : the main component observed in XPS core levels corresponds to the frozen initial f^n population (vertical arrow in Fig. 10). In addition weak components, accounting for other populations of the localized f states not present in the initial state (at least not as localized states) can also be expected. They are commonly visualized as resulting from shake-up and shake-down processes of an electron from or to E_F (arrows). As shown in Fig. 11, [50] the 3d

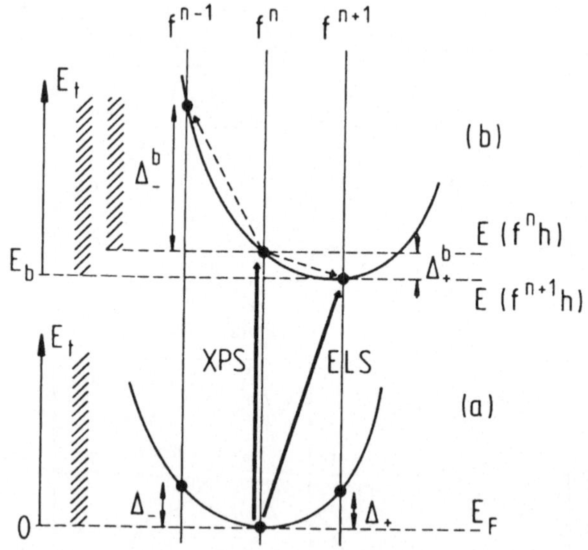

Fig. 10 Schematic diagram of the total energy of a Wigner-Seitz cell as a function of the f population for a neutral atom (a) and an atom with a core hole (b). The shaded regions indicate continuous excitations of the extended states.

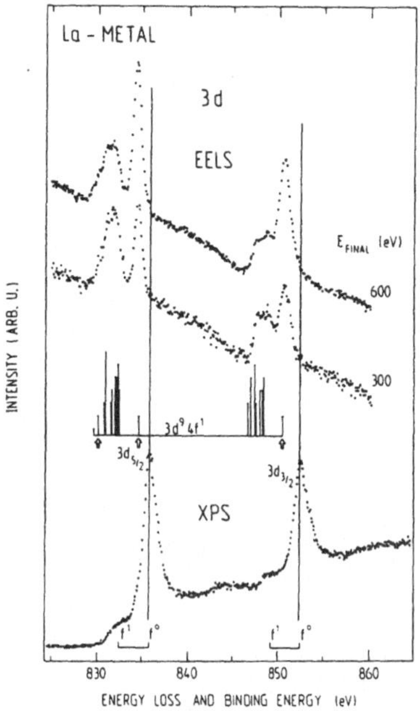

Fig. 11 XPS and EELS spectra of the 3d core level of La plotted on a common energy scale [50].

XPS spectrum of La, which has undoubtedly a $4f^0$ initial configuration, contains also a small $4f^1$ component on the low energy side of the main $4f^0$ line. In order to demonstrate the correctness of this assignment, in Fig. 11 the XPS spectrum is compared on a common energy scale to EELS spectra corresponding to the transition $3d^{10}\ 4f^0 \rightarrow 3d^9 4f^1$, (see Fig. 10). The two energy loss structures can be analyzed in terms of $3d^9 4f^1$ multiplets [50] which coincide with the tail and the shoulder of the XPS line.

Quite generally it is important to realize that in the presence of an open localized shell, the energy of the main structure observed in XPS core level spectra don't yield the true lowest binding energy defined in eq. (1). Shake-down satellites can usually be considered as safe symptoms revealing the presence of an outer localized level. In most cases they have a too small intensity to introduce an uncertainty about the initial state population.

A very exceptional situation, called mixed valence, occurs in the rare earths when the total energy of two different populations, for example f^n and f^{n+1}, are degenerate. For high-energy spectroscopies, it is quite satisfactory to visualize naively the ground state as a fluctuation between these two localized configurations. In this case the spectroscopic results display simply the superposition of the two corresponding spectra which are usually well separated in energy by the large chemical shift resulting from the different localizations of the charges [25]. It is also very easy to understand with the help of Fig. 7 that the lowest $f^n \rightarrow f^{n-1}$ XPS transition and $f^n \rightarrow f^{n+1}$ BIS transition must coincide with E_F since the fluctuation $f^n\ V_F \leftrightarrow f^{n+1}$($V_F$: electron at E_F) requires no energy ($\Delta_+ = 0$).

7. CERIUM

The behaviour of the 4f level in Ce metal and compounds has been now for more that 10 years a very enigmatical problem. It has given rise to innumerable studies, controversial models and passionate debates. No attempt will be made here to review the cerium problem, we will exclusively describe the puzzling but now accepted spectra obtained for the two phases γ and α of the pure metal. We will also quote a very recent theoretical approach which seems to reconcile the surprising and often contradictory results obtained by the different techniques. There is little doubt that at high temperature γ-Ce has a trivalent configuration with precisely one occupied atomic-like 4f orbital. The $\gamma \rightarrow \alpha$ phase transition is unique since it is isostructural and involves a very large volume reduction $\Delta V/V \approx 16\%$. The unusual character of this phase transition has always been attributed to some drastic modification of the electronic structure for which many different models have been proposed: a) the promotion model [51,52] assuming the mixed valence of the α-phase, b) the delocalization model [53] predicting in the α-phase the formation of a 4f band (Mott transition) on the basis of thermochemical considerations, c) the screening model [54] supposing final states with different d-populations, d) the two-well potential model producing two types of 4f wave functions [55], e) the "Kondo-like" model [56,57] which seems suitable for describing the thermodynamical aspects of the transition.

Each model may provide an explanation for one particular type of spectrum but the difficulty lies much deeper in the fact that before any interpretation, the conventional analysis of the different types of spectra yields contradictory results. We shall briefly present some recent experimental observations. Fig. 12 [58] shows the photoemission spectra of γ-and α-Ce (stabilized by 10% Th) at $\hbar\omega = 50$ eV and at the 4d resonance ($\hbar\omega = 122$ eV). The occurrence of two peaks showing many symptoms indicating that they originate from f-symmetry states is enigmatical in γ-Ce supposed to have a pure $4f^1$ configuration. The transition to α-Ce induces only a moderate intensity transfer from the low-lying peak to the peak just below E_F. This behaviour may help to eliminate different models but certainly not to determine the nature of the electronic transition driving the phase transition. If the unoccupied DOS is correctly revealed by BIS spectra, in Fig. 13 [59] the shape of the main structure between 3 and 6 eV is in perfect agreement with the broadened f^2 final states predicted by the fractional parentage coefficients. Assuming that the extended states form a featureless empty band, the spectrum of γ-Ce seems to demonstrate the pure trivalence ($4f^1$) of this phase. The BIS spectrum of α-Ce shows a striking difference : a narrow and intense peak emerges now just above E_F. It can be fitted to the two spin-orbit split f^1 final states and therefore seems to reveal an important tetravalent (f^0) component

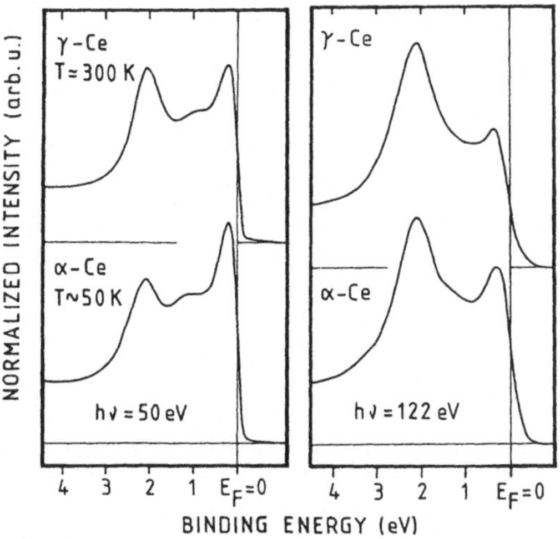

Fig. I2 Photoemission spectra of γ-and α-Ce excited at low photon energy (50 eV) and on resonance (I22 eV) [58].

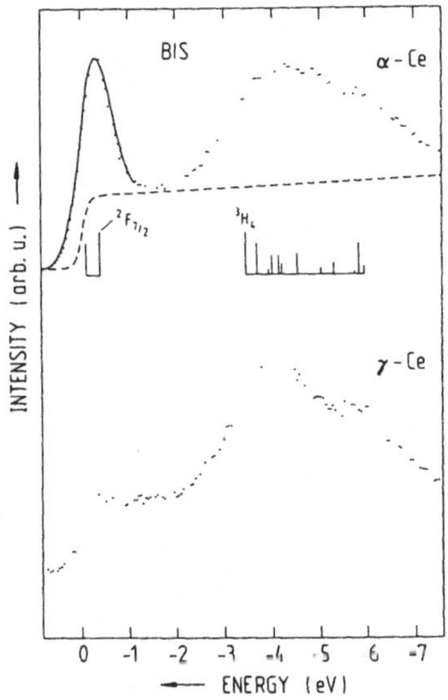

Fig. I3 BIS spectra of α- and γ-Ce [59].

in the initial state ($Ce^{4+}/Ce^{3+} \sim 30\%$). Finally we consider in Fig. 14 [59] the core level excitations analyzed in terms of multiplets corresponding to the coupling of a 3d hole with the different 4f populations. For γ-Ce, XPS shows essentially a $4f^1$ main line and a weak $4f^2$ component which can be attributed to a shake-down process (Fig. 10). The XPS spectrum of α-Ce is only slightly different : a very weak $4f^0$ component (~ 5% of the total intensity) has emerged and, much more surprisingly, the f^2 component has increased roughly by a factor of 2 ! The two EELS spectra of γ-and α-Ce can be considered to be perfectly identical within the statistical error. They seem to demonstrate that in both phases the Ce atoms have a pure $4f^1$ configuration since only $3d^94f^2$ multiplets are observed. The contradictory conclusions drawn from the different spectra (Fig. 12-14) force us to realize that the proposed analysis is no longer correct. It is impossible to compare simply these results to the different models for the electronic structure of α- and γ-Ce, but within these models the final state spectra resulting from the different transitions induced in the spectroscopies need to be calculated.

Very recently Gunnarsson and Schönhammer [60] have performed an exhaustive calculation of this type, yielding numerical predictions for the different types of spectra as a function of a limited number of parameters. They used a generalized Anderson impurity model and described the transitions within the sudden approximation. We will not attempt to describe this calculation but make only simple comments. The exceptional situation met in Ce is apparently attributable to i) the strong coupling Δ of the localized 4f level with the conduction electrons resulting from the bare energy position of the 4f level lying within the conduction band and from moderately small 4f orbital size, ii) the very large degeneracy N of the 4f level. The most spectacular consequence resulting from this calculation is that the signal strength ratios observed in the spectroscopies have no longer a linear

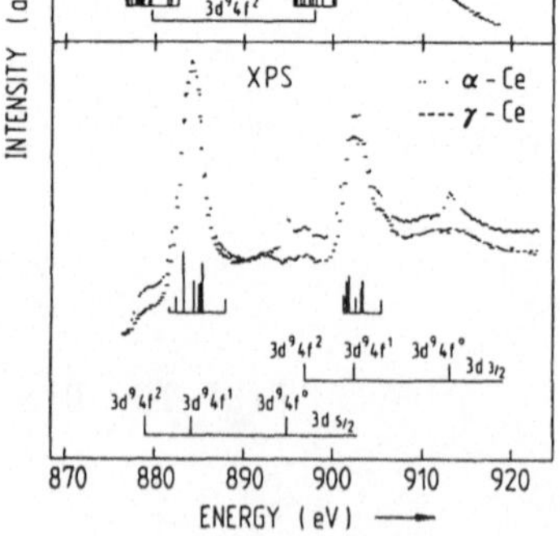

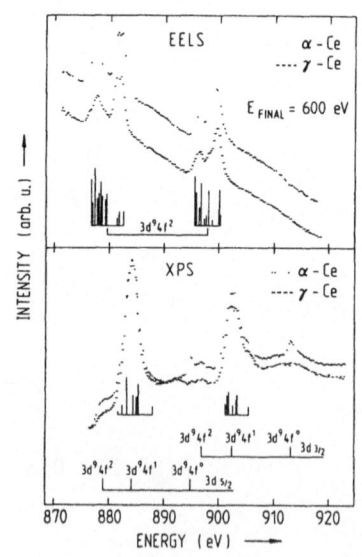

Fig. 14 Comparison of EELS and XPS spectra of the 3d level in α and γ-Ce [59].

relationship with the corresponding weight ratios of the initial configurations. Until now this possibility has never been seriously considered for different initial 4f counts. This type of formalism provides obviously an unified framework eliminating largely the contradictions between the different results [60]. The influence of the loss of orthogonality between initial and final f wave functions and the importance of the multiplets will still have to be estimated.

The two "f" photoemission peaks observed in the valence band spectra (Fig. 12) are also predicted by this theory. They might be naively understood as originating from the removal of the degeneracy of f and d states which results from their strong mixing. Nevertheless, these excited final states have a reasonable lifetime as shown by the widths of the corresponding structures and one should be able to ascribe them to some population of the localized f orbital. One can make the guess that the peak at 2 eV has a dominant $|f^0>$ character whereas the peak at E_F has rather a $|f^1>$ character. An increase of Δ in the $\gamma\to\alpha$ transition could explain the variation of the two peaks ratio and the presence of a $|f^1>$ peak above E_F in BIS. One can only hope that the theorists will find a way to go beyond the mystery of their formalisms in order to provide a physical picture of these states.

REFERENCES

[1] J.C. Phillips, Phys. Rev. 123, 420 (1961).
[2] D. Smith and O.W. Day, J. Chem. Phys. 62, 113 (1975).
[3] J.F. Janak, Phys. Rev. B18, 7165 (1978).
[4] A.R. Williams and N.D. Lang, Phys. Rev. Lett. 40, 954 (1978).
[5] L.I. Schiff, "Quantum Mechanics" (Mc Graw Hill, New York, (1955) p. 217.
[6] R. Manne and T. Åberg, Chem. Phys. Lett. 7, 282 (1970).
[7] J.W. Gadzuk and M. Sunjic, Phys. Rev. B12, 524, (1975).
[8] C. Noguera, D. Spanjaard and J. Friedel, J. Phys. F9, 1189 (1979).
[9] F. Seitz, "The Modern Theory of Solids," (Mc Graw Hill, New York, 1940) p. 313.
[10] P.W. Anderson. Phys, Rev. Lett. 18, 1049 (1967).
[11] S. Doniach and M. Sunjic. J. Phys. C3, 285 (1970).
[12] G.K. Wertheim and P.H. Citrin, Topics in Applied Physics 26, ch. 5 (1978).
[13] P. Nozieres and C.T. De Dominicis, Phys. Rev. 178, 1097 (1969).
[14] F.J. Himpsel, J.A. Knapp and D.E. Eastman, Phys. Rev. B19, 2919 (1979).
[15] W. Eberhardt and E.W. Plummer, Phys. Rev. B21, 3245 (1980).
[16] D. Chandersis, J. Lecante and Y. Petroff, Phys.Rev. B27, 2630(1983).
[17] F.J. Himpsel and Th. Fauster, Phys. Rev. B26, 2679 (1982).
[18] R. Raue, H. Hopster and R. Clauberg, Phys. Rev. B50, 1623 (1983).
[19] J. Ungis, A. Seiler, R.J. Cclotta and D. Pierce, Phys. Rev. Lett. 49, 1047 (1982).
[20] C.J. Maetz, U. Gerhardt, E. Dietz, A. Ziegler and R.J. Jelitto, Phys. Rev. Lett. 48, 1686 (1982).
[21] L.I. Johansson L.-G. Petersson, K.-F. Berggren and J.W. Allen, Phys. Rev. B22, 3294 (1980).

[22] B. Reihl, G. Hollinger and F.J. Himpsel, Phys. Rev. B28, 1490(1983).
[23] J.C. Carver, G.K. Schweitzer and T.A. Carlson, J. Chem. Phys. 57, 973 (1972).
[24] G.K. Wertheim, S. Hüfner and H.J. Guggenheimm, Phys. Rev. B7, 556 (1973).
[25] M. Campagna, G.K. Wertheim and Y. Baer, Topics in Applied Physics 27, 217 (1979).
[26] C.S. Fadley and D.A. Shirley, Phys. Rev. A2, 1109 (1970).
[27] D.R. Penn, Phys. Rev. Lett. 42, 921 (1979).
[28] A. Liebsch, Phys. Rev. Lett. 43, 1431 (1979) and Phys. Rev. B23, 5203 (1981).
[29] L.C. Davis and L.A. Feldkamp, Solid State Comm. 34, 141 (1980).
[30] N. Mårtensson and B. Johansson, Phys. Rev. Lett. 45, 482 (1980).
[31] G. Treglia, F. Ducastelle and O. Spanjaard, J. Physique 41, 281 (1980) and J. Physique 43, 341 (1982).
[32] A. Kotani and Y. Toyozawa, J. Phys. Soc. Japan 37, 912 (1974).
[33] U. Fano, Phys. Rev. 124, 1866 (1961).
[34] A. Zangwill and P. Soven Phys. Rev. Lett. 45, 204 (1980).
[35] L.A. Feldkamp and L.C. Davis, Phys, Rev.Lett. 43, 151 (1979).
[36] S.M. Girvin and D.R. Penn, Phys. Rev. B22, 4081 (1980).
[37] Y. Yafet. Phys. Rev. B21, 5023 (1980).
[38] L.C. Davis and L.A. Feldkamp, Phys. Rev. B23, 6239 (1981).
[39] J. Barth, G. Kalkoffen and C. Kunz, Phys. Lett. 74A, 360 (1979).
[40] R. Clauberg, W. Gudat, E. Kisker, E. Kuhlmann and G.M. Rothbert, Phys. Rev. Lett. 47, 1314 (1981).
[41] R.R. Turtle and R.J. Liefeld, Phys. Rev. B7, 3411 (1973).
[42] J.F. Herbst, R.E. Watson and J.W. Wilkins, Phys. Rev. B17, 3098 (1978) and ref. Therein.
[43] J.K. Lang, Y. Baer and P.A. Cox, J. Phys. F11, 121 (1981).
[44] P.A. Cox, Structure and Bonding 24, 59 (1975).
[45] P.A. Cox, J.K. Lang and Y. Baer, J. Phys. F11, 113 (1981).
[46] Y. Baer, "Electron Spectroscopy Studies," in Handbook on the Physics and Chemistry of the Actinides, A.J. Freeman, G. Lander and C. Keller, eds. (North Holland Publ. Co.) to appear.
[47] J.R. Naegele, J.C. Spirlet, L. Manes and W. Müller, Phys. Rev. Lett, to appear and ref. 46.
[48] Y. Baer and J. Schoenes, Solid State Comm. 33, 885 (1980).
[49] Y. Baer, H.R. Ott and K. Andres, Solid State Comm. 36, 387 (1980).
[50] H.R. Moser, B. Delley, W.D. Schneider and Y. Baer to be published.
[51] K.A. Gschneidner and R. Smoluchowski, J. Less-Common Met. 5, 374 (1963).
[52] B. Coqblin and A. Blandin, Adv. Phys. 17, 281 (1968).
[53] B. Johansson, Phil. Mag. 30, 469 (1974).
[54] S.H. Liu and K.-M. Ho, Phys. Rev. B26, 7052 (1982).
[55] M. Schlüter and C.M. Varma, Helvetica Physica Acta 56, 147 (1983).
[56] J.W. Allen and R.M. Martin, Phys. Rev. Lett. 49, 1106 (1982).
[57] M. Lavagna, C. Lacroix and M. Cyrot, J. Phys. F13, 1007 (1983).
[58] N. Mårtensson, B. Reihl and R.D. Parks, Solid State Comm. 41, 573 (1982).
[59] W. Wuilloud, H.R. Moser, W.-D. Schneider and Y. Baer, Phys. Rev. B, (1983) to appear.
[60] O. Gunnarsson and K. Schönhammer, this book and ref. therein.

EFFECTS OF ENVIRONMENT ON THE ELECTRONIC CONFIGURATION IN INTERMEDIATE VALENT RARE EARTH SYSTEMS

Thomas Penney

IBM Thomas J. Watson Research Center
P.O. Box 218
Yorktown Heights, New York 10598, U.S.A.

In the intermediate valence problem one must consider highly correlated localized f electrons in the presence of delocalized d and s electrons (denoted d for simplicity). Two configurations, $f^n d^m$ and $f^{n-1} d^{m+1}$, have comparable energies and must be considered. It is possible that a completely electronic theory, such as the Anderson/Kondo impurity/lattice models, may contain the essence of coupling these states and will describe the magnetic susceptibility and specific heat. However, these configurations have different sizes and bonding properties. The proper electronic configuration will depend and, conversely, be a major influence on the structure. In this paper we will discuss the relationship of valence to volume, homogeneous valence transitions, surface valence transitions, local valence transitions, and the existence of gaps in the electronic density of states.

Valence and Volume

The lattice constant is an extremely useful indicator of valence because the volume associated with the $4f^{n-1}5d^{m+1}$ configuration is significantly smaller than that of the $4f^n5d^m$. For $4f^{n-1}$ $5d^{m+1}$ there is one less 4f electron inside the 5s, 5p, 5d and 6s electrons screening the nuclear charge, z, and one more electron in the d band. The effect is exactly as Friedel[1] explained for the first half of the 3d, 4d or 5d transition metal series. There is a large decrease in volume each time z is increased and an electron is added to the bottom (bonding) half of the d band. Williams, Gelatt, and Janak[2] have shown, from first principles total energy calculations, that for transition metals the d electrons provide most of the cohesion while compression of the outer sp electrons provide the repulsion. Figure 1 shows the volume derivative of the total energy as a function of volume for molybdenum, divided up into d and sp contributions. At the equilibrium position, x=0, the 4d contraction is balanced by the 5sp repulsion. The approximate behavior of ions as hard spheres and the usefulness of ionic radii is based on the rapid increase in the kinetic energy of the outermost electrons with decreasing volume.

Pauling[3] was able to assign radii to the monovalent alkali metals and halogen ions which predict the lattice constants of the alkali halides with NaCℓ structure to about 1% accuracy. In a study of the valence of the NaCℓ type rare earth chalcogenides and pnictides, Iandelli,[4] following Templeton and Dauben,[5] developed a set of ionic radii for the rare earths. Agreement with the lattice constants is within a few percent in most cases and ≈5% in the worst cases. For reasons discussed above, metallic $Gd^{3+}S$ (f^7d) is much smaller than the semiconductor $Eu^{2+}S$ (f^7). A large ionic radius, r^{2+}, was assigned to Eu^{2+} and approximately 20% smaller values to the Gd^{3+} and Eu^{3+} trivalent radii, r^{3+}. Plotting lattice constants versus the r^{3+} provides a clear separation between divalent and trivalent rare earth valence.

In a more recent work, Iandelli and Palenzona[6] studied the systematics of lattice constant vs. ionic radius for a large number of rare earth intermetallic compounds. Surprisingly they found reasonable correlations using the ionic radii developed for the NaCℓ structure materials. Following Ref. 6, figure 2 shows the cell size for the entire rare earth cadmium (RCd) series, plotted against the r^{3+} (solid dots). In this case geometry works very well. The solid line through the points has a slope $2/\sqrt{3}$ appropriate for the CsCℓ structure (simple cubic for each constituent). Furthermore EuCd and YbCd with divalent rare earth valence are well off the line if plotted against their r^{3+} but right on the line if the r^{2+} are used. The RZn series shows the correct slope and divalent EuZn lies on the solid line if plotted with its divalent radius of 1.12Å (open diamond). In each case a dotted line is drawn at the measured cell size connecting the trivalent and divalent radii. The intersection of the solid line through the trivalent compounds with the dotted line gives the valence, by interpolation along the dotted line. In the case of YbZn, for example, the valence is about 2.3 which is a good estimate since geometry works for this series.

Also shown in Fig. 2 is the $RA\ell_2$ series with cubic $MgCu_2$ structure. The slope of 0.9 does not agree with the geometrical value $4/\sqrt{11} = 1.2$ for this structure. $EuA\ell_2$ does not lie on the line when plotted against r^{2+}. Its valence has been assumed to be 2+, but it has not been measured by L_{III} absorption. For this reason the appropriate r^{2+} for Yb in this series is not well defined. It could be taken from the usual r^{2+} value which would

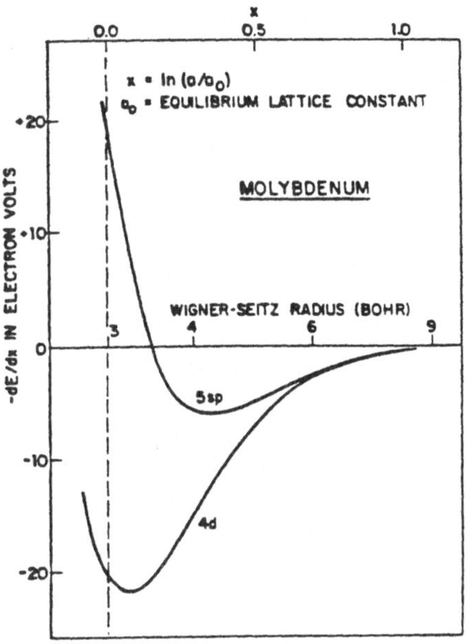

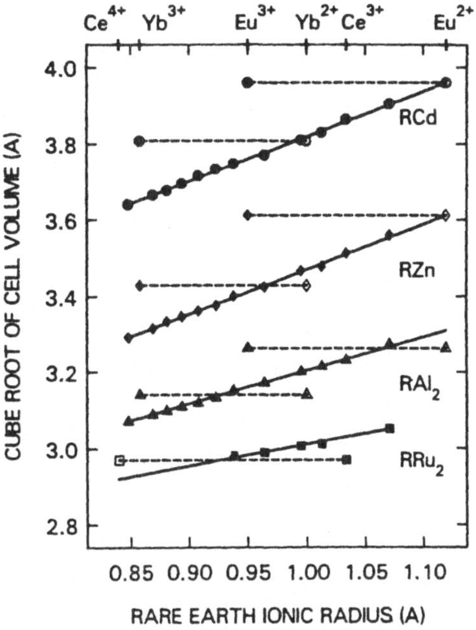

Fig. 1. Derivative of total energy with respect to lattice constant, a. From Ref. 2.

Fig. 2. Cube root of cell volume for various rare earth series v. trivalent rare earth radii (solid symbols) or vs. divalent radii for Yb and Er or tetravalent radius for Ce (open symbols). The RZn series is shifted down 0.1Å and the RAl_2 and RRu_2 are shifted down 0.8Å for clarity. Data from Ref. 6.

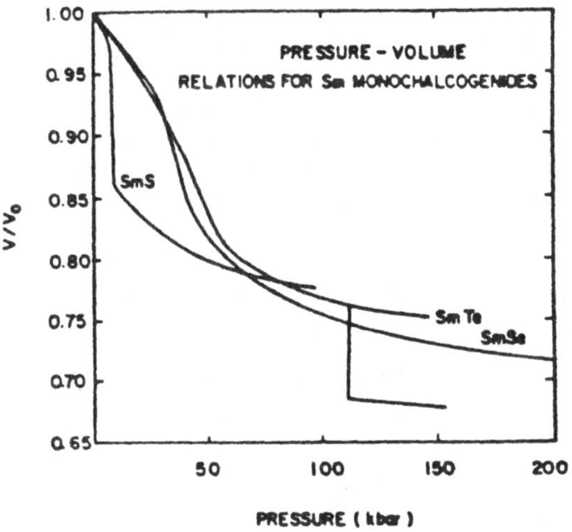

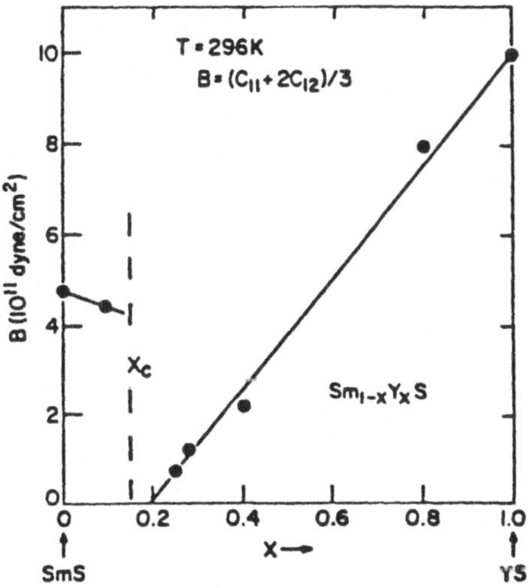

Fig. 3. Pressure volume relations for Sm chalcogenides from Ref. 9.

Fig. 4. Bulk modulus vs. composition for $Sm_{1-x}Y_xS$ from Refs. 17 and 18.

imply a Yb valence of ≈2.6. Alternatively it could be estimated by proportionally reducing the r^{2+} values of Yb and Eu to make $Eu^{2+}A\ell_2$ fall on the line, which would imply v ≈ 2.2. In this case lattice constant is a good indicator of intermediate valence for $YbA\ell_2$ but a poor estimator of the value. For comparison photo-emission[7] as well as L_{III} x-ray absorption[8] give v = 2.4.

The valence of Ce compounds is an unresolved question. On the basis of L_{III} edges, Wohlleben[8] has suggested that many nonmagnetic intermetallic compounds which were thought to be tetravalent have valences of about 3.2 or 3.3. He further suggests that the Ce^{4+} ionic radius should be about 0.84Å rather than 0.94Å. In the case of the RRu_2 (also $MgCu_2$ structure) use of the new radius (Fig. 2) for $CeRu_2$ gives a valence of 3.5 while the old gives 4.0; L_{III} edges gives 3.3.[8] For Ce compounds lattice constant is an indicator of whether a compound is in the nearly trivalent, γ, phase (e.g. $CeA\ell_2$) or the collapsed α phase (e.g. $CeRu_2$). The nature of the nonmagnetic α phase is still a subject of debate.

Homogeneous Valence Transitions and Soft Bulk Modulus

Just as the electronic configuration shows itself in the volume, so can pressure reduce the volume and produce continuous or discontinuous valence transitions with no change in crystal symmetry. Jayaraman[9] and co-workers have measured the pressure-volume behavior of the divalent rare earth chalcogenides over several hundred kilobars and found anomalously large but continuous volume decreases in most of them, which can be understood only as continuous valence changes. SmS has a discontinuous transition[9] to an intermediate valence (v ≈ 2.7) phase[10] at 6.5 kbar. SmS is semiconducting at ambient pressure with the d band slightly above the f^6 level and the Fermi level in the gap. With pressure the d-band drops,[11] f^6 to f^5d transitions take place, and the discontinuous phase transition occurs. The transition stops at an intermediate valence because the smaller more compressed phase has a larger bulk modulus so the elastic energy increases nonlinearly during the collapse.[12,13,14]

The $Sm_{1-x}Y_xS$ solid solution system has a discontinuous valence transition at x = 0.15, analogous to the pressure transition[15,16] $Y^{3+}S$ is smaller than SmS and has a d band with one electron. With increasing YS concentration, the common d band descends and the Fermi level drops into the f^6 states as before. In the intermediate valence phase, the Fermi level cuts both the f^6 level and the d band. It is possible to make f^6 to f^5d transitions with no energy cost in the electronic system. The result is that for the sample (Fig. 4) nearest the phase transition in the intermediate valence phase, the bulk modulus falls to a value about 1/10 of the value one would expect for the average of "$Sm^{2+}S$" and $Y^{3+}S$.[17,18]

It is the bulk modulus and not a phonon which tends to zero because there is no symmetry change. A simple theory[14] for this effect showed that the bulk modulus was reduced from the average value of the constituents by two effects. One is due to the redistribution of electrons under strain, either static or ultrasonic. This is exactly what happens in crystal field systems.[19,20] The second effect is due to the fact that one electronic configuration is smaller so that as the crystal is strained the reference volume changes to reduce the energy.

Pressure volume measurements[21] of SmS show that its bulk modulus in the collapsed intermediate valence phase is about half that of YS. SmS at pressures smaller than that of the transition is also soft,[17,21] (B = 475 kbar) compared with divalent EuS (B = 600 kbar).[22] L_{III} edge measurements[23] show that the valence is 2.1 which implies hybridization across the f^6-f^5d energy gap. Presumably the amount of mixing changes under pressure leading to slight softening.

$Ce_{1-x}Th_x$ goes through the γ-α transition with increasing pressure or decreasing temperature. Detailed measurements[24] of resistance and length vs. temperature showed a systematic change from continuous to discontinuous transitions as x was decreased. Although the bulk modulus was not measured the length measurements show that B must go to 0 at the critical point. Stoichiometric intermediate valent TmSe is soft but off-stoichiometric integral valent $Tm_{0.87}{}^{3+}Se$ is not.[25,26] Pressure volume measurements[27,28] of the $TmSe_{1-x}Te_x$ system show behavior similar to the Ce-Th systems or a liquid gas system. In each case as the critical concentration is approached the bulk modulus tends to zero.

The volume-valence instability leads to modification of the phonon dispersion of collapsed $Sm_{0.75}S$, Fig. 6.[29] The unusual effects are the reduction of the longitudinal optic (LO) phonons below the transverse optic (TO) modes and the reduction of the longitudinal acoustic (LA) below the TA for most of the $(\zeta\zeta\zeta)$ direction. The $(\zeta\zeta\zeta)$ LA branch of SmS shows a similar anomaly after the pressure collapse, although the magnitude is smaller.[30]

Various theoretical fits to the $Sm_{0.75}Y_{0.25}S$ data have been made.[31,32,33] All fit the unusual features and all are based on the ability of Sm to change its electronic configuration and therefore its charge and volume.

Phonon dispersion measurements on $CeSn_3$, which showed no anomalies compared to $LaSn_3$, called into question the generality of associating soft bulk moduli or phonon anomalies with intermediate valence.[34] The answer seems to be that in $CeSn_3$ the valence is close to 3 and is difficult to change with strain, so no softenings result. Another way to put it is that $CeSn_3$ is far from a valence or configurational transition.

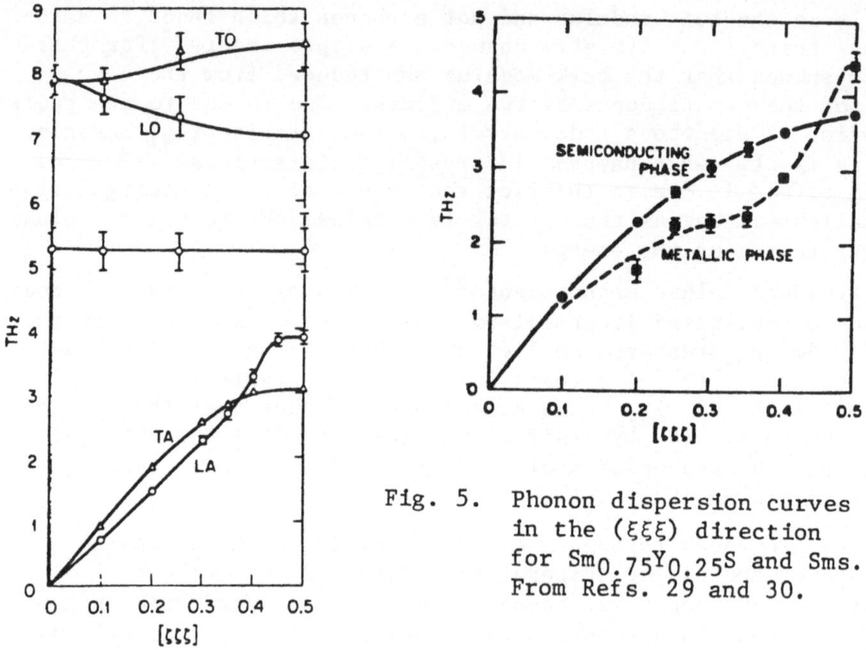

Fig. 5. Phonon dispersion curves in the (ξξξ) direction for $Sm_{0.75}Y_{0.25}S$ and Sms. From Refs. 29 and 30.

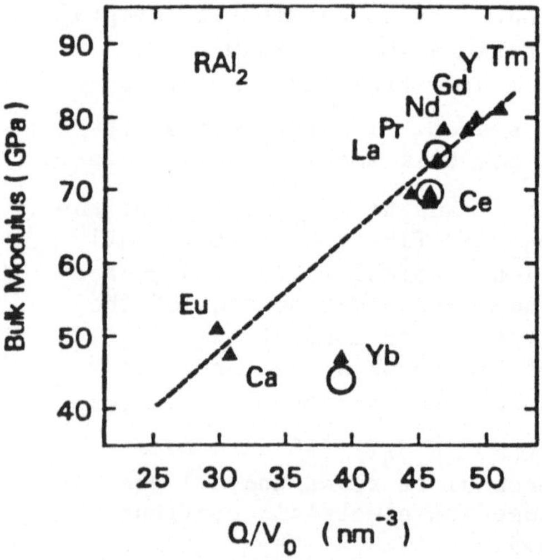

Fig. 6. Bulk modulus vs valence/ volume ratio for RAl_2 compounds. Triangles ultrasonic, circles x-ray. From Ref. 35 and Refs. therein.

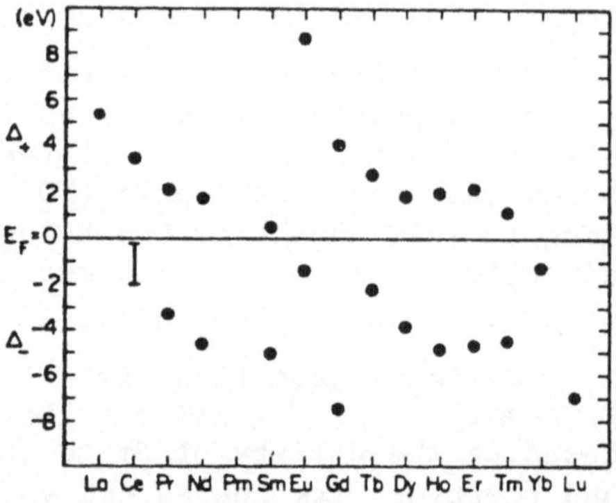

Fig. 7. Minimum energy required to increase (Δ_+) and decrease (Δ_-) by one electron, the population of the 4f shell in the rare earth metals. From Ref. 40.

The picture is supported by pressure volume and ultrasonic sound velocity measurements of the $RA\ell_2$ system.[35] As discussed above, $YbA\ell_2$ clearly has an intermediate valence with approximate value 2.4. The bulk modulus is soft compared to the average (dotted line in Fig. 7) one would expect for a mixture of divalent and trivalent compounds with the same valence-volume (Q/V_0) ratio. By comparison $CeA\ell_2$ which has a valence close to 3 and a transition (γ-α) at about 80kbar,[36] shows very slight softening.[35] Ultrasonic measurements on polycrystalline Ce[37] and Ce-Th[38] show that for a large region around the γ-α transition longitudinal (compressional) sound waves are soft and transverse (shear) are not. Therefore, even though the exact nature of the γ-α transition is still unresolved, since it involves a volume change, the bulk modulus will be soft at the transition.

Surface Valence Transitions

Just as pressure may delocalize an f electron causing a volume decrease, so can a free surface localize an f electron. $YbAu_2$ has a trivalent lattice constant but a large divalent photoemission was observed by Wertheim, Wernick and Crecelius.[39] They suggested that generally the surface valence might be different from the bulk, if the 4f level was close to the Fermi level. Using photoemission, Baer and Lang[40] have measured, for the rare earth metals, the energy Δ_- to remove an electron from the bulk 4f level and put it at the Fermi level. Using inverse photoemission (BIS), they measured the energy Δ_+ to take an electron from the Fermi level and localize it in an f-state (Fig. 7). The sum of Δ_- and Δ_+ is the Coulomb correlation energy, U. For Sm metal, Δ_+ is very small but Δ_- is large. Sm metal is therefore trivalent but very close to being divalent. Wertheim and Crecelius[41] proposed that on the surface, becaue of lower coordination, the d electron band width would be reduced. This would raise the Fermi level causing d to f transitions if the f level were close to E_F (small Δ_+). The surface layer would then expand causing more d band narrowing, accelerating the process. The result is a divalent or intermediate valent surface. This process is just the opposite of the pressure induced volume phase transition observed in SmS. It is an interesting question whether once the surface transition starts it will go all the way to divalent or whether it could stop at an intermediate valence. In the bulk, anharmonic lattice effects and/or hybridization stabilize intermediate valence. It has been argued that the surface of Sm metal is divalent[42] or that the surfaces of Sm metal and SmB_6 are mixed integral valent.[43] Ultraviolet photoemission (70 eV) on $Sm_{1-x}Y_xS$ shows only a divalent signal in both the black and gold (intermediate valent) phases.[44] These results show that the surface, up to at least the 6Å escape depth of the photoelectrons, is completely divalent.

In the bulk, TmTe is divalent, TmSe is intermediate valent or trivalent depending on composition, while TmS is trivalent. On the surface, however, TmS is divalent.[45] $YbA\ell_2$ has a bulk valence of about 2.4. The photoemission[7] valence spectrum shows four peaks (Fig. 8). The sharp peak at E_F is the $f^{14} \rightarrow f^{13}(^2F_{7/2})$ bulk transition. The second sharp peak is a replica of the first, split off by the J = 7/2, J = 5/2 spin orbit splitting of the final state. The two broad peaks are the surface contribution and its replica. Since this surface peak is completely below E_F and 0.9 eV below the bulk peak, the 2+ state is stable on the surface by that amount. The surface is not intermediate valent. In general, f states at a surface have lower energy with respect to the Fermi energy than in the bulk.

Local Valence Changes

Sherrington and von Molnar[46] considered the effect of local lattice distortions in homogeneous intermediate valence materials. They noted that the lattice should expand around a large divalent ion and contract around a small trivalent one if they occurred in their integral valence states. If the large ion was found in the small lattice site, or vice versa, there would be an elastic energy cost, E, relative to the relaxed case. In the Anderson Hamiltonian the conduction band-f electron mixing term, V_{kf}, leads to an f band width, Δ. This term favors an intermediate valence state, if Δ is larger than the energy difference between the electronic configurations. When one includes the lattice then there is a competition between Δ and the elastic energy E. For the case of Δ small compared to E, they predicted that there would be large local distortions. The f band width, Δ, would be renormalized to a smaller value because the lattice distortion inhibits the valence change. If Δ is large, on the other hand, then the tunneling between valence states will be fast and the lattice will adopt an intermediate value.

Friedel[47] has made a similar argument contrasting the width Δ of a 4f virtual bound state with the Debye energy, θ. For $\Delta \gg \theta$ one has a homogeneous situation while for $\Delta \ll \theta$ one expects each ion will have a volume depending on its valence. He suggests that in the latter case one would expect ordered rather than random valence/volume arrangements.

Kohn, Lee and Lin-Liu[48] considered the problem for an SmS cluster including the degeneracy, N, of the magnetic configurations. They calculated as a function of Δ, the difference, (δR), between the average Sm-S distances associated with the integral valent configurations. They found that with increasing Δ the difference, (δR), tended to zero. This result and the previous ones indicate that for large Δ (large electronic d-f mixing) an undistorted lttice with an average lattice constant should be expected.

Extended X-ray absorption fine structure, EXAFS, is a good way to look at local structure[49]. In the EXAFS process, an X-ray photon, $\hbar\omega$, is absorbed and a photo electron is emitted whose wave vector $\mathbf{k}$ is given by $\hbar^2k^2/2m = \hbar\omega - E_0$ where E_0 is the threshold energy for X-ray absorption. The photoelectron is reflected back from neighbouring atoms, j, at a distance r_j with a phase shift ψ_j. Since this reflected wave can be in or out of phase with the outgoing wave, the x-ray absorption is modulated by a term $\sin[2kr_j+\psi_j(k)]$ and the distance r_j can be determined.

EXAFS studies of intermediate valence TmSe[50], SmS (under pressure)[51] and $Sm_{0.75}Y_{0.25}S$[52] found only one average rare earth nearest neighbor distance in agreement with a picture of Δ larger than local elastic deformation energies, E. In the last case EXAFS was able to measure a Y-S distance of 2.77Å, different from the single Sm-S distance of 2.84. These distances can be compared with the x-ray diffraction lattice constant values ($a_0/2$) of 2.73, 2.83, 2.98 and 2.80 for YS, $Sm_{0.75}Y_{0.25}S$, SmS and fictitious "Sm^{3+}S." The single Sm-S distance is intermediate between Sm^{2+}S and Sm^{3+}S, indicating homogeneous intermediate valence. The Y-S distance is slightly over half way from the $Sm_{0.75}Y_{0.25}S$ to the YS $a_0/2$ values. Presumably this represents a compromise between the Y-S strain and the strain of the S's and their neighbors.

A similar effect is observed in Br EXAFS of 5% and 10% Br substituted in KCℓ[53]. The Br-K distance is found to be about half way between the average lattice constant and the KBr lattice constant by x-ray diffraction. In the case of $Ga_{1-x}In_xAs$, there is a bimodal distribution with the Ga-As and In-As distances from EXAFS increasing by only 1.4% while the lattice constant increases by 7% from GaAs to InAs[54]. From these studies we conclude that local distortions can be measured but that in the intermediate valence materials TmSe, SmS under pressure, and $Sm_{0.75}Y_{0.25}S$, there is only one species of rare earth ion which appears at a site of average size.

The $SmS_{1-x}As_x$ system[55] (Fig. 9) collapses from an inhomogeneous black phase to a metallic (presumably homogeneous) intermediate valence phase for As concentrations $9 < x < 25\%$. Kasuya[56] suggested that the rapid lattice constant decrease in the low As concentration black phase was due to the six nearest Sm neighbors of each As collapsing from divalent to trivalent. Preliminary As EXAFS measurements[57] show that for the 12% As sample there is only one As-Sm nearest neighbor distance and that twice this distance agrees with the lattice constant from x-ray diffraction (Fig. 9). In the black phase a 4% As sample showed only one As-Sm nearest neighbor distance but the value was much smaller than the lattice constant value, about 40% of the difference in lattice constants between the black and metallic samples. This result supports the model of an inhomogeneous local collapse of the Sm ions around each As. One might expect the same value for the As-Sm distance in the two samples since the Sm neighbors have collapsed in both bases. However as we have seen in the case of $KC\ell_{1-x}Br_x$ and $Sm_{1-x}Y_xS$ the K-Br and Y-S distances were not as small as the KBr and YS lattice constants. Just as in these cases, we suppose that local strain in the 4% As sample prevents the nearest neighbors from assuming the positions they would in a more homogeneous system.

Gaps

Intermediate valent SmB_6 has a gap of about 3-4 meV in the density of states with the Fermi level in the gap. Resistivity[58,59] optical conductivity[59], far infrared absorption[58], oxide barrier tunneling[59], GaAs Schottky barrier tunneling[60], metal point contact spectroscopy[61], and specific heat[58] are all consistent with such a gap.

Intermediate valent, stoichiometric TmSe in its antiferromagnetic phase has a gap of about 2 meV as measured by GaAs Schottky barrier tunneling[60] and metal point contact spectroscopy (MPCS)[61]. The resistivity[62] rises dramatically with decreasing temperature consistent with a gap, but does not follow a simple exponential as one would expect for a temperature independent gap.

The case for SmS favors a gap but the evidence is less clear cut. The Schottky barrier tunneling[60] shows a resistance peak at zero voltage about 50% above background and a width of about 2 meV. In SmB_6 and TmSe the peak was 5 and 3 times background respectively. The metal point contact spectroscopy[61] for SmS gives a width of about 5 meV. The difficulty may arise because of the pressure sensitivity of SmS which might smear the gap. At zero pressure bulk SmS is black. With increasing pressure it transforms to the gold intermediate valence phase at 6.5 kbar. On reduction of the pressure it transforms back to the black phase at about 1 kbar. If the surface is rubbed or scratched it turns gold and stays gold. The GaAs Schottky barrier tunneling was done on a polished surface, $2\Delta_{FWHM} = 2.1$ meV, and on a black surface transformed to gold by the pressure of the GaAs tip, $2\Delta_{FWHM} = 1.7$ meV. The SmS used was slightly Sm rich so that the bulk of the material in the black phase was not insulating at low temperature. The MPCS was done on a polished gold surface only.

GaAs Schottky barrier tunneling is a proven technique in that against Pb it has correctly measured the gap, the strong coupling Pb phonons above the gap and a GaAs phonon[63]. However zero bias anomalies are a problem for tunneling in general and Schottky barriers in particular[64,65]. For a Si Schottky barrier against SmS, we have observed a peak in the tunneling resistance about 3 times background and $2\Delta_{FWHM}$ 1.7 meV. A field of about 5kOe removes the peak. These results are similar to the previous GaAs/SmS results. We also find that for one Si Schottky barrier against another piece of the same Si material there is a peak in the tunneling resistance 2 times background with width $2\Delta_{FWHM} = 1.5$ meV and which disappears in about 5kOe. The Si is 1.6×10^{19} p-type etched with HF in atmosphere and then submerged in liquid helium where the junction is made by bringing the two pieces together.

We conclude that well-formed gaps such as those in SmB_6 and TmSe are well measured by a variety of techniques including tunneling. In SmS where the gap may be smeared, the existance of a gap of order a

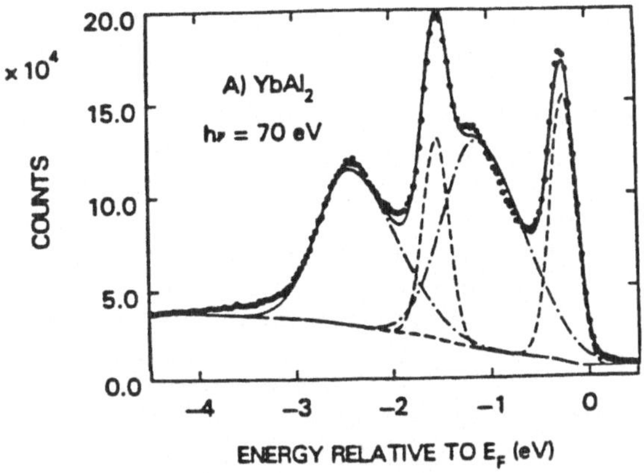

Fig. 8. Photoemission spectrum of $YbAl_2$ from Ref. 7.

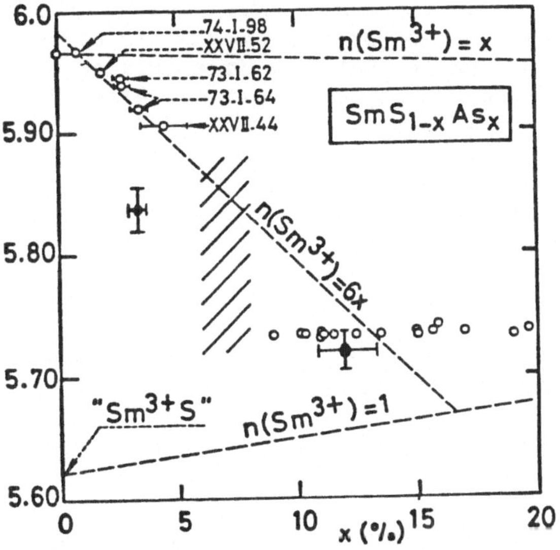

Fig. 9. Lattice constant (open circles) vs. composition for $Sm_{1-x}As_xS$, from Ref. 55. EXAFS measurements (dots) of twice the As-S distance, from Ref. 57.

few meV is likely but the value is somewhat uncertain.

Conclusion

In systems containing Ce, Sm, Eu, Tm and Yb, the electronic configuration of the rare earth is closely coupled with its environment. Composition, pressure, surfaces, impurities and phonons change the electronic system. For Ce systems there is a magnetic γ-like trivalent, f^1, phase and/or a nonmagnetic α-like phase with part of the f electron less well localized in some way, but not promoted to a d state. For the other rare earths mentioned, the two electronic configurations $f^n d^m$ and $f^{n-1}d^{m+1}$ are related by a promotion of an f electron to a d state. These transitions, as a function of pressure or composition, can be continuous or discontinuous, near or far from ambient conditions. The sensitivity of the rare earth configurations to their environment and the effects on phonons will depend on how close a particular system is to a valence transition at ambient. Since TmTe is divalent and TmS is trivalent, intermediate ($v \approx 2.6$) TmSe can be thought of as being in the middle of a phase transition between the other two. Similarly, $Sm_{0.75}Y_{0.25}S$, $YbA\ell_2$ and α-Ce are caught in a phase transition. In all of these materials the electronic configurations can change with pressure so the bulk modulus is soft and the valence depends strongly on environment. The divalent state of Tm, Sm and Yb is stable on the surfaces of these materials. On the other hand, nearly integral valent materials such as $Tm_{0.87}Se$, $CeA\ell_2$ and $CeSn_3$ have electronic configurations which depend much less sensitively on pressure and so are not drastically soft. A gross environmental change such as a free surface can drive a valence change as shown by the divalent surface of trivalent $YbAu_2$ and TmS. If configurational mixing were weaker than found in homogeneous intermediate valence systems, polaronic distortions around nearly integral valent ions might occur. It is apparent that for a full treatment of intermediate valence the local environment as well as the bare electronic system must be considered.

Acknowledgement

We thank G.S. Cargill III, G. Guntherodt S. von Molnar, R.A. Pollak, W.A. Thompson and D. Wohlleben for valuable discussions.

References

1. J. Friedel in the Phys of Metals 1, Electrons, J.M. Ziman ed., Cambridge University Press, 1969, p. 340.
2. A.R. Williams, C.D. Gelatt Jr., J.F. Janak, in Theory of Alloy Phase Formation, L.H. Bennett ed., Proc. 108th AIME Annual Meeting New Orleans, 1979.
3. L. Pauling, The Nature of the Chemical Bond, Cornell, 1960.
4. A. Iandelli in Rare Earth Research, E.V. Kleber ed., MacMillan, New York (1961) p. 135.
5. D.H. Templeton and C.H. Dauben, J. Am. Chem. Soc. 76, 5237 (1954).
6. A Iandelli and A. Palenzona in Handbook of the Physics and Chemistry of Rare Earths, K.A. Gschneidner Jr., and L. Eyring, North Holland (1979) Vol. 2, p.1.

7. G. Kaindl, B. Reihl, D.E. Eastman, R.A. Pollak, N.Martensson, B. Barbara, T. Penney and T.S. Plaskett, Sol. State Comm. 41, 157 (1982).
8. K.R. Bauchspiess, W. Boksch, E. Holland-Moritz, H. Launois, R. Pott and D. Wohlleben, Valence Fluctuations in Solids, L.M. Falicov, W. Hanke, M.B. Maple eds., North-Holland (1981) p. 417 and private communication.
9. A. Jayaraman, Handbook on the Physics and Chemistry of Rare Earths, op cit, Vol. 2, p. 575 and A. Jayaraman, A.K. Singh, A. Chatterjee and S. Usha Devi, Phys. Rev. B9, 2513 (1974).
10. M.B. Maple and D.K. Wohlleben, Phys. Rev. Lett. 27, 511 (1971).
11. B. Batlogg, E. Kaldis, A. Schlegel, P.Wachter, Phys. Rev.14B, 5503(1976).
12. L.L. Hirst, J. Phys. Chem. Solids 35, 1285 (1974).
13. C.M. Varma and V. Heine, Phys. Rev. B11, 4763 (1975).
14. T. Penney and R.L. Melcher, J. Physique 37, Colloq C4-275 (1976).
15. L.J. Tao and F. Holtzberg, Phys. Rev. B11, 3842 (1975).
16. T. Penney and F. Holtzberg, Phys. Rev. Lett. 34, 322 (1975).
17. R.L. Melcher, G. Guntherodt, T. Penney and F. Holtzberg, 1975 Ultrasonics Symposium Proceedings, IEEE Cat. #75 CHO 994-4811, p. 16.
18. T. Penney, R.L. Melcher, F. Holtzberg and G. Gentherodt, AIP Conf. Proc. 29, 392 (1975).
19. B. luthi in Dynamical Properties of Solids, G.K. Horton and A.A. Maradudin eds., North Holland (1980), Vol. 3, p. 245. B. Luthi, AIP Conf. Proc. 34, 7 (1976).
20. See chapter by P. Fulde.
21. R. Keller, G. Guntherodt, W.B. Holzapfel, M. Dietrich and F. Holtzberg, Sol. State Comm. 29, 753 (1979).
22. A. Jayaraman, B. Batlogg, R.G. Maines, H. Bach, Phys. Rev. B26, 3347 (1982).
23. J. Roehler, G Krill, J.P. Kappler, M.F. Ravet and D. Wohlleben in Valence Instabilities, P. Wachter and H. Boppart eds., North Holland (1982), p. 215.
24. J.M. Lawrence, M.C. Croft and R.D. Parks, Phys. Rev. Lett. 35, 289 (1975).
25. B. Batloggs, E. Kaldis, H.R. Ott, Physics Letts 62A, 220 (1977).
26. H. Boppart, A. Treindl, P. Wachter in Valence Instabilities (1982), op cit. p. 103.
27. H. Boppart, W. Rehwald, E. Kaldis and P. Wachter, Valence Instabilities (1982), op cit, p. 81.
28. H. Boppart, Thesis; see also chapter by P. Wachter and H. Boppart.
29. H.A. Mook, R.M. Nicklow, T. Penney, F. Holtzberg and M.W. Shafer, Phys. Rev. B18, 2975 (1978). H.A. Mook and R.M. Nicklow, Phys. Rev. B20, 1656 (1979).
30. H.A. Mook, D.B. McWhan, F. Holtzberg, Phys. Rev. B25, 4321 (1982).
31 N. Wakabayashi, Phys. Rev. B22, 5833 (1980).
32. H. Bilz, G. Guntherodt, W. Klepmann and W. Kress, Phys. Rev. Lett. 43, 1998 (1979).
33. P. Entel, N. Grewe, M. Sietz, K. Kawalski, Phys. Rev. Lett. 43, 2002 (1979).
34. L. Pintschovius, E. Holland-Moritz, D. Wohlleben, S. Stahr, J. Liebertz, Sol. State Comm. 34, 953 (1980).
35. T. Penney, B. Barbara, T.S. Plaskett, H.E. King, Jr. and S.J. LaPlaca, Sol. State Comm. 44, 1199 (1982).

36. M.C. Croft and A. Jayaraman, Sol. State Comm. 29, 9 (1979).
37. F.F. Voronov, V.A. Goncharova and O.V. Stal'garova, Sov. Phys. JETP 49, 687 (1979).
38. H. Wehr, K. Knorr and R. Feile, Sol. State Comm. 40, 507 (1981).
39. G.K. Wertheim, J.H. Wernick and G. Crecelius, Phys. Rev. B18, 875 (1978).
40. Y. Baer and J.K. Lang, J. Appl. Phys. 50, 7485 (1979).
41. G.K. Wertheim and G. Crecelius, Phys. Rev. Lett. 40, 813 (1978).
42. W. Gudat, M. Campagna, R. Rosei, J.H. Weaver, W. Eberhardt, F. Hulliger and E. Kaldis, J. Appl. Phys. 52, 2123 (1981).
43. J.W. Allen, L.I. Johansson, I. Lindau and S.B. Hagstrom, Phys. Rev. B21, 1335 (1980).
44. B. Reihl, G. Kaindl, F. Holtzberg, R.A. Pollak, G. Hollinger, N. Martensson, in Valence Instabilities (1982), op cit, p. 287.
45. N. Martensson, B. Reihl, R. A. Pollak, F. Holtzberg and G. Kaindl, Phys. Rev. B25, 6522 (1982).
46. D. Sherrington and S. von Molnar, Sol. State Comm. 16, 1347 (1975).
47. J. Friedel, J. Mag. Mag. Mat. 17-19, XXXVIII (1980).
48. W. Kohn, T.K. Lee and Y.R. Lin-Liu, Phys. Rev. B25, 3557 (1982).
49. P.A. Lee, P.H. Citrin, P. Eisenberger and B.M. Kincaid, Rev. Mod. Phys. 53, 769 (1981).
50. H. Launois, M. Rawiso, E. Holland-Moritz, R. Pott and D. Wohlleben. Phys. Rev. Lett. 44, 1271 (1980).
51. K.N. Franf, G. Kaindl, J. Feldhaus, G. Wortman, W. Krone, G. Materlik and H. Bach in Valence Instabilities (1982), op cit, p. 189.
52. J.B. Boyce, R.M. Martin and J.W. Allen in Valence Fluctuations in Solids (1981), op cit, p. 427.
53. T. Murata, P. Lagarde, A. Fontaine, M. Raoux, EXAFS and Near Edge Structure, A. Bianconi, L. Incoccia and S. Striprich eds., Springer (1983) p. 271.
54. J.C. Mikkelsen, Jr. and J.B. Boyce, Phys. Rev. Lett. 49, 1412 (1982).
55. F. Holtzberg. O. Pena, T. Penney and R. Tournier, in Valence Instabilities and Related Narrow Band Phenomena, R.D. Parks ed., Plenum (1977). p. 507.
56. T. Kasuya, J. Physique 37, Colloq. C4-261 (1976).
57. G.S. Cargill, III, F. Holtzberg and T. Penney, to be published.
58. S. von Molnar, T. theis, A. Benoit, A. Briggs, J. Flouquet, J. Ravex and Z. Fisk, in Valence Instabilities (1982), op cit, 0. 389, and refs. therein.
59. B. Batlogg, P.H. Schmidt and J.M. Rowell in Valence Fluctuations in Solids (1981), op cit, p. 267.
60. G. Guntherodt, W.A. thompson, F. Holtzberg and Z. Fisk, Phys. Rev. Lett. 49, 1030 (1982), and Valence Instabilities (1982), op cit, p. 313.
61. I. Frankowski and P. Wachter, Valence Instabilities (1982), op cit, p. 309.
62. P. Haen, F. LaPierre, J.M. Mignot and R. Tournier, Phys. Rev. Lett. 43, 304 (1979).
63. W.A. thompson and S. von Molnar, J. Appl. Phys. 41, 5218 (1970).
64. E.L. Wolf in Inelastic Electron Tunneling Spectroscopy, T. Wolfram ed., Springer (1978), p. 220.
65. E.L. Wolf, D.L. Losee, D.E. Cullen and W.D. Compton, Phys. Rev. Lett. 26, 438 (1971).

LOCALIZED MOMENTS IN METALS: CRYSTAL FIELD EXCITATIONS AND THEIR COUPLING TO ELECTRONS AND PHONONS

Peter Fulde

Max-Planck-Institut für Festkörperforschung
7000 Stuttgart 80
Fed. Republic of Germany

I. INTRODUCTION

In these lectures we want to consider metallic systems with well localized magnetic moments. In particular we are interested in effects which are caused by the crystalline electric field (CEF) prevailing at the locations of the moments. Such a field is set up by the surroundings of the ion containing the local moment and it acts on the orbital parts of the wave function describing the moment. When it is larger than the spin-orbit coupling it leads to a quenching of the orbital moment (transition metal ions), but when the spin-orbit coupling is larger (rare earth ions) the CEF leads to lifting of the ground state degeneracy (Stark splitting). It is the coupling of these CEF split energy levels with conduction electrons and phonons which leads to a large number of interesting effects.

From the above it is clear that we shall concentrate on rare earth (RE) systems. RE ions are characterized by an incomplete 4f shell within which the electrons couple according to the Russell-Saunders case. It applies when the Coulomb interactions among the electrons are larger than the spin-orbit energy. In that case the individual spins $\underline{s}_i$ and angular momenta $\underline{\ell}_i$ are added to $\underline{S} = \sum_i \underline{s}_i$ and $\underline{L} = \sum_i \underline{\ell}_i$, respectively. The total angular momentum $\underline{J} = \underline{L} + \underline{S}$ characterizes the incomplete 4f shell. The addition of the $\underline{s}_i$ and of the $\underline{\ell}_i$ is done according to Hund's rules. For electrons occupying the $2(2\ell+1)$ states of a shell these rules state:

(1) add the $\underline{s}_i$ so as to maximize $|\underline{S}|$ and
(2) add the $\underline{\ell}_i$ as to maximize $|\underline{L}|$ under the constraints that rule (1) and the Pauli principle are obeyed.

The addition of $\underline{S}$ and $\underline{L}$ is determined by the spin-orbit coupling which implies that $\underline{S}$ and $\underline{L}$ are antiparallel for a shell less than half-filled and parallel for a shell more than half-filled. The Hund's rules ensure that the Coulomb repulsion between electrons in a shell is minimized since configurations are avoided in which it would be particularly large, i.e., when electrons are in the same orbital state. All further considerations will be restricted to the lowest J multiplet of the 4f shell since energy separations between different multiplets are typically of the order of 10^3K. For example for Pr^{3+} with two 4f electrons the ground state multiplet is $^3H_4 (S=1, L=5, J=4)$.

II. DESCRIPTION OF THE CEF

Consider a RE ion in a solid. The 4f-electrons experience an electrostatic potential $V(\underline{r})$ due to the surrounding charge distribution. If there is no overlap between this charge distribution and the 4f wave functions the potential fulfills Laplace's equation $\Delta V(\underline{r}) = 0$. Then $V(\underline{r})$ can be expanded in terms of multipoles

$$V(\underline{r}) = \sum_{\ell,m} \gamma_{\ell m} r^{\ell} Y_{\ell m}(\Theta, \phi). \tag{1}$$

The summation is limited to $\ell \leq 6$ since higher multipoles can not cause transitions between $\ell_{4f}=3$ electronic states. The CEF Hamiltonian is written as

$$H_{CEF} = e \sum_i V(\underline{r}_i) \tag{2}$$

where i runs over all 4f-electrons. As discussed above for RE ions the CEF energy is much smaller than the spin-orbit energy. This results in important simplifications. One can show that the sum of polynomials in x_i, y_i, z_i appearing in Eq. (2) can be replaced by a sum of polynomials of J_x, J_y, J_z operators having the same transformation properties as the original expression. They act on the 4f shell as a whole and not on a particular 4f electron as do the $x_i (y_i, z_i)$ polynomials. The "Stevens operator equivalents" allow for expressing H_{CEF} in terms of the J_x, J_y, J_z as

$$H_{CEF} = \sum_{\ell,m} B_\ell^m O_\ell^m . \qquad (3)$$

The B^m_ℓ are the CEF parameters usually determined by experiments and the O_ℓ^m are polynomials in J_α like, e.g., $O_2^2 = J_x^2 - J_y^2$, $O_4^4 = [J_+^4 + J_-^4]/2$, $O_6^6 = [J_+^6 + J_-^6]/6$ etc. In particular for a CEF with cubic point symmetry one has

$$H_{CEF} = B_4^0 [O_4^0 + 5O_4^4] + B_6^0 [O_6^0 - 21O_6^4] \qquad (4)$$

with the two CEF parameters B_4^0 and B_6^0. For a given point symmetry group theory provides us with the representations and degeneracies of the different CEF levels. For example in a cubic CEF the ground multiplet of Pr^{3+} with J = 4 splits into a Γ_1(singlet), Γ_3(doublet), Γ_4(triplet), Γ_5(triplet). Energy splittings are typically within the range of 0.1-10meV.

We list in the following a few experimental methods which allow for a measurement of the CEF energy levels.

(a) Inelastic neutron scattering: the magnetic scattering-cross section measures $\mathrm{Im}\chi(\underline{q},\omega)$, where $\chi(\underline{q},\omega)$ is the dynamical magnetic susceptibility. Magnetic dipolar transitions between levels show up as peaks.
(b) Specific heat measurements: the CEF excitation energies give rise to Schottky anomalies which can be measured.
(c) Magnetic susceptibility measurements: especially CEF singlet ground state systems can be often identified by a constant, instead of a Curie like behavior at low temperatures T.
(d) Other methods include paramagnetic resonance, Mössbauer effect, superconducting tunneling etc.

In the following we will assume a given RE system with known CEF energy levels. The RE ions can be impurities in a metallic matrix or they can form a lattice. We will specify from case to case which system we are considering. The CEF levels will interact with conduction electrons and phonons and we shall study a number of effects resulting from those interactions.

III. INTERACTIONS WITH CONDUCTION ELECTRONS

When a conduction electron is scattered by the 4f shell of a RE ion it changes its quantum numbers from $|k\ell m\sigma\rangle$ into $|k'\ell'm'\sigma'\rangle$. There is a total of 14 different interactions which can take place between the conduction electron and the 4f shell. This is so since ℓ and ℓ' can combine to $\Lambda = 0, \ldots, 2\ell_{4f} = 6$ and the spins can combine to $\Sigma = 0,1$ and there is a different interaction for each combination (Λ,Σ). Prominent examples are the spherical Coulomb scattering $(\Lambda=0,\Sigma=0)$, the isotropic exchange interaction $-2J_{ex}\underline{s}\cdot\underline{S}_{4f}(\Lambda=0,\Sigma=1)$, the aspherical Coulomb charge scattering $A^{\ell}_{Q}O_2^{m}(\underline{J})O_2^{m}(\underline{\ell})$, $(\Lambda=2,\Sigma=0)$ with $O_2^{m}(\underline{J})$ and $O_2^{m}(\underline{\ell})$ denoting quadrupolar operators acting on the RE ion and on the conduction electron, respectively, and the orbital exchange interaction $B_s\underline{\ell}\cdot\underline{L}$ $(\Lambda=1,\Sigma=0)$ giving rise to skew scattering. A detailed investigation of the various interactions has been given by Hirst. Without doubt the isotropic exchange interaction is the dominant interaction in most cases (excluding the trivial spherical Coulomb scattering) but also the other interactions can become important and responsible for particular effects.

1. Effective mass in RE metals

The virtual excitation of CEF levels by conduction electrons can lead to an effective mass m* in a similar way as phonons or paramagnons. Recent low temperature experiments by Forgan on Pr have shown that $m^*/m = 4$ due to this effect. It is

$$\frac{m^*}{m} = 1 - \frac{\partial \Sigma(p_F,\omega)}{\partial\omega}\Bigg|_{\omega=0} \qquad (5)$$

where $\Sigma(\underline{p},\omega)$ is the electron self-energy. To lowest order it is calculated from the diagram shown in Fig. 1. The excitations are described by a Boson type of propagator $R(\underline{q},\omega)$ which in the case of the exchange interaction is the magnetic susceptibility $\chi(\underline{q},\omega)$. One finds for T=0

$$\frac{m^*}{m} = 1-(g_L-1)^2 J_{ex}{}^2 \frac{N(0)}{2p_F{}^2}\int_0^{2p_F} dq\; q\chi(q,0) \qquad (6)$$

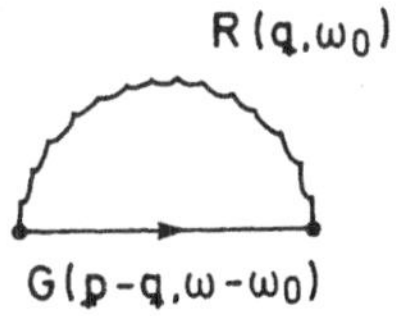

Fig. 1: Lowest order contribution for $\Sigma(\underline{p},\omega)$. $G(\underline{p},\omega)$ is the electron Green's function.

where g_L is the Lande factor and N(O) is the density of states per spin at the Fermi surface. It is seen that only the static susceptibility enters m^* at T=O. For illustration consider a lattice of RE ions with two singlet CEF levels per site separated by an energy δ and coupled via $\underline{J}$ through a matrix element M. Then Eq. (6) reduces to

$$\frac{m^*}{m} = 1+2(g_{L-1})^2 J_{ex}{}^2 N(O)\frac{|M|^2}{\delta}. \qquad (7)$$

For small δ, m^*/m can become very large. Pr metal is an ideal candidate for studying this effect because it is a singlet ground state system with small excitation energies. It is found that within an estimated error of 10 % the effective mass is given by

$$\frac{m^*}{m} = 1+(g_L-1)^2 J_{ex}{}^2 N(O)\frac{1}{2}\mathrm{Tr}[\chi_0^{hex}(O)+\chi_0^{cub}(O)] \qquad (8)$$

where $\mathrm{Tr}\chi_0^{hex}(O)=11.4\,\mathrm{meV}^{-1}$ and $\mathrm{Tr}\chi_0^{cub}(O)=4.8\ \mathrm{meV}^{-1}$ are traces of the zero field and temperature susceptibilities of the hexagonal and cubic sites of the dhcp Pr metal, respectively. Furthermore $N(O)=1.06\ \mathrm{eV}^{-1}$ per atom and spin and $J_{ex}=0.092$ eV. This results in $m^*(T=O)/m=4$. A test of the theory is the behavior of m^*/m in an applied magnetic field. Due to the changes of the CEF levels and energies in an applied field m^*/m becomes strongly field dependent. A comparison between theory and experiments is shown in Fig. 2 As far as $m^*(H)/m$ is concerned, the theory contains no adjustable parameters. It should be mentioned that it is no problem to extend the calculation to finite temperatures, i.e., to find $m^*(T)/m$.

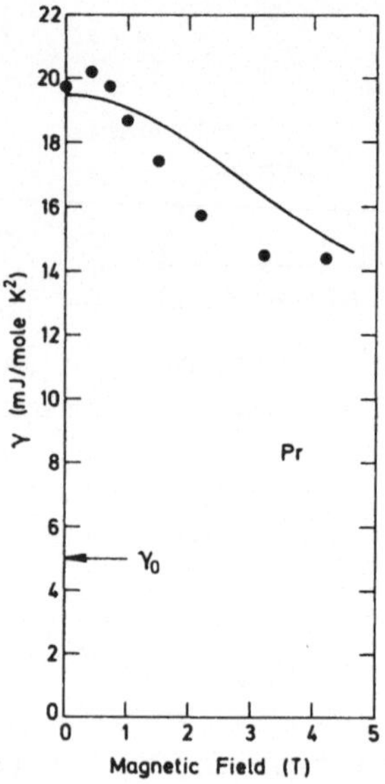

Fig. 2: Magnetic field dependence of the Sommerfeld constant $\gamma(H)$ for Pr metal as measured by Forgan (dots) and as calculated (solid line).

In the heavy RE metals the effects are somewhat smaller with the mass enhancement factor lying between 1.5-2. However, here the situation is more uncertain since band-structure calculations indicate substantial reductions in the density of states in the magnetically ordered state. They must be properly included in the theory for m^*/m.

2. Thermoelectric power

The thermoelectric power Q is a particulary interesting transport coefficient since it is very sensitive to details of the conduction electron scattering mechanism. This is seen by writing

$$Q = -\frac{ne}{m\sigma T} \int_{-\infty}^{+\infty} d\omega \left(-\frac{\partial f}{\partial \omega}\right) \omega \left(1+\frac{a\omega}{\varepsilon_F}\right) \tau(\omega) . \tag{9}$$

Here σ is the conductivity, n is the electron density, a is a constant of order unity, $f(\omega)$ is the Fermi function and $\tau(\omega)$ is the frequency dependent lifetime of the conduction electrons. The latter is related to the self-energy by $\tau(\omega)=-(2\mathrm{Im}\Sigma(\omega))^{-1}$. The term $a\omega/\varepsilon_F$ comes from the electron-hole asymmetry in the density of states near the Fermi surface. For $\tau(\omega)=\tau_0$=const. it is this asymmetry which gives rise to the small thermopower in a metal, $Q_0 = a\pi T/3e\varepsilon_F$, since $(-\partial f/\partial \omega)$ is an even function of ω and the integral would vanish otherwise. Therefore any term in $\tau(\omega)$ with odd powers of ω should strongly influence Q. It is therefore of interest that the virtual excitation of CEF levels leads in third order perturbation theory to a term $\sim\omega$ in $\tau(\omega)$. For RE impurities of concentration n_I with two CEF singlet levels it is found that, for $\omega, T \ll \delta$,

$$\mathrm{Im}\Sigma(\omega) = -\pi n_I N(O) M_e^2 [1 - 4\frac{M_i}{M_e}\tanh\frac{\delta}{2T}\cdot\frac{\omega}{\delta}]. \qquad (10)$$

Here M_e and M_i are the matrix elements for elastic (Coulomb) and inelastic (e.g.isotropic exchange) scattering, respectively. The second term on the r.h.s. is a true many body effect since it arises from an integral of the form

$$h(x) = P\int_{-D}^{+D} \frac{d\varepsilon}{\varepsilon - x}(1 - f(\varepsilon)) \qquad (11)$$

as in the Kondo problem (D is the bandwidth and P is the principal value). However, the physics is different in the two cases since here the effect vanishes when the two CEF states are equally populated. The additional contribution Q_1 to the thermopower due to the term $\sim\omega$ in Eq. (10) is

$$Q_1 = -\frac{4N(O)M_i^2}{eM_e}\frac{\delta}{2T}\tanh\frac{\delta}{2T}[1 + \frac{\delta}{2\pi T}\mathrm{Im}\psi'(\frac{i\delta}{2\pi T})]. \qquad (12)$$

It has been found experimentally in $La_{1-x}Tb_xAl_2$ and a comparison with the theory is shown in Fig. 3. The effect is also present in systems such as $PrPb_3$ and TmCd where the RE ions form a regular lattice. Thereby an interesting new result is obtained. In a regular lattice the Bloch electrons are eigenstates to the periodic potential and therefore there is no elastic scattering from the isotropic Coulomb interaction. However, such a scattering is required for Q_1 to occur. This implies that some of the other interactions such as the aspherical Coulomb scattering must be appreciable. They can lead to different elastic scattering potentials V_g and V_e for the ground and excited state and then only the thermal average

$$\langle V \rangle = n_g V_g + n_e V_e \qquad (13)$$

is then taken into account by the introduction of Bloch electrons (n_g and n_e are the corresponding thermal population factors). In the cases of $PrPb_3$ and TmCd the experimental results imply the presence of an aspherical Coulomb scattering and an orbital exchange interaction.

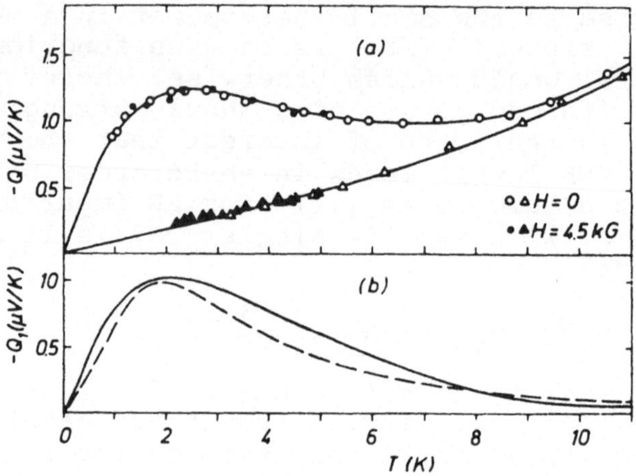

Fig. 3: (a) Thermopower for $La_{1-x}Tb_xAl_2$ for x=0.02 (circles) and x=0 (triangles) as measured by Umlauf, (b) experimental (solid line) and theoretical (dashed line) curve.

3. Linewidths of CEF levels

The interaction of CEF levels with the conduction electrons will lead to a finite linewidth of the former. This broadening usually dominates the one due to the interaction with phonons. Of interest is not only the T dependence of the linewidth of the CEF levels of single ions but also the quasielastic linewidth which appears close to the magnetic transition of induced moment systems such as Pr_3Tl. We start out from the Hamiltonian

$$H=H_{CEF}+H_{el}-2(g_L-1)J_{ex}\sum_i \underline{s}_i\cdot\underline{J}_i \tag{14}$$

where H_{el} denotes the conduction electron Hamiltonian and i runs over the RE ions. These can be either dilute impurities or ions forming a regular lattice as in Pr_3Tl. The line spectrum is obtained by computing $\mathrm{Im}\chi(\underline{q},\omega)$ for the RE system. This is done by means of a projection operator or memory function technique as developed by Mori and Zwanzig. This method is always very appropriate for calculating dynamical response or correlations when the static functions are known. When the memory function is calculated to lowest order in J_{ex}, the following results are found.

Single ion case: The linewidth for elastic transitions is proportional to $[(g_L-1)J_{ex}N(0)]^2T$ (Korringa law). For well separated inelastic transitions the linewidth at low T is found to be $\sim[(g_L-1)J_{ex}N(0)]^2\delta$, where δ is the excitation energy. At higher temperatures when the broadened lines overlap they interact strongly with each other and the spectrum can no longer be described by a superposition of single lines. Finally all oscillator strength appears in a quasielastic line which, with increasing T narrows before it broadens again (Korringa law for degenerate level scheme). This is shown in Fig. 4 for a Ce^{3+} ion. An interesting additional effect occurs when the metallic matrix becomes superconducting. If $\delta<2\Delta(0)$, where $\Delta(0)$ is the superconducting energy gap at T=0, then the number of electronic states, which can be excited by the CEF excitation decreases exponentially with T. This results in a noticeable decrease in the CEF linewidth for $T<T_c$. The effect has been observed for Tb in $LaAl_2$ ($T_c=3.24K$, $\Gamma_1-\Gamma_4$ splitting $\delta=8K$) and is shown in Fig. 5.

Lattice of RE ions: Here two different mechanisms contribute to damping. One is due to the excitation of conduction electrons while the other results from the interaction of delocalized CEF excitations and is also present in a Heisenberg magnetic system. A theory has been developed which contains both contributions. For that purpose one expands the relaxation function with the help of the projection technique into a continued fraction

$$\phi(\omega)=\frac{1}{\omega}[\chi(\underline{q})-\chi(\underline{q},\omega)] = \cfrac{\chi(\underline{q})}{\omega-\cfrac{\delta_1(\underline{q})}{\omega-\sum(\underline{q},\omega)}} \quad . \tag{15}$$

Here $\delta_1(\underline{q})$ is the excitation energy in RPA when for simplicity a two-singlet level system is considered. The damping effects are contained in $\sum(\underline{q},\omega)$ which is again a relaxation function but in a space orthogonal to that of the original dynamical variables. $\sum(\underline{q},\omega)$ can be separated into

$$\sum(\underline{q},\omega) = \sum_{el}(\underline{q},\omega)+\sum_{eff}(\underline{q},\omega). \tag{16}$$

$\sum_{el}(\underline{q},\omega)$ is the contribution from the electronic excitations

$$\sum_{el}(\underline{q},\omega) = \frac{A}{\omega+i\frac{2}{\pi}qv_F} \tag{17}$$

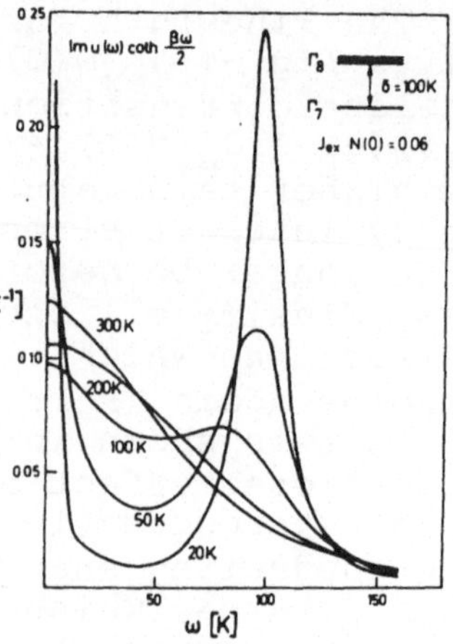

Fig. 4: Spectral distribution for Ce^{3+} in $LaAl_2$. $u(\omega)$ is the single ion susceptibility.

Here $A = \delta^2 u J^{eff}_{\underline{k}=0}$ where u is the single-ion susceptibility and $J^{eff}_{\underline{k}}$ is the Fourier transform of the RKKY interaction. The second term $\sum_{eff}(\underline{q},\omega)$ is due to the interaction between delocalized excitations and is

$$\sum\nolimits_{eff}(\underline{q},\omega) = \frac{B}{\omega + i(\frac{\pi}{2}B)^{1/2}} \tag{18}$$

with

$$B = \frac{\delta^2}{2N} \sum_{\underline{k}} J^{eff}_{\underline{k}} [\delta_1(\underline{k}) \tanh \delta_1(\underline{k})/2T]^{-1} . \tag{19}$$

$\delta_1(\underline{k})$ denotes the energy of the delocalized CEF excitations (magnetic excitons). From Eqs. (15) and (17),(18) it is seen that $\phi(\omega)$ has a 3-pole structure consisting of a quasielastic line and inelastic Stokes and anti-Stokes lines. When the electronic damping mechanism do-

Fig. 5: Linewidth of Tb in $LaAl_2$ as measured by Feile, Kjems, Loewenhaupt and Hoenig above and below the superconducting T_c.

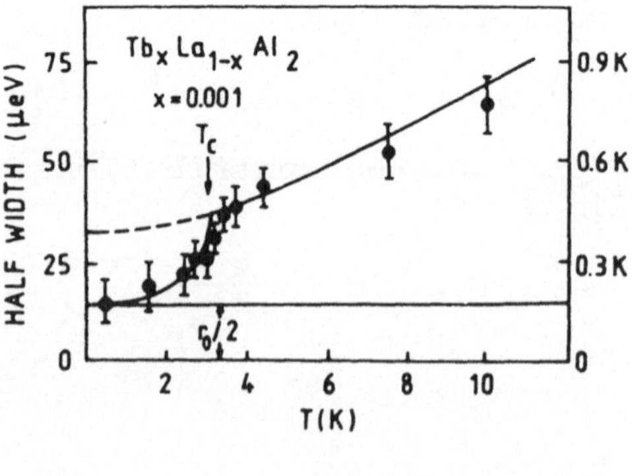

minates the linewidth changes according to q^{-1}. This will be the case of $J^{eff}_k \leq J_{crit}$ where J_{crit} is the critical coupling constant for induced magnetism to occur in a two-singlet level system, and/or when N(O) is sufficiently large. The effect has been observed in Pr metal. When $J^{eff}_k > J_{crit}$ then $k_B T_c$ becomes of the order of δ and the magnetic excitation interactions become important for damping. This is somewhat surprising since the electronic damping is $\sim J_{ex}^2$ while the damping due to magnetic exciton interactions is $\sim J_{ex}^4$. For small q the linewidth of the inelastic transitions becomes independent of q and the magnetic excitons become less soft as T approaches T_c from above than a molecular field theory would predict Finally close to T_c all intensity is obtained in the diverging quasielastic line. This behavior will be discussed more in Section V.

IV. INTERACTIONS WITH PHONONS

Lattice deformations in the vicinity of a RE ion change the CEF and therefore also the CEF energy levels. This results in a coupling between CEF levels and phonons (magnetoelastic coupling). It is well known that the deformation tensor $v^m_{\alpha\beta}$ (m=site index) can be decomposed into a symmetric, pure strain part $\varepsilon^m_{\alpha\beta}$ and an antisymmetric rotational part $\omega_{\alpha\beta}$, i.e.,

$$\begin{aligned} \varepsilon^m_{\alpha\beta} &= (v^m_{\alpha\beta}+v^m_{\beta\alpha})/2 \\ \omega^m_{\alpha\beta} &= (v^m_{\alpha\beta}-v^m_{\beta\alpha})/2 \,. \end{aligned} \tag{20}$$

Since both parts can lead to magnetoelastic interactions the total Hamiltonian reads

$$H = H_{phon} + H_{CEF} + H_{str} + H_{rot} \tag{21}$$

where H_{phon} is the Hamiltonian of the unperturbed phonons and H_{str} and H_{rot} describe the strain and rotational interactions, respectively. We shall discuss a number of effects due to these interactions. Thereby we distinguish between effects which are also present in the long wavelength limit of the phonons ("static effects") and effects which vanish in that limit ("finite frequency effects"). To the first group belong the temperature dependence of the elastic constants and the Voigt (or Cotton-Mouton) effect for phonons in an applied magnetic field. To the second group belong the Faraday rotation of phonons and the nonreciprocity of surface waves in an applied magnetic field. Furthermore we want to show that in materials like $CeAl_2$ the magnetoelastic interactions are so strong that a bound state is formed between a CEF excitation and an optical phonon.

1. Static effects

Let us consider a cubic system in the absence of an external magnetic field. The free energy F_{el} due to an elastic, homogeneous deformation takes a particularly simple form when the three irreducible representations Γ_1, Γ_3 and Γ_5 of the six $\varepsilon_{\alpha\beta}$ components are introduced, i.e.,

$$\begin{aligned} \Gamma_1(\text{singlet}):\ & \varepsilon_v = \varepsilon_{xx} + \varepsilon_{yy} + \varepsilon_{zz} \\ \Gamma_3(\text{doublet}):\ & \varepsilon_2 = 2^{-1/2}(\varepsilon_{xx}-\varepsilon_{yy}) \\ & \varepsilon_3 = 6^{-1/2}(2\varepsilon_{zz}-\varepsilon_{xx}-\varepsilon_{yy}) \\ \Gamma_5(\text{triplet}):\ & \varepsilon_{xy}, \varepsilon_{xz}, \varepsilon_{yz}. \end{aligned} \tag{22}$$

In that case

$$F_{el} = \frac{1}{6}(c_{11}^o+2c_{12}^o)\varepsilon_v^2+\frac{1}{2}(c_{11}^o-c_{12}^o)(\varepsilon_2^2+\varepsilon_3^2)+ + 2c_{44}^o(\varepsilon_{xy}^2+\varepsilon_{xz}^2+\varepsilon_{yz}^2), \tag{23}$$

where the $(c_{11}^o+2c_{12}^o)$ etc. are the corresponding elastic constants. We want to study the changes in the elastic constants due to the magnetoelastic interactions. For that purpose we consider

$$c_{44}^o = \frac{1}{4}\frac{\partial^2 F_{el}}{\partial \varepsilon_{xz}^2} \tag{24}$$

and assume that the only non-vanishing strains are ε_{xz}. In that case H_{str} reduces to

$$H_{str} = -g\sum_i (J_xJ_z+J_zJ_x)_i\varepsilon_{xz} \tag{25}$$

where the index i refers to different RE sites. This form can be understood by realizing that $(J_xJ_z+J_zJ_x)_i$ has the same transformation behavior as ε_{xz}. Alternatively one can consider a point charge model with displaced charges eZ as indicated in Fig. 6. The change in the potential δV due to the charge displacements which correspond to $\varepsilon_{xz} \neq 0$ is given by

$$\delta V(\underline{r}) = \frac{3Ze}{a^3}(xz+zx)\varepsilon_{xz}. \tag{26}$$

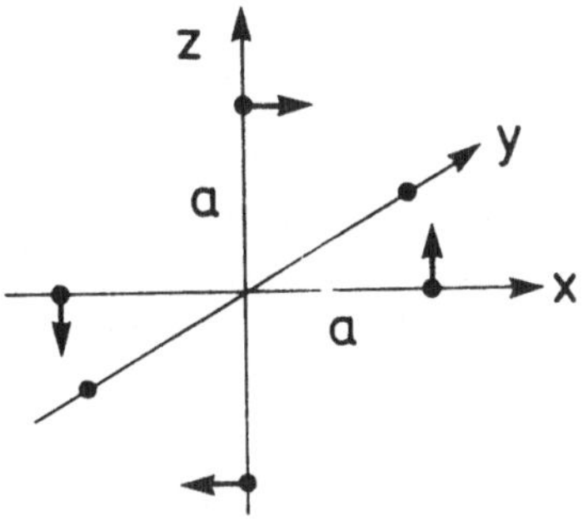

Fig. 6: Charge displacements corresponding to $\varepsilon_{xz} \neq 0$

When the Stevens operator equivalents are used this expression goes over into Eq. (25) but with an explicit expression for g. Consider N RE-ions which couple to ε_{xz} according to Eq. (25). Then the free energy is

$$F = F_{el} - TN\ln\left(\sum_{j}^{2J+1} \exp[-\delta_j(\varepsilon_{xz})/T]\right) \tag{27}$$

where $\delta_j(\varepsilon_{xz})$ denotes the CEF energy levels in the presence of H^{xz}_{str}. For a better understanding of the consequences of this change in F we consider instead the influence on the CEF levels of an external magnetic field $\underline{H}$. The interaction energy is then

$$H_{mag} = \mu_B g_L \sum_i \underline{J}_i \underline{H} \,. \tag{28}$$

By comparing the two forms (25) and (28) one notices that ε_{xz} corresponds to $\underline{H}$ and the quadrupole operator $(J_x J_z + J_z J_x)/2$ to the magnetic moment operator $\mu_B g_L \underline{J}_i$. In analogy to the magnetic susceptibility per site

$$\chi = -\frac{1}{N}\frac{\partial^2 \Delta F}{\partial H^2} = -\left\langle\frac{\partial^2 E}{\partial H^2}\right\rangle + \frac{1}{T}\left\langle\left(\frac{\partial E}{\partial H}\right)^2\right\rangle - \frac{1}{T}\left\langle\frac{\partial E}{\partial H}\right\rangle^2 , \tag{29}$$

where E is the energy of a RE ion in a field $\underline{H}$, one can define a quadrupolar susceptibility

$$\chi_Q = -\frac{1}{N}\frac{\partial^2 \Delta F}{\partial \varepsilon_{xz}^2} = -\left\langle\frac{\partial^2 E}{\partial c_{xz}^2}\right\rangle + \frac{1}{T}\left\langle\left(\frac{\partial E}{\partial \varepsilon_{xz}}\right)^2\right\rangle - \frac{1}{T}\left\langle\frac{\partial E}{\partial \varepsilon_{xz}}\right\rangle^2 . \tag{30}$$

The renormalized elastic constant is then given by

$$c_{44} = \frac{1}{4}\frac{\partial^2 F}{\partial \varepsilon_{xz}^2} = \frac{1}{4}\left[\frac{\partial^2 F_{el}}{\partial \varepsilon_{xz}^2} + \frac{\partial^2 \Delta F}{\partial \varepsilon_{xz}^2}\right] \tag{31}$$

$$= c_{44}^{o} - \frac{N}{4}\chi_Q(T) .$$

This results in a T dependence of c_{44} through $\chi_Q(T)$. Its form will strongly depend on the CEF ground state. We distinguish between two cases:

(a) The ground state has a vanishing quadrupolar moment. Then $\chi_Q(T)$ shows Van Vleck behavior, i.e., it remains constant for $T \to 0$. To this group belong, e.g., $TmSb(\Gamma_1)$, $SmSb(\Gamma_7)$ or $DySb(\Gamma_6)$, with the ground state shown in brackets.

(b) The ground state has a finite quadrupolar moment. Then $\chi_Q(T)$ shows Curie type of behavior, i.e. $\sim T^{-1}$ for $T \to 0$. In that case c_{44} necessarily vanishes at a temperature T_c at which a phase transition will take place. To this group belong, e.g., $HoSb(\Gamma_3)$ and $ErSb(\Gamma_8)$.

When in addition a magnetic field is applied a number of interesting effects can occur. First of all the rotational interactions come into play. In the absence of a field and in the static limit the $\omega^m_{\alpha\beta}$ describe just a rotation of the crystal as a whole which does not have any consequences. But when a field is present then due to the Zeeman energy the relative orientation of the crystal and the field becomes of importance and rotational interactions lead to measurable effects. For example the velocities of otherwise degenerate elastic waves can become different. Consider a cubic crystal with a magnetic field applied in the principal z direction. The two waves shown in Fig. 7 have the same strain deformations but different rotational parts. Due to H_{rot} the two waves acquire different velocities which have been measured.

Also the strain interactions can lead to interesting effects when a field is applied. As an example we consider the Voigt or Cotton-Mouton effect for phonons. We choose again a cubic crystal with a field in the z direction and consider two waves propagating along the x direction with polarizations along the y and z direction, respectively. Their velocities are given by

$$\begin{aligned} v_\perp &= v_0(1-(4c^0_{44})^{-1}\ \chi_Q(J_xJ_y)) \\ v_\parallel &= v_0(1-(4c^0_{44})^{-1}\ \chi_Q(j_xJ_z))\ . \end{aligned} \qquad (32)$$

In the absence of a field $\chi_Q(J_xJ_y)=\chi_Q(J_xJ_z)$, and both waves have the same velocity, but when a field is present $\chi_Q(J_xJ_y) \neq \chi_Q(J_xJ_z)$ and $v_\perp \neq v_\parallel$. The 4f shell is slightly deformed by the field as indicated by the ellipsoid in Fig. 8 and the response is different when the operators J_xJ_y and J_xJ_z act on it. Cotton-Mouton phase shifts have been observed in $CeAl_2$.

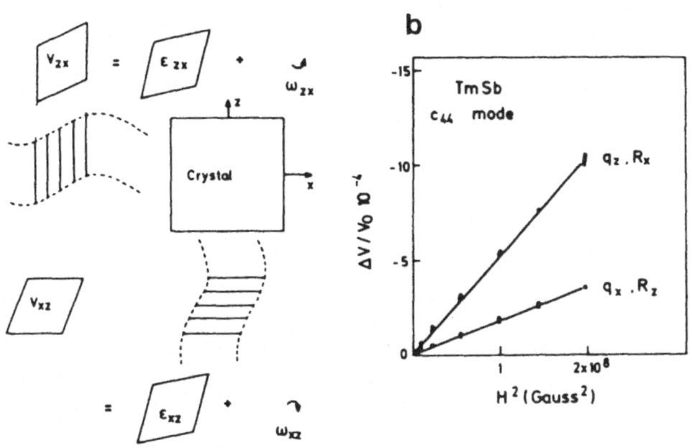

Fig. 7: (a) Transverse sound waves in a cubic crystal. Both waves have the same strain deformations but opposite rotational deformations.
(b) Experimental results (Wang and Lüthi)

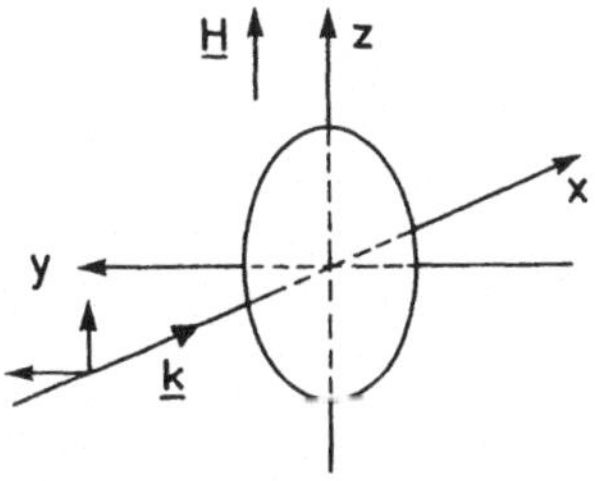

Fig. 8: Voigt geometry. The ellipsoid indicates the slightly deformed 4f shell in an applied magnetic field.

2. Finite frequency effects

In addition to the static effects there exist a number of finite frequency effects which vanish in the long wavelength limit. We want to consider two examples, namely the Faraday rotation of phonons and the non-reciprocity of surface waves in an applied magnetic field. The simplest way of demonstrating these effects is to start from the equations of motion for the displacements $\underline{u}$

$$-\rho\omega^2 u_\alpha = \sum_\beta \frac{\partial \sigma_{\alpha\beta}}{\partial x_\beta} \qquad (33)$$

$$\sigma_{\alpha\beta} = \frac{\partial F}{\partial v_{\alpha\beta}} .$$

Here ρ is the mass density and $\sigma_{\alpha\beta}$ is the stress tensor. By calculating the free energy F by second order perturbation theory in H_{str} one finds for finite frequencies

$$\sigma_{\alpha\beta} = \frac{\partial H_{el}}{\partial v_{\alpha\beta}} + N\left\langle \frac{\partial H_{str}}{\partial v_{\alpha\beta}}\right\rangle_0 - N\chi\left(\frac{\partial H_{str}}{\partial v_{\alpha\beta}};H_{str};\omega\right). \qquad (34)$$

Here $\langle .....\rangle_0$ refers to a thermal average in the absence of magnetoelastic interactions. $\chi(A;B;\omega)$ is the generalized isothermal susceptibility of the Hermitean operators A and B.

Let us consider again a cubic crystal with $\underline{H}$ along the z axis and furthermore we want to assume that the only non-vanishing strains are $\varepsilon_{xy}, \varepsilon_{xz}$ and ε_{yz}. The corresponding operators of the 4f shell are $O_1=J_xJ_y+J_yJ_x$, $O_2=J_yJ_z+J_zJ_y$, $O_3=J_xJ_z+J_zJ_x$. One finds for the susceptibility $\chi=\chi'+i\chi''$ at low frequencies

$$\chi'(O_i;O_j) = \begin{pmatrix} S_{11} & 0 & 0 \\ 0 & S_{22} & 0 \\ 0 & 0 & S_{22} \end{pmatrix}$$

$$\chi''(O_i;O_j) = i\omega \begin{pmatrix} 0 & 0 & 0 \\ 0 & 0 & A_{23} \\ 0 & -A_{23} & 0 \end{pmatrix} \qquad (35)$$

where $S_{\nu\nu}$ ($\nu=1,2,3$) and A_{23} remain finite in the limit $\omega\to 0$. For $\underline{H}=0$ it is $S_{11}=S_{22}$ and $A_{23}=0$. We want to consider transverse phonons propagating along the z axis. The equations of motion then take the form

$$-\rho\omega^2 u_x = \frac{\partial}{\partial z}\left[2\left(c^o_{44}-\frac{NS_{11}}{4}\right)\varepsilon_{xz}+i\omega^2 NA_{23}\varepsilon_{zy}\right]$$

$$-\rho\omega^2 u_y = \frac{\partial}{\partial z}\left[2\left(c^o_{44}-\frac{NS_{11}}{4}\right)\varepsilon_{yz}-i\omega^2 NA_{23}\varepsilon_{xz}\right]. \qquad (36)$$

The eigenstates are circularly polarized states $u_x \pm iu_y$ with wave vectors

$$k_+^2 = \frac{\omega^2}{c_{44}} \rho[1-A_{23}\rho/c_{44}]$$
$$k_-^2 = \frac{\omega^2}{c_{44}} \rho[1+A_{23}\rho/c_{44}] \quad . \tag{37}$$

The turn in the phase angle per unit length is $\psi=(k_+-k_-)/2$ and therefore $\sim\omega^2$. A Faraday rotation of phonons has been observed in $CeAl_2$ but there are problems with a quantitative comparison between theory and experiment.

Next we consider a surface in the x-z plane and a wave propagating along the x direction. Its amplitude decreases as it penetrates into the crystal which is in the y direction. The displacements are assumed to be along the z direction. The equation of motion becomes then

$$-\omega^2 \rho u_z = c_{44}\left[\frac{\partial^2 u_z}{\partial x^2} + \frac{\partial^2 u_z}{\partial y^2}\right] \tag{38}$$

where $c_{44}=c_{44}^o-Ng^2S_{22}$. This equation is supplemented by the condition that the surface is stress free. This implies

$$\left[c_{44}\frac{\partial u_z}{\partial y} - \frac{iNg^2}{4}\omega A_{23}\frac{\partial u_z}{\partial x}\right]\bigg|_{y=0} = 0 \ . \tag{39}$$

For the u_z the following ansatz is made

$$u_z = u_o \exp[-i\omega t+ikx-\alpha y] \tag{40}$$

which leads to

$$\alpha = Ng^2 A_{23}\omega k/4c_{44} \tag{41}$$

when set into Eq. (39). The surface wave has then the dispersion

$$\omega^2 = c_{44}k^2\rho^{-1}[1-(\alpha/k)^2] \ . \tag{42}$$

Since $\alpha>0$ it follows from Eq. (41) that depending on the sign of A_{23} only one direction of k is possible and the corresponding -k is forbidden. In piezoelectrics such a mode is called Bleustein-Gulyayev mode. Although it should be difficult to measure it demonstrates in a transparent way how non-reciprocity comes about. The similar situation occurs for Rayleigh waves where the effect should be large enough to be observable but for which the explanations are more lengthy.

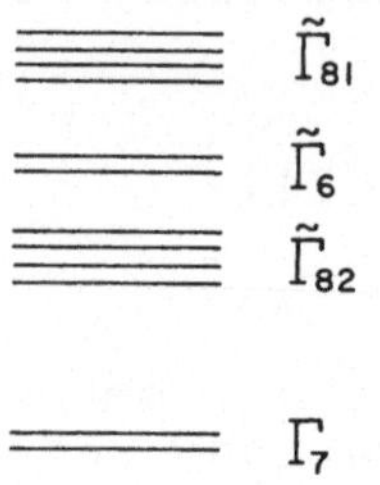

Fig. 9: Excitations of Ce^{3+} in the presence of strong magnetoelastic coupling

3. Bound state between a CEF excitation and a phonon

In $CeAl_2$ the magnetoelastic coupling is particularly large. The CEF ground state of Ce^{3+} is a Γ_7 doublet with a Γ_8 quartet excited state at $\Delta \simeq 100$ K. Measurements of the phonon density of states have revealed that there are phonons present of approximately Γ_5 symmetry with energy $\hbar\omega_0 \simeq 140$ K. Therefore the ground state Γ_7 with one of these phonons μ, i.e., $|\Gamma_7,\mu\rangle$ has an energy which is of the same order as that of the state $|\Gamma_8,0\rangle$. Let us write the interaction in the form

$$H_{str} = -g_0 \sum_{\mu=1}^{3} (a_\mu + a_\mu^+) O_\mu \qquad (43)$$

where a_μ^+, a_μ are the operators for the Γ_5 phonons and $O_1 = J_xJ_y + J_yJ_x$ etc. as before. Then one must diagonalize $H_{CEF} + H_{str}$ in the space spanned by the ten states $|\Gamma_7,\mu\rangle$ and $|\Gamma_8,0\rangle$. By observing that $|\Gamma_7,\mu\rangle = \Gamma_7 \otimes \Gamma_5 = \Gamma_6 \oplus \Gamma_8$ one can predict that the vibronic eigenstates will consist of a doublet $\tilde{\Gamma}_6$ and two quartets $\tilde{\Gamma}_{81}$, $\tilde{\Gamma}_{82}$. One obtains then an energy level scheme as shown in Fig. 9. One checks easily that there are no magnetic dipole transitions between $|\Gamma_7\rangle$ and $|\tilde{\Gamma}_6\rangle$ so that one expects a two-peak structure in an inelastic neutron scattering experiment which has been indeed observed. From the measured splitting of $\Delta_1 - \Delta_2 = 80$K one deduces $g_0 = 6.3$K which is of the same size as the corresponding coupling constant found in ultrasound experiments. When the damping due to electron-hole excitations is taken into account the spectral function looks as shown in Fig. 10.

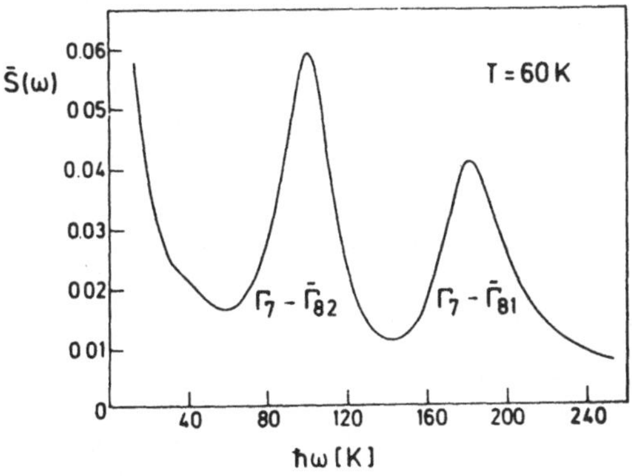

Fig. 10: Spectral function for $CeAl_2$ due to the presence of a bound state between a phonon and a CEF excitation.

When the dispersion of the phonons is taken into account the wave function takes the form

$$|\psi\rangle = \sum_i v_i |i,0\rangle + \sum_{\underline{k}\mu j} u_j(\underline{k},\mu)|j,\underline{k}\mu\rangle \tag{44}$$

where i,j are CEF eigenstates and $\underline{k},\mu$ denotes a phonon with momentum $\underline{k}$ and polarization μ. For weak coupling or small g_o one finds only a shift and a finite lifetime of the CEF Γ_8 excited state. But when g_o exceeds a threshold value a bound (and antibound) state appear. The situation resembles the exciton-phonon bound state known in semiconductor physics. The above theory does not only explain the two-peak structure observed in neutron scattering but it can also provide an explanation of the very strong temperature variation of a number of phonon branches in $CeAl_2$ which was recently observed by Reichardt and Nücker.

V. COOPERATIVE PHENOMENA

When the RE ions form a lattice there will be generally interactions taking place among them. They can be mediated through conduction electrons or phonons but in metals mediation through electrons will be dominating. It has been discussed in Section II that the different interactions of a RE ion with conduction electrons can be classified by specifying (Λ,Σ). Let us consider a system in which the RE ions have a singlet ground state. When the ion-ion interaction becomes sufficiently large then a phase transition will take place the character of which depends on the interaction which is causing it. For example when $(\Lambda,\Sigma)=(0,1)$ is the dominant interaction then the effective ion-ion interaction is of the form

$$H_{int} = -\frac{1}{2}\sum J_{ij}^{ex}\underline{J}(i)\underline{J}(j) \qquad (45)$$

and the system can undergo a magnetic phase transition. This is called an induced magnetic moment system and an example is Pr_3Tl.

When the interaction $(\Lambda,\Sigma)=(2,0)$ is dominating then

$$H_{int} = -\frac{1}{2}\sum_{ij\alpha n} K_{ij}(\alpha)O_\alpha^n(i)O_\alpha^n(j) \qquad (46)$$

where $O_\alpha^n(i)$ is a quadrupolar operator of representation α acting on site i and n is a degeneracy index. Then the system can undergo a quadrupolar phase transition and one is speaking of an induced Jahn-Teller or quadrupole moment system. An example is $PrCu_2$. Similarly other interactions (Λ,Σ) lead to various other H_{int}. Since there is a large amount of literature available on cooperative phenomena in RE systems we shall be comparatively brief here.

In order to calculate the excitations in such interacting systems one must compute the proper dynamical susceptibility $\chi_{ij}(\omega)$ (i.e. magnetic or quadrupolar, depending on the type of interaction). For the magnetic case one finds in a molecular field type of approximation

$$\chi_{ij}(\omega) = u(\omega)\delta_{ij}+u(\omega)\sum_\ell J_{i\ell}^{ex}\chi_{\ell j}(\omega) \qquad (47)$$

while in the quadrupolar case $J_{i\ell}^{ex}$ is replaced by $K_{ij}(\alpha)$. $u(\omega)$ is the single ion susceptibility which for a given CEF level scheme has the form

$$u(\omega,T) = \sum_{i,j=1}^{2J+1} |M_{ij}|^2 \frac{\delta_{ij}(n_j-n_i)}{\delta_{ij}^2-(\omega+i0^+)^2} \quad . \qquad (48)$$

The M_{ij} are the matrix elements between CEF states i and j of the operator $\underline{J}$ in the magnetic and O_α^n in the quadrupolar case and $\delta_{ij}=\delta_i-\delta_j$. The n_i are the thermal population factors of the CEF levels. Taking the Fourier transform of Eq. (48) leads to

$$\chi(\underline{q},\omega) = \frac{u(\omega)}{1-J(\underline{q})u(\omega)} \qquad (49)$$

where $J(\underline{q})$ is the transform of the interaction, i.e. J_{ij}^{ex} or $K_{ij}(\alpha)$. The poles of Eq. (49) yield the excitations of the system. For the Hamiltonian (45) these are the magnetic excitons and there is excellent neutron scattering work available where the q and T dependences of the excitations have been measured. When the static susceptibility $\chi(\underline{q},0)$ diverges a phase transition takes place. The transition temperature is then obtained from

$$J(q)u(0,T_c) = 1 . \tag{50}$$

In order to fulfil this equation for a singlet ground state system J(q) must exceed a threshold value J_{crit}. There exists a large number of investigations on the two singlet system and the case of a singlet ground state and a triplet excited state $(\Gamma_1-\Gamma_4)$ near such a phase transition. Of particular interest is the behavior of the quasielastic line. This problem has been discussed in Section II.3 where it was shown that a theory can be developed which is based on the projection operator method. In particular it is found experimentally and theoretically that in the induced ferromagnet Pr_3Tl the energy of the magnetic excitons remains finite for $q \to 0$ as $T \to T_c$ from above. The central peak intensity (CPI) has the property

$$\begin{aligned} CPI(q) &= k_B T \int_{-\omega_c}^{\omega_c} \frac{d\omega}{\omega} \mathrm{Im}\chi(q,\omega) = \pi k_B T \chi(q) \\ &= \frac{c}{q^2+\kappa_1^2(T)} . \end{aligned} \tag{51}$$

Here $\kappa_1^{-1}(T)$ is a correlation length which diverges as $T \to T_c$. For a quantitative theory one should not eliminate the conduction electrons, which mediate the RE ion interaction, by an effective ion-ion interaction. Instead their dynamical properties should be taken into account. This amounts to replacing J(q) in Eq. (49) by $J(q,\omega)$ which results in additional damping and changes in the dispersions of the excitations as discussed in Section III.3.

As mentioned above with $PrCu_2$ an induced quadrupole system has been found. The transition temperature is T_c=7.6K. It can take place because the two lowest lying CEF levels are connected only via a quadrupolar matrix element but not a magnetic one. Only the next higher excited state couples magnetically with the ground state. In this way the quadrupolar interaction becomes the dominating interaction.

ACKNOWLEDGMENT

I would like to thank Dr. R. Camley and Dr. P. Thalmeier for a number or discussions. Dr. Camley provided the particular derivation of the Faraday rotation of phonons given here.

REFERENCES

1) Recent review articles

Newman, D.J., 1971, Adv. Phys. 20, 197

Cooper, B.R., 1972, in "Magnetic Properties of Rare-Earth Metals", R.J. Elliott, ed., Plenum Press, London and New York

Fulde, P., 1978, Chapter 17 in "Handbook on the Physics and Chemistry of Rare Earths", K.A. Gschneidner and L. Eyring, ed., North-Holland Publishing Company, Amsterdam, New York, Oxford

Lüthi, B., 1980, Chapter 4 in "Dynamical Properties of Solids," Vol. 3, G.K. Horton and A.A. Maradudin, ed., North-Holland Publ. Company, Amsterdam, New York, Oxford.

2) In the following we list only a few selected references which can not be found in any of the review articles. The list is certainly not complete.

To Section:

II.1 Forgan, E.M., 1981, Physica (Utrecht), B107:65

Fulde, P.,and Jensen, J., 1983, Phys. Rev.,B 27:4085

II.3 Becker, K., Fulde, P., and Keller,J.,1977, Z. Physik B 28:9

Becker, K., 1980, Habilitationsschrift, Universität Regensburg

Becker, K., and Keller, J., 1982 in "Crystalline Electric Field Effects in f-Electron Magnetism", R.P. Guertin, W.Suski and Z.Zolnierek, ed. Plenum Press, New York and London

Zevin, V. and Barbay, E., 1980, Z.Phys., B 39:173

Feile, R., Kjems, J., Loewenhaupt, M., and Hoenig, H.E., 1981, Phys. Rev. Lett., 47:610

III.3 Lüthi, B., and Lingner, C., 1979, Z. Phys. B 34:157

Camley, R., and Fulde, P.,(to be published)

III.3 Loewenhaupt, M., Rainford B.R. and Steglich,F., 1979, Phys. Rev. Lett., 42:1709

Reichardt, W., (to be published)

Thalmeier, P., and Fulde, P., 1982, Phys.Rev.Lett., 49:1588

FUNDAMENTALS OF THE INTERMEDIATE VALENCE PROBLEM

C.M. Varma

Bell Laboratories
600 Mountain Avenue
Murray Hill, New Jersey, U.S.A.

The intermediate valence problem presents a serious challenge to theorists. So far we understand well only the single intermediate valence impurity in a conduction band and the interactions between a pair of them. For the real problem, only qualitative ideas exist. After discussing the conditions in which intermediate valence comes about, I will discuss the single impurity and the pair interaction problems, draw qualitative conclusions about the lattice, point out the difficulties in the lattice problem and discuss some approximate treatments.

Occurrence of Intermediate Valence

Let us first recapitulate briefly the origin of intermediate or homogeneous mixed-valence. Consider the electronic configurations f^n and f^{n+1} of an isolated ion with respective ground states $|f^n{:}J_a\rangle$ and $|f^{n+1}{:}J_b\rangle$. We suppose the ion to be in weak contact with an electron reservoir in which the chemical potential (i.e. the Fermi level) is at energy E_F. There will exist some energy boundary E_{ab} such that at T=0 the ion will be in the $f^n{:}J_a$ state if $E_F < E_{ab}$ and in the $f^{n+1}{:}J_b$ state if $E_F > E_{ab}$ (Fig. 1). Let each neutral atom contain n+r electrons and let us form a crystal out of a large number of them. The non-f electrons go into a set of (mostly) d-bands which constitutes the electron reservoir. Let E_{Fa} be the Fermi level when it contains r electrons per atom and E_{Fb} for r-1 per atom. If now E_{ab} should happen to lie between E_{Fa} and E_{Fb} then the mixed valence situation arises (Fig. 1). It requires no remarkable accident of nature for this to happen. The state of the system in which all ions have n f-electrons is impossible because E_{Fa} is too high to be consistent with the $f^n{:}J_a$ ionic states, and

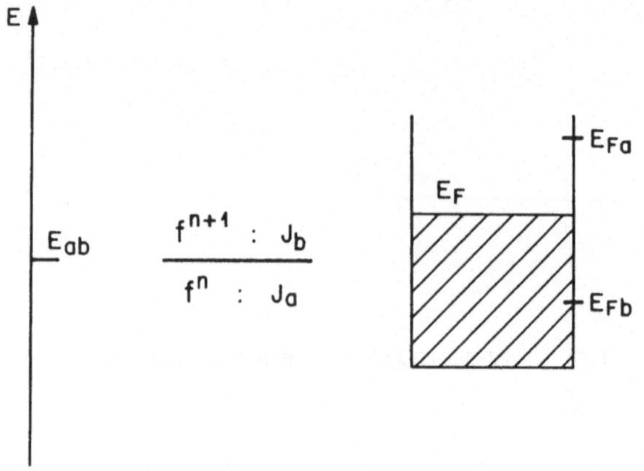

Fig. 1. Illustrates conditions of occurrence of intermediate valence.

similarly having all ions in $f^{n+1}:J_b$ is impossible because the corresponding E_{Fb} lies below E_{ab}. Thus the Fermi level in the reservoir will lie at the boundary E_{ab} with some fraction of ions in the f^{n+1} and the rest in the f^n states as demanded by having the correct total number of electrons. This consideration also reveals that about 10% of rare earth compounds should belong to the intermediate valence category. The above description, which developed in a discussion with V. Heine, only involves the rigorous concept of chemical potetial and relies on no approximations.

The phenomena are found most frequently in compounds of cerium, thulium and ytterbium as well as in those of samarium and europium. In the former, the underlying closed shell makes the 4f and the 5d atomic energy levels close in energy, and in the latter two, Hund's-rule-coupling's preference for the f^7 configuration makes different configurations close in energy. Praseodymium and terbium salts with two different valences are known to exist. It would be interesting if metallic intermediate valence compounds can be made with them.

Although Fig. 1 is good enough to give a condition for intermediate valence, the remarkable experimental properties of these materials arise from what is neglected in Fig. 1. This is that the boundary line E_{ab} gets blurred by quantum-mechanical contact with the reservoir. The point is that there are [N!/(N-Nx)!(Nx)!] configurations of the system (not counting the spin and orbital degeneracy at each site) depending on which particular Nx ions have the $f^{n+1}:J_b$ state. This degeneracy is lifted primarily using the contact with the reservoir i.e., though the hybridization of the f-orbitals with those of the itinerant electrons. The fluctuations between these configurations must be highly correlated, and correlated with the spin and orbital fluctuations. These fluctuations are at the root of the remarkable properties (and of the theoretical difficulties).

Principal Questions Posed by Intermediate Valence

These were already posed by the experiments in the early 1970's by Geballe and collaborators[1] on SmB_6 and by Jayaraman and collaborators,[2] and Maple and Wohlleben[3] on SmS. In rare earth metals where the ionization/affinity boundaries in Fig. 1 lie far from the chemical potential, the f-electrons are essentially in the Mott insulating regime due to the large ratio of the correlation energy U (about 10 eV) to the the hopping energy t (less than 0.1 eV) and undergo magnetic transitions at temperatures of the order of t^2/U. In mixed valence compounds, on the other hand, the f-electrons behave as if they are delocalized as revealed by their contribution to the linear specific heat and the Pauli susceptibility. The effective mass from these measurements is 10^2-10^3. How does such a massive particle propagate as effectively free?

Of the compounds investigated, only the TmSe behaves differently, undergoing a magnetic transition at low temperatures. TmSe is different in that, ignoring the small crystal field splitting compared to the hybridization energy, both valence states are magnetic, whereas in all other compounds investigated one of the valence states has no magnetic moment. Why does this change the behavior completely?

Experiments at energies high compared to the characteristic blurring energy clearly reveal the existence of two charge states and the multiplet structure within a given charge state. At high temperatures, the Curie law with almost the full localized moment is observed. The low energy excitations are dramatically different. They show only fluctuations from one charge to the other with accompanying spin fluctuations indicative of a Fermi liquid. No excitations characteristic of localized magnetic moments are seen. Further, there is no phase transition in going from the high energy to the low energy regime. This behavior is reminiscent of the problem of quarks in high energy physics which are supposed to be asymptotically free at high energies and confined at low energies.

It was realized very early[4] that the ground state and low lying excitations are characteristic of Fermi liquids but how this behavior comes about just because of the relative positions of the levels in fig. 1 has by and large remained a mystery. It was also clear that the properties of these materials are similar to those of dilute localized magnetic moments in metals (the Kondo problem). But why should interactions between the moments not completely change the behavior as they do in ordinary rare earth metals?

These are some of the basic questions which have to be answered before detailed comparison of any particular experiment with any theory are meaningful. In the typical intermediate valence compound, the characteristic hybridization energy is about 100K. It appears that there is another class of compound in which $CeA\ell_3$, $CeCu_2Si_2$, UBe_{13} fall, where the characteristic energy is only about 10K. The other properties reveal that the valence in this case is very close to but not exactly integral. These have been called the Kondo lattice compounds. At least two of them even undergo a superconducting transition at low temperatures! Is the physics of such material only quantitatively different from the others or are these fundamental differences?

Single Impurity Problem

Although the asymmetric Anderson Hamiltonian does not obey the Friedel sum rule, it probably gives qualitatively the same answers for matters of our interest as a modified Anderson Hamiltonian which does obey the Friedel sum rule. We stick therefore with

$$H = \sum_{\sigma,i=0} \left(\varepsilon_0 a^+_{i\sigma} a_{i\sigma} + U\, n_{i\sigma} n_{i-\sigma}\right)$$

$$+ \sum_{\sigma i<j} t_{ij}\left(b^+_{i\sigma} b_{j\sigma} + c.c.\right) + V \sum_{\sigma,i=0} a^+_{i\sigma} b_{i\sigma} + c.c. \qquad (1)$$

The local orbital (at i=0) is denoted by the a's and the conduction electron operators by the b's. $n = a^+a$. Note that in the lim $U\to\infty$, (1) is equivalent to

$$H' = \sum_{\sigma,i=0} \varepsilon_0 a^+_{i\sigma} a_{i\sigma} + \sum_{\sigma i<j} t_{ij}(b^+_{i\sigma} b_{j\sigma} + c.c.)$$

$$+ V \sum_{\sigma,i=0} (1-n_{i,-\sigma}) a^+_{i\sigma} b_{i\sigma} + c.c. \qquad (2)$$

This may be seen from the fact that

$$\left[n_{i\sigma} n_{i-\sigma}, H'\right] = 0 \ , \qquad (3)$$

which implies that the subspace with zero or single occupation in which $n_{i\sigma}n_{i-\sigma} = 0$ is unconnected to the doubly occupied subspace, $n_{i\sigma}n_{i-\sigma} = 1$. If we start with the former, we stay in it with H', thus realizing the $U \to \infty$ limit. This trick, originally due to Blandin would appear to be equivalent to Coleman's trick in which $(1-n_{i,-\sigma})$ is regarded as a Boson.

Varma and Yafet[5] considered a variational solution to the single impurity problem. To this end it is convenient to transform (1) or (2) to momentum space, k, with spherical symmetry about the impurity. The allowed states of the Hamiltonian can have total spin S = 0, 1/2, 1, •••. The possible singlet states with S = 0 and with $n_\sigma n_{-\sigma} = 0$ can be constructed in many different ways: Fermi-sea : $|\psi_0\rangle$; Fermi sea with excited particle-hole pair:

$$\sum_\sigma b^+_{k'\sigma} b_{k\sigma} |\psi_0\rangle ,$$

with total angular momentum of k and k' = 0; a hole (doublet) in the Fermi-sea plus an electron (doublet) in the local orbital;

$$\sum_\sigma a^+_\sigma b_{k\sigma} |\psi_0\rangle ;$$

two holes and an electron in the Fermi-sea and an electron in the local orbital:

$$\sum_{\sigma\sigma'} a^+_\sigma b^+_{k''\sigma'} b_{k'\sigma} b_{k\sigma} |\psi_0\rangle ,$$

with appropriate angular momentum restrictions and so on.

Similarly the doublet (S = 1/2) states with $n_\sigma n_{-\sigma} = 0$ are $a^+_\sigma|\psi_0\rangle$, $b^+_{k\sigma}|\psi_0\rangle$, and so on.

The triplet states, S = 1, with $n_\sigma n_{-\sigma} = 0$ are

$$b_{k'\sigma} b_{k\sigma} |\psi_0\rangle, \quad \sum_\sigma b_{k'-\sigma} b_{k\sigma} |\psi_0\rangle ,$$

$$\left(a^+_{-\sigma} b_{k-\sigma} - a^+_\sigma b_{k\sigma}\right) |\psi_0\rangle, \quad \text{and so on.}$$

Note that all the states have either one or no electrons in the local orbital and various number of particle-hole pairs with their angular momentum and spins phased appropriately. For a given total S, the various possible states are coupled by the mixing term of H or H'. Thus we can take their linear combination as variational wave-functions

$$|\psi(S=0)\rangle = \Big\{a_0 + \sum_{k,\sigma} \alpha_k a^+_\sigma b_{k\sigma} + \sum_{kk'\sigma} \alpha_{kk'} b^+_{k\sigma} b_{k'\sigma} + \cdots\Big\} |\psi_0\rangle, \qquad (4)$$

$$|\psi_\sigma(S=1/2)\rangle = \Big(\beta_0 a^+_\sigma + \sum_k \beta_k b^+_{k\sigma} + \cdots\Big) |\psi_0\rangle , \qquad (5)$$

and so on.

Note that the doublet (S = 1/2) energy is only a renormalization of the initial energy ε_0. Therefore to compare the singlet and doublet energies to any order, the singlet wave-function considered should always be consistently an order higher. The singlet will then always be lower in energy because it uses both the $\sigma = 1/2$ and $\sigma = -1/2$ states of the doublet together with the spin 1/2 states of the Fermi-sea to form a singlet. To lowest order (Varma and Yafet) the doublet energy is simply $-\varepsilon_0$ and the singlet energy is obtained by $\delta\langle H\rangle/\delta\alpha_0 = \delta\langle H\rangle/\delta\alpha_k = 0$ giving

$$\varepsilon\, \alpha_0 = -\varepsilon_0\alpha_0 + \sum_k V_k a_k , \tag{6}$$

$$\varepsilon\, \alpha_k = V_k\alpha_0 - \varepsilon_k\alpha_k , \tag{7}$$

from which the singlet state energy is determined to be

$$\varepsilon = -\varepsilon_0 + \sum_k |V_k|^2/(\varepsilon + \varepsilon_k) , \tag{8}$$

i.e.,

$$\varepsilon_{\text{singlet}} = \varepsilon_{\text{doublet}} - \frac{\Gamma}{\pi}\left(\ln\frac{4W}{|\varepsilon|} - 2\right) , \tag{8a}$$

where $\Gamma = \pi V^2\rho$, ρ is the conduction electron density of states with a bandwidth 2W. For $|\varepsilon_0| \gg \Gamma$, the binding energy is the Kondo value $\approx W\exp(-1/J\rho)$, where J is the Schrieffer-Wolff exchange integral for $U \to \infty$, $J \approx V^2/|\varepsilon_0|$. For $\varepsilon_0 \approx 0$, the intermediate valence regime, the binding energy is

$$E_{\text{sing}}(\varepsilon_0 \approx 0) \approx \Gamma/\pi \ln(4W/\Gamma) . \tag{9}$$

In the latter case a similar calculation including a magnetic field yields a magnetic susceptibility

$$\chi \approx \mu^2/E_{\text{singlet}} .$$

In the intermediate valence limit, the answers obtained by the lowest order variational method are correct in the 'scaling regime' (this will be classified below) and have been reobtained by Bringer and Lustfeldt,[6] Haldane,[7] Ramakrishnan[8] and Keiter and Grewe[9] and others. It is worth realizing that the Brillouin-Wigner perturbation theory used by Ramakrishnan, Keiter and Grewe etc. is identical, order by order, to the wave function expansion considered above. Haldane's poor man scaling method is very instructive and gives the reason why the simple approach works. The idea is to write the equations for $\psi(S=0)$ and $\psi(S=1/2)$ to lowest order exactly as above with the coefficients α_k and β_k determined by second-order perturbation theory. In the expression for the energy of the singlet and the doublet states, the intermediate particle-hole states are eliminated gradually starting with the highest energy states and the difference between the singlet and doublet energy considered as a function of the energy-cut-off. By contrast in the above all the intermediate states are eliminated at once. If there is only one scaling parameter in the problem with respect to which the scaling is logarithmic both procedures will yield similar answers. This may be illustrated graphically. In lowest order (Varma and Yafet), only the self-energy diagram shown in Fig. 2 is calculated in which an f^0 states absorbs and re-emits a conduction electron becoming f^1 and f^0 again. This diagram gives a logarithmic contribution. In the scaling theory this term is considered for the highest energy electron-holes. For lower energy states successively more and more nested diagrams of the type shown in Fig. 3 are considered. They are all higher order in V and logarithmic in W. Thus the lowest order gives the right answer for small Γ. The contribution of their nested diagrams to all orders can be written down as a continued fraction; Inagaki[10] has found that the eigenvalue then is given by the integral equation

$$\varepsilon = \Sigma_o(\varepsilon) \tag{11}$$

where

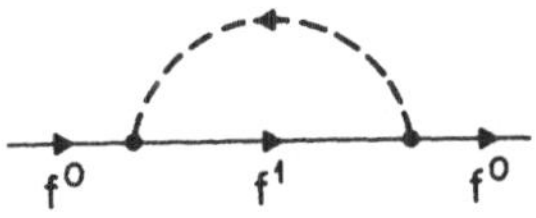

Fig. 2. Lowest order process (not a Feynman diagram) for a non-degenerate intermediate valence impurity.

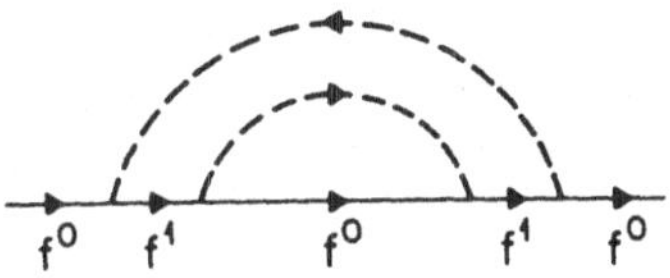

Fig. 3. Next order process.

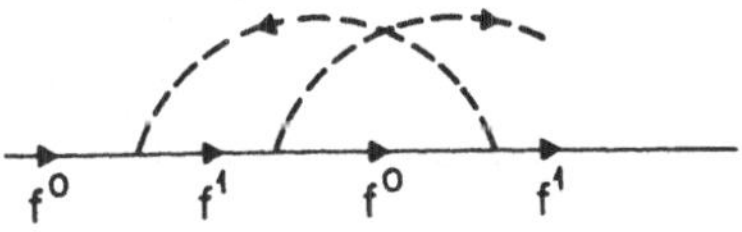

Fig. 4. Vertex correction to Fig. 2.

$$\Sigma_0(\varepsilon) = V^2 \sum_k \frac{1}{\varepsilon+\varepsilon_0-\varepsilon_k+\Sigma_1(\varepsilon-\varepsilon_k)} , \tag{12}$$

and

$$\Sigma_1(\varepsilon) = V^2 \sum_k \frac{1}{\varepsilon+\varepsilon_k+\Sigma_0(\varepsilon+\varepsilon_k)} . \tag{13}$$

These must be solved numerically to get corrections over Eq. (8a).

The scaling method does not sum crossed-diagrams of the type Fig. 4, i.e., it neglects vertex corrections. They are non-logarithmic but become as important as the logarithmic terms when the scale for the latter approaches Γ, i.e. in the strong coupling limit. The only methods giving quantitatively correct answers then are the numerical methods (Krishnamurthy, Wilkins and Wilson[11]) or exact analytical methods employing special techniques (Andrei,[12a] Wiegmann,[12b] Schlottman[12c]).

Orbital Degeneracy: Anderson has suggested that there may be some simplifications in the problem in the large orbital degeneracy limit. This has been further investigated by Ramakrishnan[8], and by Coleman.[13] Recently Gunnarsson and Schönhammer[14] have suggested that the lowest order variational wavefunctions, eqs. (4) and (5) etc. become exact for infinite orbital degeneracy. In the scaling regime, this may be seen from consideration of orbital angular momentum conservation. For impurity orbital angular momentum ℓ, (ℓ=1,$\cdots$,N) the mixing provides coupling only to conduction electron states of local angular momentum ℓ. The state $|\psi_0\rangle$ is connected to N different combinations $\sum_{k_\ell,\sigma} \alpha_{k\ell}\, a^+_{\ell\sigma}\, b_{k_\ell\sigma}|\psi_0\rangle$ in the lowest order. In the next or any higher order nested diagram, for a given initial ℓ, the same ℓ conduction electron must be absorbed and re-emitted due to angular momentum conservation as shown in Fig. 5. Therefore the lowest order contribution is proportional to N but any subsequent order proportional to 1.

Even though a singlet state appears both in the local moment (Kondo) regime and in the intermediate valence regime, there are important physical differences between the two. This may be observed from the value of the coefficient α_0 in Eq. (4), when normalized. From (6) and (7) for the local moment regime $\alpha_0 \approx |\Gamma/\varepsilon_0|$ so that charge fluctuations are negligible, and the spin fluctuations dominate. In the intermediate valence regime $\alpha_0 \approx 0(1/2)$. Therefore charge fluctuations are important - the spin fluctuations are of the same order.

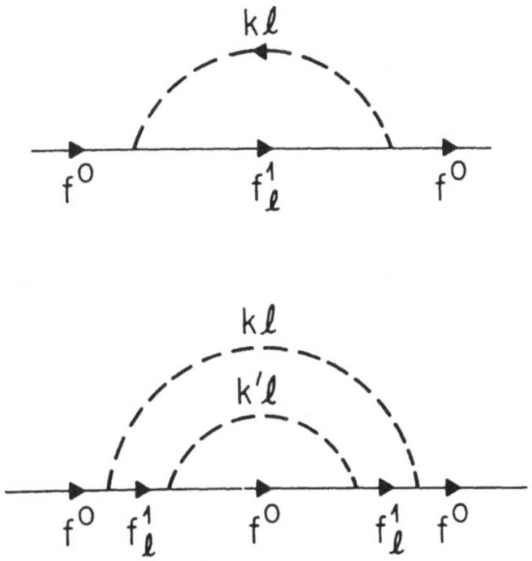

Fig. 5. Processes for intermediate valence impurity with orbital degeneracy.

Single M.V. Impurity with Both States Magnetic [15]

TmSe is the only known mixed-valence compound in which both the valence states, $f^{12}(3^+)$ and $f^{13}(2^+)$, (ignoring the crystal field splitting for 2^+, which is small compared to the hybridization) have a magnetic moment in the ground state. It is therefore of interest to examine whether this behavior follows from the property of a single impurity. Although there are quantitative differences from the case of the last section, the ground state of the single impurity is found to be a singlet in this case also.

As a simple model, let f^1 and f^2 be the two valence states of the impurity. Any other occupation states are assumed so far removed in energy that they don't need to be considered. Since f^2 has spin 1, there must be at least two distinct localized orbitals, $|a\rangle$ and $|b\rangle$, differing in their orbital magnetic quantum number, say $m_\ell(a) = 1$ and $m_\ell(b) = 0$. In the simplest model, an f level with only two orbitals need be considered. (Generalization to the orbitally degenerate case will briefly be mentioned.) The mixing interaction is assumed to be spherically symmetric (and spin-independent), so it conserves m_ℓ. Use the radial representation, $|k,\ell,m_\ell\rangle$, for the free electron band, and drop the ℓ index. Because of spherical symmetry, $|a\rangle$ mixes only with $|k,1\rangle$ and $|b\rangle$ mixes only with $|k,0\rangle$, i.e.,

$$\langle a|H_{mix}|k,1\rangle = \langle b|H_{mix}|k,0\rangle = V . \qquad (14)$$

Assume V and the density of states ρ to be constant, independent of k.

The most general wave function, just as before, is a sum of terms, each having the following form: The f level is occupied by one or two electrons; there is a corresponding number of holes in the Fermi sea so as to conserve the total number of electrons; and there are any number 0,1,2,••• of electron-hole excitations.

Since the impurity states have respectively $S_f = 1/2$ and $S_f = 1$, the possible states of the system are a singlet, a doublet, or a triplet. First, the simplest wave functions are considered.

a) **Spin doublet.** Let $|\psi_0\rangle$ be the ground state of the singlet Fermi sea. Take as a variational wave function:

$$|\psi_D\rangle_\uparrow = \alpha c^+_\uparrow|\psi_0\rangle + \sum_k (\beta(k)/\sqrt{6}) \; [(c^+_{a\uparrow}c^+_{b\downarrow} + c^+_{a\downarrow}c^+_{b\uparrow})c_{k0\downarrow} + 2c^+_{a\uparrow}c^+_{b\uparrow}c_{k0\uparrow}]|\psi_0\rangle \; . \qquad (15)$$

(The similar wave function with ↓ spin down is degenerate with (15)).

Let the total energy be:

$$E = E_0 + \varepsilon_f - \omega \qquad (16)$$

where ω is the binding energy. On varying $\langle H\rangle$ with respect to α and $\beta(k)$, the binding energy for the doublet state is found to be:

$$\omega_D = \frac{3\Gamma}{2\pi} \, \ell n \, \frac{W+\omega_D}{\omega_D} \; . \qquad (17)$$

b) **Spin singlet.** Consider a singlet state obtained by transferring an electron from states $(k,1,\sigma)$ to the f states (a,σ). Let now H_{mix} act and transfer electrons. As was done in Eq. (5) this can be represented by:

$$(a;k1) \leftarrow V \rightarrow (ab;k1,k'0) \leftarrow V \rightarrow (b;k'0) \leftarrow V \rightarrow (ab;k''1,k'0) \leftarrow V \rightarrow (a;k''1) \leftarrow V \rightarrow \cdots \qquad (18)$$

Note that, in contrast to the doublet case where the singly occupied a state did not mix with the singly occupied b state, here these same states are connected in order V^2. Hence they are both present in the wave function, which we take to be of the form:

$$|\psi_0\rangle = \sum_k (\alpha_1(k)/\sqrt{2})\,(c^+_{a\uparrow}c_{k1\uparrow} + c^+_{a\downarrow}c_{k1\downarrow})|\psi_0\rangle$$

$$+ \sum_{k'} (\alpha_0(k')/\sqrt{2})\,(c^+_{b\uparrow}c_{k'0\uparrow} + c^+_{b\downarrow}c_{k'0\downarrow})|\psi_0\rangle$$

$$+ \sum_{k,k'} (\beta(k,k')/\sqrt{3})\,[c^+_{a\uparrow}c^+_{b\uparrow}c_{k'0\uparrow}c_{k1\uparrow} + c^+_{a\downarrow}c^+_{b\downarrow}c_{k'0\downarrow}c_{k1\downarrow}$$

$$+ \frac{1}{2}(c^+_{a\uparrow}c^+_{b\downarrow} + c^+_{a\downarrow}c^+_{b\uparrow})\,(c_{k'0\uparrow}c_{k1\downarrow} + c_{k'0\downarrow}c_{k1\uparrow})]|\psi_0\rangle \tag{19}$$

The last term in $\beta(k,k')$, is formed by coupling the triplet ab state to a triplet 2 hole state, so as to form a singlet. The binding energy is found to be

$$\alpha(k)(\omega+\varepsilon_k) = \frac{3\Gamma}{2\pi}\int_0^W \frac{\alpha(k)+\alpha(k')}{\omega+\delta+\varepsilon_k+\varepsilon_{k'}}\,d\varepsilon_{k'} \tag{20}$$

where the variational equations require $\alpha_0 = \alpha_1 \equiv \alpha$ for compatibility and $\delta = \varepsilon_f+U$.

That we should have obtained an integral equation is not surprising in view of the fact that the two kinds of terms $c^+_{a\sigma}c_{kl\sigma}$ and $c^+_{b\sigma}c_{k'0\sigma}$ interact in order V^2 (or $\tilde{\Gamma}$). The first term in the integral in Eq. (20) alone would give the same binding energy as the doublet case. The interaction responsible for the binding energy of the doublet is of order V (the mixing strength) while that responsible for the doublet-singlet difference is of order V^2. Thus we expect this difference to be small compared to ω_D. Numerical results are given in Yafet-Varma.[15]

c) **Spin triplet.** We start as we did for the singlet, by transferring an electron from state $kl\sigma$ to the f state $a\sigma$. As in the singlet case, H_{mix} connects this state in second order to the state where the orbital b is singly occupied, but now we have two kinds of states in which both a and b are occupied: First, two holes in the singlet state coupled to the triplet (ab), and second, two holes in the triplet state coupled to the triplet (ab) to form a triplet.

The wave function of the triplet for $M_s = 0$ is taken to be:

$$|\psi_T\rangle_0 = \sum_k (\alpha_1(k)/\sqrt{2})(c^+_{a\downarrow}c_{k1\downarrow} - c^+_{a\uparrow}c_{k1\uparrow})|\psi_0\rangle$$

$$+ \sum_{k'} (\alpha_0(k')/\sqrt{2})\ (c^+_{b\uparrow}c_{k'0\downarrow} - c^+_{b\uparrow}c_{k'0\uparrow})|\psi_0\rangle$$

$$+ \frac{1}{2}\sum_{kk'} \beta(kk')(c^+_{a\downarrow}c^+_{b\uparrow} + c^+_{a\uparrow}c^+_{b\downarrow})(c_{k'0\uparrow}c_{k1\downarrow} - c_{k'0\downarrow}c_{k1\uparrow})|\psi_0\rangle$$

$$+ \frac{1}{\sqrt{2}}\sum_{kk'} \gamma(kk')(c^+_{a\downarrow}c^+_{b\downarrow}c_{k'0\downarrow}c_{k1\downarrow} - c^+_{a\uparrow}c^+_{b\uparrow}c_{k'0\uparrow}c_{k1\uparrow})|\psi_0\rangle \quad (21)$$

The binding energy is given by the eigenvalue equation:

$$\alpha(k)(\omega+\varepsilon_k) = \frac{3\Gamma}{2\pi}\int \frac{\alpha(k)+(1/3)\alpha(k')}{\omega+\delta+\varepsilon_k+\varepsilon_{k'}}\, d\varepsilon_{k'} \quad (22)$$

This differs from Eq. 20 for the singlet in having the factor (1/3) in front of $\alpha(k')$. Thus the effective potential that couples $\alpha(k)$ to $\alpha(k')$ is only 1/3 of what it is for the singlet, and hence the triplet lies above the singlet but below the doublet.

Interaction Between Two Impurities

Consider two impurities at sites A and B embedded in the conduction electron sea. For simplicity, we take the model in which each site has two configurations in which either one (f^1) or two (f^2) localized electrons are close in energy in the sense specified in the last section. In f^2 we will consider that, for the isolated ion, one of the states S = 0 or S = 1 is of much lower energy than the other so that only one of them need be considered. In this way we will simulate the situation when both or one of the valence states have magnetic moments.

If the interactions between the two sites are ignored, the results of the previous sections for the interaction between each of the impurity sites and the conduction electrons may be summarized by the Hamiltonian

$$H_o = \sum_S H_{AS} + H_{BS} \quad (23)$$

where S denotes the total spin. H_{AS} can be written in terms of the projections $P_1\psi_A$ and $P_2\psi_A$, where ψ_{AS} is the eigenstate at A and $P_1\psi_A$, $P_2\psi_A$ project on to the space in which there are 1 and 2 electrons in the localized orbital respectively.

$$H_{AS} = H_{AS}^{(1)} |P_1\psi_{AS}\rangle \langle P_1\psi_{AS}| + H_{AS}^{(2)} |P_2\psi_{AS}\rangle\langle P_2\psi_{AS}|$$

$$+ V_{AS}^{(1,2)}(|P_1\psi_{AS}\rangle \langle P_2\psi_{AS}| + \text{H.C.}) \tag{24}$$

The last term gives the mixing between the configurations with the same total spin. $H_{AS}^{(1)}$, $H_{AS}^{(2)}$ contain all the effects of renormalizations due to excitations of electron-hole pairs to any order. It is understood that the total number of electrons (local and Fermi-sea) is the same in all terms. Similarly H_{BS} is defined.

The particle-hole pairs in the Fermi-sea contained in ψ_A, ψ_B can also travel between sites A and B, thereby coupling them. It is convenient to classify the conduction electron states by their angular momentum about sites A and B and by their parity about the mid-plane between the sites. Also, for considering the interactions, it is necessary to have as basis states the product space of ψ_A and ψ_B.

The interaction can only be of two kinds:[16]

(1) Spin interactions between the two sites without any charge exchange. These may be written as

$$H_{AB}^{R}(S_A,S_B) = \sum_{\alpha,\beta=1,2} H_{AB,\alpha\beta}|P_\alpha\psi_A S_A, P_\beta\psi_B S_B\rangle\langle P_\alpha\psi_A S'_A, P_\beta\psi_B S'_B| , \tag{25}$$

with the requirement that $\vec{S}_A + \vec{S}_B = \vec{S}'_A + \vec{S}'_B$. A typical term of this interaction is illustrated in Fig. 6. It is clear that to leading order this interaction is of order $V^4\rho^3$; it is merely like the RKKY interaction with a spin structure $\vec{S}_A \cdot \vec{S}_B$. In the context of valence fluctuation problems it has been discussed before.[16,17] Higher orders (more electron-hole pairs) than of the type illustrated in Fig. 5 have necessarily the same spin structure.

Orbital Degeneracy: For orbital degeneracy N, the RKKY interactions become of O(1/N) compared to the leading contribution, i.e., $O(V^2)$, to the singlet-triplet energy difference from single-site interactions with the conduction electrons. This is seen from the fact (compare figs. 5 and 2) that RKKY, which gives a contribution proportional to V^4, is similar to the V^4 contribution to the single-site energies.

(2) Spin interactions with charge exchange. These may be written as

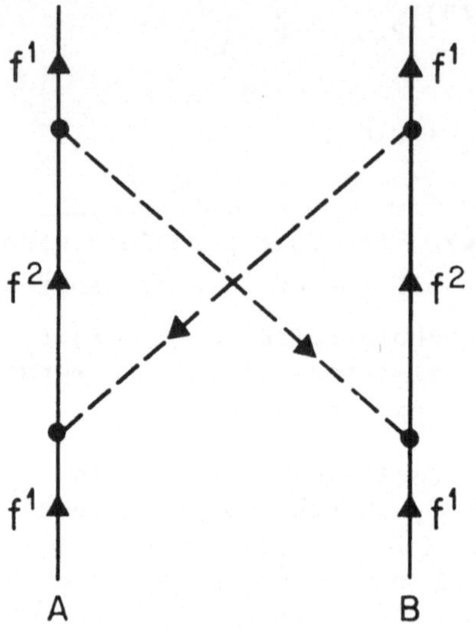

Fig. 6. Lowest order process for interaction between two intermediate valence impurities without charge exchange.

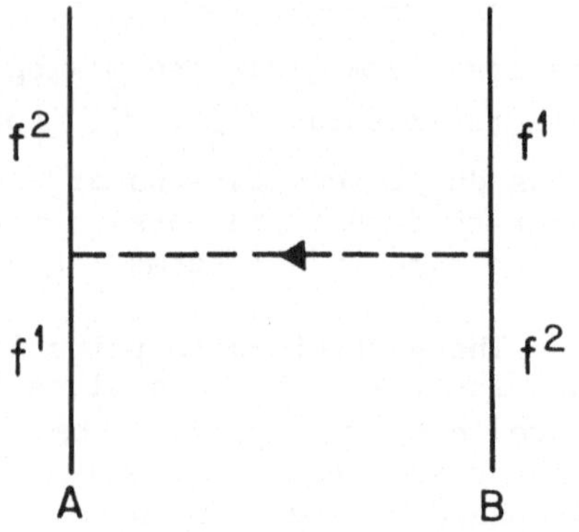

Fig. 7. Lowest order process for interaction with charge exchange.

$$H^C_{AB}(S_A,S_B) = \sum_{\alpha\neq\beta=1,2} H_{AB,\alpha\beta}\left|P_\alpha\psi_A S_A,\ P_\beta\psi_B S_B\right\rangle\left\langle P_\beta\psi_A S'_A,\ P_\alpha\psi_B S'_B\right| \ , \qquad (26)$$

again with $\vec{S}_A + \vec{S}_B = \vec{S}'_A + \vec{S}'_B$.

This process is illustrated in Fig. 7. It is evident that for the fluctuating valence case, when the energy of the f^1 configuration plus the Fermi energy is degenerate with the energy of the f^2 configuration, this process is of order $V^2\rho$ and therefore much more important than spin interactions without charge exchange. In the so-called Kondo lattice case, when the difference in energy of the two charge configurations is ε_0, the charge-exchange interaction is reduced and becomes of order $V^2\rho(V^2\rho/\varepsilon_0)^2$. This however means that for $\varepsilon_0\rho \lesssim 1$, this interaction is larger in magnitude than the RKKY-type interactions.

Orbital Degeneracy: In a two-site effective Hamiltonian, the coefficient of this process **is of the same order** as the singlet-triplet energy difference at a site. This is because this process is real, (not virtual as in the case of RKKY) and of order V^2 because the original configurations get switched around.

This kind of interaction was first proposed in a different context by Zener,[18] and investigated further by Anderson and Hasegawa[19] and by deGennes[20] and has been termed "Double-exchange". The physics of this process is simply that the energy gained by delocalizing the f-electrons depends upon the relative spin configuration of the initial and the final sites. Due to the large Hund's rule coupling, the energy gain is maximized when the spins at the two sites are aligned parallel. The spin dependence of this interaction can be shown to be of the form:

$$\left|\vec{S}_A + \vec{S}_B\right| \; . \tag{27}$$

There is an interesting difference, however, if one of the valence states has no magnetic moment. In this case the kinetic energy reduction is independent of the direction of the moment of the other valence state, as is physically obvious and also as given by Eq. 27. I believe this point is important in understanding the difference between TmSe and the other known mixed-valence materials. This may be seen from the following argument.

Physical Consequences of the Form of the Interactions

Consider a material where one of the states has no magnetic moment and the ratio of the two valence states is x. The f-electron delocalization energy in the paramagnetic phase is of order $V^2\rho x(1-x)$. The gain in energy if it orders magnetically is only due to the RKKY type interactions, which are $O(V^4\rho^3)$. A simple application of Stoner-like criteria for magnetism, $I\chi = 1$, where I is the interaction energy which is of $O(V^4\rho^3)$, and χ is the itinerant electron susceptibility, equal roughly to $O([V^2\rho x(1-x)]^{-1})$, leads to the qualitative conclusion that an ordered magnetic state is not possible unless x or $1 - x \approx O(V^2\rho^2) < 10^{-4}$!

For a material like TmSe, however, the dominant interaction tends to align the moment. On this qualitative basis, TmSe was predicted[21] to order ferromagnetically for suitable x. This was indeed found to be the case.[22] Additional signatures that the "Double-exchange" interaction is the dominant effect are that the resistivity drops by about two orders of magnitude on ferromagnetic order[22] and that the ordered moment changes with x.[23]

Remarks on the Lattice Problem

On the basis of the solution to the single-site and the two-site problem, one may write down an effective Hamiltonian for the lattice, including only f electron spin and charge operators and effectively eliminating the wide band. The solution of this quite non-trivial problem would be a worthwhile exercise but has not been attempted.

In the literature there are several CPA solutions to the problem, which regard different spin and charge configurations as components of an alloy and determine their relative probabilities self-consistently. Such single-site approximations to the problem beg the real issues.

A refinement of this is the Bethe-Peierls approximation attempted by Schlüter and me.[24] This approximation cannot give low temperature properties accurately, since long wavelength spin and charge excitations are not properly considered in this method.

Another alternative is real space numerical methods[25,26] with use of some renormalization scheme. Because at least 12 different states must be considered at each site for the minimal mixed-valence problem, it does not appear feasible for more than 8 sites on the largest computer presently available.

A variational method for the lattice is presented below for the specific purpose of answering the question of why the f-electrons are not localized when the resonance due to them is close to the Fermi-level.

A calculation using the multiplicate renormalization group to obtain the solution to the one-dimensional mixed-valence lattice in the leading logarithmic approximation has recently been done by Fred Zawadowski and me and will be published elsewhere.

Our principal results are that for $U \to \infty$, an effective Hamiltonian can be generated in which only the subspace (0,1) for the number of correlated particles per site enters in the calculations. With this Hamiltonian, a self-consistent Green's function of Fermi-liquid character can be defined for the heavy (f) particles, and a renormalization group scheme defined systematically. The renormalization group results lead to the weak-coupling fixed point demonstrating the stability of the Fermi-liquid. We have also examined the response functions of this Fermi-liquid and find the triplet superconducting response to be the most divergent.

This last result is very interesting in view of the experimental discovery of superconductivity in some of the heavy fermion solids ($CeCu_2Si_2$, UBe_{13}). I find it hard to understand them as ordinary phonon-induced s-wave superconductors, since the energy scale of the Coulomb repulsions is, if anything, smaller than that of the phonon-induced interactions. Triplet superconductivity is possible with short-range repulsions. The usual argument against triplet superconductivity - that any impurity scattering is pair-breaking - is not very strong for these materials because of their very small correlation lengths.

There is a fond hope in some quarters[27] that the problem is only that of a collection of isolated impurities. This is justified by considering infinite orbital degeneracy and ignoring the RKKY interactions. I believe that the physics at orbital degeneracy of order 10 is different from that at infinity, and that interaction effects due to charge fluctuations as well as the RKKY interactions are important to consider. It should be realized that the arguments presented so far that lead to the isolated impurity picture would rule out magnetic order in all rare-earth materials.

A few remarks about the so-called 'Kondo lattice': A possibility mentioned by me[4] briefly was that the special properties of the intermediate valence compounds might arise because the moments at a given site might renormalize to zero due to the interactions with the conduction electrons at a temperature higher than the interactions between the moments. Independently Doniach[28] applied this idea for the case that there are no significant f-charge fluctuations, i.e., the 90% of the cases that are not intermediate valence, as discussed in connection with fig. 1. Then, the temperature at which the moment renormalizes to zero is the Kondo temperature $T_K \approx W \exp(-W/2J)$ where W is the conduction electron bandwidth and J the exchange integral of the localized electrons, assumed s-like, with the conduction electrons that are s-wave about a given site. The interaction among the moments is $E_{int} \approx J^2/W$. So if $T_K > E_{int}$, the ground state may remain non-magnetic. The problem with this is that the cross-over occurs at J of the order of W/10 and this high value of J is impossible. Secondly in careful renormalization group calculations on the one-dimensional Kondo lattice, Jullien[25] finds that there is a gap in the quasiparticle spectrum for all values of J indicating the argument may not be quite right.

The problem that a large value of J is necessary disappears, however, if large orbital degeneracy, N, is considered. In this case $T_K \approx W \exp(-W/NJ)$, but E_{int} remains as above. For $W \approx 1$ eV, $N \approx 10$, $J \approx O(10$ meV) now satisfy the criteria $T_K > E_{int}$. Such values for J are not too unreasonable. Moreover this calculation yields $T_K \approx O(1$ meV). For materials like $CeCu_2Si_2$, which are probably integral valent, if the measured low-temperature specific heat is fitted to T/T_K, T_K of O(1 meV) is found! Should this idea therefore be considered seriously?

The Gutzwiller Method

The Gutzwiller variational method[29] was used by Brinkman and Rice to study the metal-insulator transition for the Hubbard model. I will use it here (work done in collaboration with Lisa Randall and Werner Weber) to glean some insight as to why f-electrons remain delocalized in the mixed-valence materials despite large correlation energies.

Consider a mixed-valence Hamiltonian

$$H = \sum_{i,\sigma} t_a(a^+_{i\sigma}a_{i+1,\sigma} + c.c.) + U \sum_{i,\sigma} n_{i\sigma}n_{i-\sigma}$$

$$+ \sum_{i,\sigma} t_b(b^+_{i\sigma}b_{i+1,\sigma} + c.c.) + V \sum_{i,\sigma} a^+_{i\sigma}b_{i\sigma} + c.c. \quad (28)$$

where $n_{i\sigma} = a^+_{i\sigma}a_{i\sigma}$. For $t_b = V = 0$, this is just the Hubbard model, for which at 1/2 filling, m* (the effective electron mass) = q^{-1}, (the quasi-particle renormalization), behaves according to Brinkman-Rice as

$$m^* = m/[1 - (U/U_c)^2], \quad (29)$$

where U_c, the critical U for the metal-insulator transition, is equal to $8\bar{t}$ where $\bar{t}$ is the average one-electron kinetic energy.

In order to eliminate the possibility of a semiconducting gap, for the mixed-valence case, we consider t_a sufficiently large (but $t_b >> t_a$) so that for U = 0 and finite V, a gap does not appear at the chemical potential. We have instead (in one dimension) two Fermi-wavevectors, k_α and k_β, where α and β are the band-indices. As U is gradually increased we must have $k_\alpha + k_\beta$ = constant, by Luttinger's theorem, until a phase transition intervenes. k_α/k_β, however, can be used as a variational parameter. Throughout we deal with the half-filled situation, i.e., two electrons per site.

As a generalization of Gutzwiller to the case of two orbitals, a and b, per site but with only one of them, a, correlated, construct a variational wave-function in the following manner:

$$\Phi_{G_a,\Gamma_a,G_b,\Gamma_b} = \prod_{G_a} a^+_{g\uparrow} \prod_{\Gamma_a} a^+_{\gamma\downarrow} \prod_{G_b} b^+_{g\uparrow} \prod_{\Gamma_b} b^+_{g\downarrow}|0\rangle$$

is a wave-function in which $G_a = (g_1,\cdots,g_{m_a})$ are a set of sites occupied by ↑ spin electrons in the a orbital, $\Gamma_a = (\gamma_1,\cdots,\gamma_{\mu_b})$ the set of sites occupied by ↓ spin electrons in the b orbitals, and similar definitions for G_b and Γ_b. The variational wavefunction chosen is

$$\psi = \sum_{G_a,G_b,\Gamma_a,\Gamma_b} A(G_a,G_b,\Gamma_a,\Gamma_b)\Phi \quad (30)$$

with

$$A(G_a, G_b, \Gamma_a, \Gamma_b) = \eta^{\nu} T_{G_a} G_{G_b} T_{\Gamma_a} T_{\Gamma_b} , \tag{31}$$

where the T's are the determinants given in Gutzwiller's notation by

$$T_{G_a} = \left(L^{-1/2} e^{ik_a g_a} R(k) \left| \begin{matrix} k_1 \alpha k_1 \beta & k_2 \alpha k_2 \beta \cdots k_{m_a} \beta \\ g_1 a g_1 b & g_2 a g_2 b \cdots g_{m_a} b \end{matrix} \right. \right) . \tag{32}$$

R(k) is the 2 × 2 rotation matrix which transforms from orbital to band space. η is calculated by considering the fraction of sites which are doubly occupied in the a-orbital and ν is a variational parameter.

For R(k) we choose

$$R(k) = r(k)\mathbf{R}_o(k)$$

where $\mathbf{R}_o(k)$ is the matrix for the uncorrelated problem and r(k) has

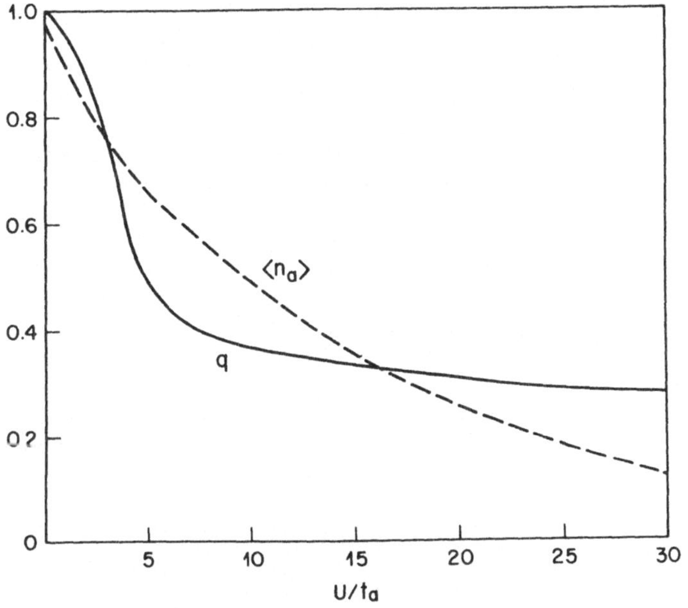

Fig. 8. Quasi particle renormalization and the occupation of the correlated orbitals vs. U/t_a calculated by the Gutzwiller method.

three different values r_1, r_2 and r_3 for $k < k_a$, $k_a < k < k_b$, $k > k_b$ respectively. The most important of these is r_2 and in the calculations r_2 is used as another variational parameter with $r_1 = r_3 = 1$. After some extensive calculations, the result of minimization of the energy with respect to all the variational parameters yields the effective mass in the "heavy" band as a function of U/t_a shown in fig. 8. The remarkable result is that with the Gutzwiller variational method, this model has no delocalized to localized transition for the a-type electrons even for arbitrarily large correlation energies. Also plotted is the number of electrons in the a-type orbital. For large U/t_a this decreases. This means that for increasing correlation energy localization is prevented by transferring electrons to the wider band and paying a cost in increased kinetic energy. I believe this physical point will persist (and perhaps manifest itself even more strongly) in methods which treat correlations better.

References

1. J.C. Nickerson, R.M. White, K.N. Lee, R. Bachman, T.H. Geballe and G.W. Hull, Jr., Phys. Rev. 3:2030(1971).
2. A. Jayaraman, V. Narayanamurti, E. Bucher and R.G. Maines, Phys. Rev. Letters 25:1430(1970).
3. M.B. Maple and D. Wohlleben, Phys. Rev. Letters 27:511(1971).

4a. C.M. Varma, Rev. Mod. Phys. 48:219(1976).

4b. C.M. Varma, in "Valence Instabilities and Related Narrow Band Phenomena", Edited by R.D. Parks, Plenum, New York (1977).

5. C.M. Varma and Y. Yafet, Phys. Rev. B13:295(1975).
6. A. Bringer and H. Lustfeldt, Z. Phys. B28:213(1977).
7. F.D. Haldane, Phys. Rev. Letters 40:416(1978).
8. T.V. Ramakrishnan in "Valence Fluctuations in Solids", Edited by L.M. Falicov, W. Hanke & M.B. Maple, North Holland (1981).
9. H. Keiter, in "Valence Fluctuations in Solids", Edited by L.M. Falicov, W. Hanke & M.B. Maple, North Holland (1981).
10. S. Inagaki, Prog. Theor. Phys. 62:1441(1979).
11. H.R. Krishnamurthy, K.G. Wilson and J.W. Wilkins, Phys. Rev. Letters 35:1101(1975).

12a. N. Andrei, Phys. Rev. Letters 45:379(1980).

12b. P.B. Wiegmann, Pis'ma Zh. Eksp. Teor. Fiz. 31:393(1980).

12c. See the chapter by P. Schlottman.

13. See the chapter by P. Coleman.
14. O. Gunnarsson and K. Schönhammer, Phys. Rev. Letters 50:604(1983), and also their chapter in this book.
15. Y. Yafet and C.M. Varma, Phys. Rev. (submitted).
16. Reference 4b.
17. See the articles by T.V. Ramakrishnan, H. Keiter, and C.M. Varma and M. Schlüter, in Valence Fluctuations in Solids, Edited by L.M. Falicov, W. Hanke, and M.B. Maple, North Holland (1981).

18. C. Zener, Phys. Rev. 82:403(1951).
19. P.W. Anderson and H. Hasegawa, Phys. Rev. 100(1955)675.
20. P.G. de Gennes, Phys. Rev. 118(1960)141.
21. C.M. Varma, Solid State Comm. 30:537(1979).
22. B. Battlogg, H.R. Ott and P. Wachter, Phys. Rev. Letters 42:282(1979).
23. C. Vettier, J. Flouquet, J.M. Mignot and F. Holtzberg, J. Magn. Magn. Mat. 15-18:987(1980).
24. C.M. Varma and M. Schlüter in "Valence Fluctuations in Solids", Edited by L.M. Falicov, W. Hanke & M.B. Maple, North Holland (1981).
25. R. Jullien in "Valence Instabilities", Edited by P. Wachter and H. Boppart, North Holland (1982).
26. C.M. Varma, M. Schlüter and Y. Yafet, in "Valence Instabilities", Edited by P. Wachter & H. Boppart, North Holland (1982).
27. See for example the chapter by D. Newns.
28. S. Doniach, Physica 91B:231(1977).
29. W.F. Brinkman and T.M. Rice, Phys. Rev. B2:4302(1970).
M.C. Gutzwiller, Phys. Rev. 137:A1726(1965).

"BAND BASED" METHODS FOR MAGNETIC RESPONSE

D. M. Edwards

Department of Mathematics, Imperial College
London SW7 2BZ
U.K.

1. INTRODUCTION

These lectures are concerned with methods of treating magnetic response within the framework of band theory. These methods have mostly been developed with application to transition metals and alloys in mind, and such applications form the central core of the present material. However, we shall attempt to discuss the extent to which such methods may be applied to other systems, such as actinide and even intermediate valence compounds. An approach based on band theory is certainly called for when the electrons responsible for the formation of magnetic moments are strongly itinerant, moving rapidly from atom to atom. In this case, as with the 3d electrons in transition metals or 5f electrons in some actinide systems, the magnetic electrons contribute importantly to bonding and to the Fermi surface. However, the application of band theory to such systems cannot be too naive. The existence of magnetic moments indicates that the interaction between electrons is important so that correlation effects are strong. Treatment of correlation has proceeded along various lines:

1. Assume that long range correlation reduces the Coulomb interaction to a screened interaction which is predominantly intra-atomic. One then writes down a model tight-binding Hamiltonian which contains a Hubbard interaction parameter U or, in the realistic many-band case, a set of such U parameters representing strong on-site interactions between electrons in the various orbitals. One of the following approaches is then adopted.

(a) Treat the strong interactions U by many-body methods which explicitly introduce short-range correlation by going beyond the Hartree-Fock (HF) and random phase approximation (RPA) level of treatment.

(b) Assume that short-range correlation is approximately taken care of by writing down a new model Hamiltonian in which the strong U interactions are replaced by weaker effective ones, U_{eff}. This model is then treated at the HF-RPA level.

2. Use spin density functional theory. Method 1(a) is extraordinarily difficult to apply to real systems and has been largely restricted to discussions of many-body effects at $T = 0$, for example the large body of work [1-4] on the so-called 6eV satellite in photemission from Ni. The great importance of SDFT for magnetism is that it essentially justifies method 1(b) and gives a practical method for determining all the parameters of the model Hamiltonian. This is discussed in §2. At the same time it is clear that this approach is only rigorous for static properties at $T = 0$. However, for the systems of interest, the range of frequencies probed by inelastic neutron scattering is small compared with electronic excitation energies, characterized by the band width. Thus the extension of method 1(b)-2 to calculate the dynamical susceptibility $\chi(\underline{q},\omega)$ at $T = 0$ is very reasonable. Another important extension of this approach is to discuss finite temperature properties within an adiabatic approximation. For a magnetic system, even above its ordering temperature, T_c, the direction of the local magnetic moment fluctuates slowly on the electronic energy scale. At a given instant the electronic system may then be regarded as in the ground state appropriate to a certain disordered exchange potential. The development of this idea is described in §3. It is so far restricted to static properties at finite T although a more phenomenological extension has been made to discuss $\chi(\underline{q},\omega)$. To give a unified treatment of correlation, including many-body effects such as satellites in the one-particle spectrum and a microscopic theory of $\chi(\underline{q},\omega)$ at finite T, it is necessary to return to method 1(a). A recent idea for solving this difficult problem is mentioned in §4.1. A practical solution to this problem for arbitrary interaction strength would, of course, lead to the Utopia of a unified theory of systems ranging from transition metals, with relatively broad 3d bands, to intermediate valence materials with very narrow 4f bands. However, much useful information, over a wide range of materials, is to be gained from unrestricted HF theory with spin-orbit coupling included. This should become clear from the discussion of uranium-based compounds in §2.5 and cerium-based compounds in §4.2

2. MAGNETIC RESPONSE AT T = 0

2.1 Spin Density Functional Theory (SDFT)

The important result of SDFT is that the ground-state energy, together with the density $\rho_\sigma(\underline{r})$ of electrons of spin $\sigma(\sigma = \pm 1$ or $\uparrow,\downarrow)$, may be determined exactly, in principle, by solving a self-consistent one-electron problem. For simplicity we neglect spin-orbit coupling until further notice. The fundamental Hartree-type equations are

$$[-\frac{\hbar^2}{2m}\nabla^2 + V_\sigma(\underline{r})]\, u_{i\sigma}(\underline{r}) = \varepsilon_{i\sigma}\, u_{i\sigma}(\underline{r}) \tag{1}$$

where

$$V_\sigma(\underline{r}) = v(\underline{r}) - \mu_B \sigma B(\underline{r}) + e^2 \int \frac{\rho(\underline{r}')}{|\underline{r}-\underline{r}'|} d^3r' + V^\sigma_{XC}[\rho_\uparrow(\underline{r}), \rho_\downarrow(\underline{r})] \tag{2}$$

and

$$\rho_\sigma(\underline{r}) = \sum_{\varepsilon_{i\sigma}<E_F} |u_{i\sigma}(\underline{r})|^2, \qquad \rho = \rho_\uparrow + \rho_\downarrow . \tag{3}$$

Here $v(\underline{r})$ is the external field due to the nuclei, $B(\underline{r})$ is an applied magnetic field in the direction of spin quantization and $V^\sigma_{XC}(\underline{r})$ is a universal functional of the spin densities. The equations are readily generalised to include a transverse magnetic field. In the original derivation of these equations [5,6] it was assumed from the outset that the exact $\rho_\sigma(\underline{r})$ for the interacting electron system can be realized by a non-interacting system in some suitable external potentials $V_\sigma(\underline{r})$. Following subsequent work [7-9] one can avoid this assumption initially and still obtain essentially the Kohn-Sham equations (1) - (3). However, in general it does not follow that the true ground state corresponds to putting the N electrons into the N <u>lowest</u> one-electron states, as implied by the summation condition $\varepsilon_{i\sigma} < E_F$ in (3). Nevertheless an argument of Janak [10] shows that if a self-consistent solution of the Kohn-Sham equations does exist with the N lowest states occupied, then this solution corresponds to the ground state. The condition that such a self-consistent solution of the Kohn-Sham equations exists appears to be a fundamental limiting factor in the useful applicability of SDFT and to restrict the range of many-electron systems whose ground state has a close correspondence with a non-interacting system. In view of the successful applications of SDFT to transition metals (see recent reviews [11,12]) there is no reason to suppose that the condition breaks down in these systems. However, it may do so in more exotic systems such as intermediate valence materials. A practical limitation on the accuracy of SDFT is the approximation which must be made to the functional V^{σ}_{XC}, this being defined as the functional derivative $\delta E_{XC}/\delta\rho_{\sigma}(\underline{r})$ of the exchange-correlation energy $E_{XC}[\rho_{\uparrow}(\underline{r}), \rho_{\downarrow}(\underline{r})]$. The latter is usually derived from theories of the uniform electron gas (local density approximation, LDA).

Since the spin density $m(\underline{r}) = \rho_{\uparrow} - \rho_{\downarrow}$ may in principle be calculated exactly for any external magnetic field we may calculate spin susceptibilities exactly. It is easily shown [11] that for a paramagnet the spin density induced at $\underline{r}$ by a field at $\underline{r}'$ is given by

$$\chi(\underline{r},r') = \chi^{0}(\underline{r},\underline{r}') + \iint d^{3}r_{1}\, d^{3}r_{2}\, \chi^{0}(\underline{r},\underline{r}_{1})\, I_{XC}(\underline{r}_{1},\underline{r}_{2})\, \chi(\underline{r}_{2},\underline{r}') \tag{4}$$

where

$$I_{XC}(\underline{r},\underline{r}') = -\frac{\delta^{2} E_{XC}}{\delta m(\underline{r})\, \delta m(\underline{r}')} \tag{5}$$

and χ^{0} is the non-selfconsistent susceptibility derived from (1). In the LDA E_{XC} is of the form $\int d^{3}r\rho(\underline{r})\varepsilon_{XC}(\rho(\underline{r}), m(\underline{r}))$, where $\varepsilon_{XC}(\rho,m)$ is the exchange-correlation energy per particle of a uniform electron gas with density ρ and spin density m. Hence

$$I^{LD}_{XC}(\underline{r},\underline{r}') = f_{XC}(\rho(\underline{r}))\, \delta(\underline{r}-\underline{r}') \tag{6}$$

where $f_{XC}(\rho) = -[\partial^{2}(\rho\, \varepsilon_{XC}(\rho,m))/\partial m^{2}]_{m=0}$. At sufficiently low frequencies ω it is reasonable to use (4) as an approximate equation for the dynamical susceptibility $\chi(\underline{r},\underline{r}';\omega)$, simply replacing $\chi^{0}(\underline{r},\underline{r}')$ by $\chi^{0}(\underline{r},\underline{r}';\omega)$. An equation similar to (4) holds for the transverse susceptibility of a ferromagnet, with $I_{XC}(\underline{r},\underline{r}')$ replaced by a function proportional to $f(\underline{r})\,\delta(\underline{r}-\underline{r}')$, in the LDA, with

$$f(\underline{r}) = [V_{\downarrow}(\underline{r}) - V_{\uparrow}(\underline{r})]/[\rho_{\uparrow}(\underline{r}) - \rho_{\downarrow}(\underline{r})]\,. \tag{7}$$

Here $V_{\uparrow}$, $V_{\downarrow}$ are the self-consistent ground-state potentials of (1). Clearly, as pointed out previously [13], the transverse susceptibility in the LDA is exactly equivalent to the RPA susceptibility for the many-body Hamiltonian

$$H = \sum_{i}[\underline{p}_{i}^{2}/2m + v_{o}(\underline{r}_{i})] + \sum_{i<j}\sum \int d^{3}r\, f(\underline{r})\, \delta(\underline{r}-\underline{r}_{i})\, \delta(\underline{r}-\underline{r}_{j})\,. \tag{8}$$

Here $f(\underline{r})$ is defined by (7) and $v_{o}(\underline{r})$ is chosen so that the Hartree-Fock potentials $v_{o} + f\rho_{\downarrow}$, $v_{o} + f\rho_{\uparrow}$ coincide with $V_{\uparrow}$, $V_{\downarrow}$ respectively. We shall return to this equivalence to relate existing theories based on methods 1(b) and 2 of §1. The longitudinal susceptibility of a ferromagnet is more complicated because it is coupled to the density response. Thus the interaction corresponding to I_{XC} is a 2 x 2 matrix with elements of the form

$$K_{\sigma\sigma'} = \frac{e^2}{|r - r'|} + \frac{\delta^2 E_{XC}}{\delta\rho_\sigma(\underline{r})\,\delta\rho_{\sigma'}(\underline{r}')} \qquad (9)$$

We may obtain an approximation to the uniform spin susceptibility χ of a paramagnet as follows. The spin density in the non-interacting system arising from a uniform magnetic field is proportional to

$$\chi^0(\underline{r}) = \int \chi^0(\underline{r},\underline{r}')\, d^3 r' \propto \sum_i |u_i(\underline{r})|^2\, \delta(\varepsilon_i - E_F) \qquad (10)$$

and $\int\chi^0(r)\, d^3r \propto N(E_F)$, the electronic density of states per atom. If we assume that approximately $\int\chi(\underline{r}, r')d^3r' \propto \chi^0(\underline{r})$, we find by integrating (4) with respect to $\underline{r}$ and $\underline{r}'$ and using (6), that

$$\chi = \frac{\chi^0}{1 - U_{eff}\, N(E_F)} \qquad (11)$$

where

$$U_{eff} = \int d^3r\, \gamma^2(\underline{r})\, f_{XC}(\rho(\underline{r})) \qquad (12)$$

with

$$\gamma(\underline{r}) = \sum_i |u_i(\underline{r})|^2\, \delta(\varepsilon_i - E_F)/N(E_F). \qquad (13)$$

Vosko and Perdew [14] have shown that the approximation in going from (4) to (11) yields a lower bound for χ. Janak [15] has evaluated (11) for a large number of metals and finds that the "Stoner criterion" $U_{eff}\, N(E_F) > 1$ is indeed only satisfied for the ferromagnetic elements Fe, Co and Ni. U_{eff} increases with atomic number as one ascends the 3d series.

2.2 Magnetic Ordering in the 3d metals

On taking the Fourier transform of (4) we have an intractable matrix equation for $\chi(\underline{q}+\underline{K}_s, \underline{q}+\underline{K}_t)$ where $\underline{K}_s$, $\underline{K}_t$ are reciprocal lattice vectors. To investigate the occurrence of orderings other than ferromagnetic we are interested in the wave-vector dependent susceptibility $\chi(q) \equiv \chi(\underline{q}, \underline{q})$. We shall describe a general method of approximately solving an equation like (4) for $\chi(\underline{q})$, or rather $\chi(q,\omega)$, in the next section . Here, however, we proceed more simply. The Stoner-like result at the end of the last section suggests that SDFT gives limited support to a molecular field picture in which a magnetic moment μ on an atom gives rise to a local molecular field $\alpha\mu$ where $\alpha = U_{eff}/(2\mu^2_B)$. This picture is consistent with the fact that in band calculations for Fe and Ni based on SDFT the exchange splitting of the 3d band is found to be fairly constant, although slightly larger at the top of the band than at the bottom. The molecular field picture implies the simple result $\chi(\underline{q}) = \chi^0(\underline{q})/(1-\alpha\chi^0(\underline{q}))$, where χ, χ^0 are now actual susceptibilities instead of the reduced quantities implied by (4). If $\chi^0(\underline{q})$ has a maximum value for a non-zero $\underline{q}$, and $\alpha\chi^0(\underline{q}) > 1$, the system will order antiferromagnetically. If the maximum occurs at a $\underline{q}$ which is incommensurate with the lattice we will find a spin density wave as in Cr.

The local (site-diagonal) susceptibility $\chi^0_\ell = N_a^{-1}\Sigma_{\underline{q}}\chi^0(\underline{q})$, N_a being the number of atoms, is a quantity of interest for two reasons. First of all, if $\chi^0_\ell > \chi^0 (= 2\mu^2_B N(E_F))$, the maximum value of $\chi^0(\underline{q})$ must occur for $\underline{q} \neq 0$. Hence this inequality, together with $\alpha\chi^0_\ell > 1$, is a sufficient condition for antiferromagnetic order. If, however, $\chi^0 > \chi^0_\ell$ and $\alpha\chi^0 > 1$ ferromatnetism may occur. The second point of interest is that $\alpha\chi^0_\ell > 1$ is the condition for a moment to occur on a site when the molecular field on all other sites is taken to be zero. More significantly it is an approximate condition for moments to exist on all atoms in a disordered state, that is, pointing in random directions; in this case the molecular fields on sites other than a central one considered average to zero. Hence if $\chi^0 > \chi^0_\ell > \alpha^{-1}$ we may expect ferromagnetism at $T = 0$ and atomic magnetic moments may persist in the disordered state above T_C. (We assume that T_C is well below the electronic degeneracy temperature so that the T dependence of χ^0_ℓ may be neglected). If, on the other hand, $\chi^0 > \alpha^{-1} > \chi^0_\ell$ we expect ferromagnetism at $T = 0$ but the atomic moments cannot persist in a state of complete disorder. We return to this point in §3. In Fig. 1 we show a schematic plot of χ^0, χ^0_ℓ and α^{-1} as a function of E_F for the 3d series. Detailed calculations showing this behaviour have been made for fcc and bcc 3d bands by Heine et al.[16]. We see that for Ni, Co, Fe $\chi^0 > \chi^0_\ell$ so that ferromagnetism is expected, whereas for Mn and Cr we expect antiferromagnetism as observed. To the left of Cr $\alpha\chi^0_\ell < 1$ so that the observed paramagnetism is expected. For Fe $\chi^0 > \chi^0_\ell > \alpha^{-1}$ so that moments can persist in the totally disordered state whereas in Ni, with $\chi^0 > \alpha^{-1} > \chi^0_\ell$, this is not the case.

2.3 The RPA Dynamical Susceptibility

As pointed out in §2.1 the equations which determine $\chi(\underline{q})$ or $\chi(\underline{q},\omega)$, in both RPA and SDFT have the same structure, that of equation (4). The problem of solving such an equation, involving in principle the inversion of an infinite matrix, is well-known in connection with the dielectric function where it is called the problem of "local-field corrections". It is convenient to introduce a tight-binding orbital basis and here we follow method 1(b) of §1, starting with a general model Hamiltonian. The analysis follows that of Cooke[17,18]. The

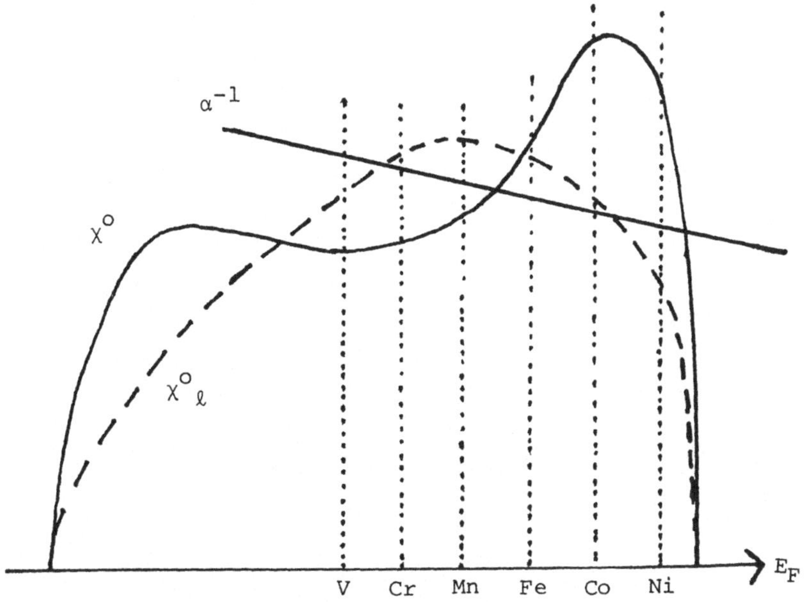

Fig. 1: Schematic plot of χ^0, χ^0_ℓ and α^{-1} as functions of E_F for the 3d series.

Hamiltonian takes the form

$$H = \sum_{\mu\nu\underline{k}} T_{\mu\nu}(\underline{k})\, c^\dagger_{\mu\underline{k}} c_{\nu\underline{k}} + \tfrac{1}{2} \sum_{\substack{\xi\eta\mu\nu \\ \underline{kpq}}} \langle \xi\underline{k}, \eta\underline{p}+\underline{q} | U | \mu\underline{p}\ \nu\underline{k}+\underline{q} \rangle\, c^\dagger_{\xi\underline{k}} c^\dagger_{\eta\underline{p}+\underline{q}} c_{\mu\underline{p}} c_{\nu\underline{k}+\underline{q}} \tag{14}$$

where $c^\dagger_{\mu\underline{k}}$ creates an electron in a Bloch state of wave-vector $\underline{k}$ formed from an atomic orbital μ, this label including a spin index. We shall assume that the interaction U is of short range (on site), this being reasonable for a paramagnet or for the transverse susceptibility of a ferromagnet in which cases the long-range Coulomb interaction plays no role. The longitudinal susceptibility is also covered for a model Hamiltonian in which the long-range forces are omitted. We use such a model in §2.5 where the inclusion of spin-orbit coupling mixes the longitudinal and transverse responses.

The first step in the RPA treatment is to form the one-electron HF Hamiltonian corresponding to (14) and to diagonalize it. Suppose that the transformation from the original $\mu \underline{k}$ basis to the HF eigenstates $n\underline{k}$ is defined by

$$c_{\mu \underline{k}} = \sum_n a_{n\mu}(\underline{k})\, c_{n\underline{k}} \tag{15}$$

with corresponding eigenvalues $E_{n\underline{k}}$. The latter quantities define the band structure. The susceptibility $\chi^{ab}(\underline{q},\omega)$ is proportional to the two-particle Green-function $\langle\langle S^a_{\underline{q}};\ S^b_{-\underline{q}} \rangle\rangle_{\omega+i\eta}$, this being the time Fourier transform of $\langle\langle S^a_{\underline{q}}(t);\ S^b_{-\underline{q}}(0)\rangle\rangle$. Here $S^a_{\underline{q}}$ is a spin operator $(a = z,\pm)$ which is a linear combination of operators $c^{\dagger}_{\nu \underline{k}+\underline{q}}\, c_{\mu \underline{k}}$, or $c^{\dagger}_{n\underline{k}+\underline{q}}\, c_{m\underline{k}}$. Thus we consider the equation of motion of the two-particle Green function involving these operators and find, using the RPA linearization and the HF equations,

$$
\begin{aligned}
(E_{n\underline{k}+\underline{q}} - E_{m\underline{k}} - \hbar\omega)\, &\langle\langle c^{\dagger}_{n\underline{k}+\underline{q}}\, c_{m\underline{k}};\ c^{\dagger}_{h\underline{p}}\, c_{\ell \underline{p}+\underline{q}} \rangle\rangle_{\omega} \\
&= \delta_{\underline{kp}}\, \delta_{mh}\, \delta_{\ell n} (f_{m\underline{k}} - f_{n\underline{k}+\underline{q}}) + (f_{n\underline{k}+\underline{q}} - f_{m\underline{k}}) \\
&\times \sum_{\underline{k}'m'n'} (\langle m\underline{k},\, n'\underline{k}'+\underline{q}\,|U|\,m'\underline{k}',\, n\underline{k}+\underline{q}\rangle \\
&\qquad - \langle m\underline{k},\, n'\underline{k}'+\underline{q}\,|U|\,n\underline{k}+\underline{q},\, m'\underline{k}'\rangle) \\
&\times \langle\langle c^{\dagger}_{n'\underline{k}'+\underline{q}}\, c_{m'\underline{k}'};\ c^{\dagger}_{h\underline{p}}\, c_{\ell\underline{p}+\underline{q}} \rangle\rangle_{\omega} .
\end{aligned} \tag{16}
$$

Here f_{nk} is the occupation number of the HF state nk.
The kernel of this equation, the bracket involving U, is separable when U is an on-site interaction. This bracket may be written

$$\sum_{\mu\nu\eta\xi} W_{\xi\mu,\nu\eta}\, a^{*}_{m\mu}(\underline{k})\, a_{n\xi}(\underline{k}+\underline{q})\, a^{*}_{n'\nu}(\underline{k}'+\underline{q})\, a_{m'\eta}(\underline{k}') \tag{17}$$

where

$$W_{\xi\mu,\nu\eta} = N_a^{-1}(\langle\mu\nu|U|\eta\xi\rangle - \langle\mu\nu|U|\xi\eta\rangle) \tag{18}$$

is an on-site matrix element between orbitals. Hence, defining

$$G_{\alpha\beta,\gamma\delta}(\underline{q},\omega) = \langle\langle \sum_{\underline{k}} c^{\dagger}_{\alpha\underline{k}+\underline{q}}\, c_{\beta\underline{k}};\ \sum_{\underline{p}} c^{\dagger}_{\gamma\underline{p}}\, c_{\delta\underline{p}+\underline{q}} \rangle\rangle , \tag{19}$$

we find

$$G_{\alpha\beta,\gamma\delta} = \Gamma_{\alpha\beta,\delta\gamma} - \sum_{\mu\nu\eta\xi} \Gamma_{\alpha\beta,\xi\mu} W_{\xi\mu,\nu\eta} G_{\nu\eta,\gamma\delta} \tag{20}$$

where

$$\Gamma_{\alpha\beta,\delta\gamma} = \sum_{\underline{k}mn} a^*_{n\alpha}(\underline{k}+\underline{q})\, a_{m\beta}(\underline{k})\, a^*_{m\gamma}(\underline{k})\, a_{n\delta}(\underline{k}+\underline{q}) \frac{f_{m\underline{k}} - f_{n\underline{k}+\underline{q}}}{E_{n\underline{k}+\underline{q}} - E_{m\underline{k}} - \hbar\omega} \,. \tag{21}$$

Cooke et al.[17,18] include 3d, 4s and 4p orbitals in their calculations of the transverse susceptibility and assume that U only has non-zero matrix elements between the 5 3d orbitals. In general the matrix $1 + \Gamma W$ to be inverted would then be 25 x 25. However, Cooke et al. make a simplifying assumption concerning W. For the transverse susceptibility the orbitals μ, η are of $\uparrow$ spin and ξ, ν of $\downarrow$ spin. Then only the second term of (18) is non-zero and Cooke et al. assume that the only non-zero elements of W are the diagonal ones $W_{\xi\xi\xi\xi}$. These take values $-U(t_{2g})$ or $-U(e_g)$ depending on whether ξ is a t_{2g} or e_g orbital. It is further asserted that, owing to the form factors, the dominant contribution to χ comes from terms of the form $G_{\xi\xi,\eta\eta}$ where ξ and η are d orbitals. The solution of (20) then only requires inversion of a 5 x 5 matrix. Some further approximations are made which, as discussed elsewhere [19], do not seem to be too serious. In general the parameterization of W must be such that the interaction term has spin rotational symmetry and, in a cubic crystal, it should also have cubic symmetry under orbital rotation. Actually the parameterization of Cooke et al. does not satisfy the latter condition. The e_g orbitals $x^2 - y^2$, $3z^2 - r^2$ are linear combinations of the corresponding orbitals $z^2 - x^2$, $3y^2 - r^2$ with respect to rotated axes, so that an interaction W which is completely diagonal with respect to the first pair of functions has off-diagonal terms in the second representation. A more symmetric parameterization is $W_{\xi\mu,\nu\eta} = -N_a^{-1} U \delta_{\mu\xi} \delta_{\nu\eta}$, in which case (20) is immediately soluble without matrix inversion. This leads to the simple molecular field picture of §2.2, as used by Lowde and Windsor [20] in their pioneering work on $\chi(\underline{q},\omega)$ in Ni. This parameterization means that a Coulomb interaction U within each orbital is supplemented by "Hund's rule" exchange interactions of the same magnitude between different orbitals. The on-site interaction term is then $U(n_i - \frac{1}{4}n_i^2) - U\underline{S}_i^2$, where n_i is the total number of electrons on the atom and $\underline{S}_i$ is the spin operator for the atom.

A less arbitrary form of W may be obtained from the LDA of density functional theory. As pointed out in §2.1 the calculation of the transverse susceptibility of a ferromagnet in this theory is equivalent to an RPA treatment of the Hamiltonian (8). Hence

$$W_{\xi\eta,\nu\mu} = -N_a^{-1} \int d^3r f(\underline{r})\, \phi_\xi(\underline{r})\, \phi^*_\eta(\underline{r})\, \phi_\mu(\underline{r})\, \phi^*_\nu(\underline{r}) \tag{22}$$

where ϕ_ξ is an atomic-orbital wave function and $f(\underline{r})$, defined by (7), is determined by the self-consistent band calculation. This interaction is easily shown to be equivalent to that which appears in the transverse susceptibility recently derived in the LDA by Callaway et al.[21]. The formulation of these authors is therefore exactly equivalent to that of Cooke at al. We wish to emphasize that SDFT, which includes all correlation effects in principle, gives a fundamental justification to calculations at the HF-RPA level provided the band structure and interaction parameters are correctly determined. This strictly applies only to static properties, but it may be hoped that calculations of the dynamical $\chi(\underline{q},\omega)$ are still meaningful. We note that the correspondence between SDFT and RPA goes beyond the LDA since $I_{XC}(\underline{r},\underline{r}')$ in (4) can be defined, and parameterized, for a non-local functional.

For small $\underline{q}$ and ω the transverse susceptibility $\chi(\underline{q},\omega)$ of a ferromagnet is dominated by the spin-wave pole and takes the form $(N_\downarrow - N_\uparrow)/(\hbar\omega - Dq^2)$ where $N_\uparrow$ and $N_\downarrow$ are the total number of $\uparrow$ and $\downarrow$ spin electrons in the ground state. Hence

$$D^{-1} = (N_\uparrow - N_\downarrow)^{-1} \lim_{\underline{q}\to 0} q^2 \chi(\underline{q},0) \tag{23}$$

so that the spin-wave stiffness constant D can be calculated rigorously in SDFT as a static quantity. By examining (20) in the limit $q \to 0$ we obtain [19] the following formula for D in a cubic crystal, the spin dependence of the one-electron states now being made explicit:

$$(N_\uparrow - N_\downarrow)D = \frac{1}{3}\sum_{n\underline{k}} (B_{n\downarrow}(\underline{k}) f_{n\underline{k}\downarrow} + B_{n\uparrow}(\underline{k}) f_{n\underline{k}\uparrow})$$

$$-\frac{1}{3}\sum_{mn\underline{k}} C_{mn}(\underline{k}) \frac{f_{n\underline{k}\uparrow} - f_{m\underline{k}\downarrow}}{E_{m\underline{k}\downarrow} - E_{n\underline{k}\uparrow}} \tag{24}$$

where

$$B_{n\sigma}(\underline{k}) = \tfrac{1}{2}\sum_{\mu\mu'} \nabla^2 T_{\mu\mu'}(\underline{k})\, a^*_{n\mu\sigma}(\underline{k})\, a_{n\mu'\sigma}(\underline{k})$$

$$C_{mn}(\underline{k}) = \sum_{\alpha=x,y,z} |A^\alpha_{mn}(\underline{k})|^2 \tag{25}$$

and

$$A^\alpha_{mn}(\underline{k}) = \sum_{\mu\mu'} \frac{\partial T_{\mu\mu'}(\underline{k})}{\partial k_\alpha} a^*_{m\mu\downarrow}(\underline{k})\, a_{n\mu'\uparrow}(\underline{k}) . \tag{26}$$

This is independent of the precise form of the interaction, as long as it is on-site and has the correct spin rotational symmetry, i.e. the interaction term of H commutes with the spin operators $S^a_{\underline{q}}$. To evaluate D one only needs a suitable tight-binding parameterization of the band structure of the system. Without resorting to tight-binding Edwards and Rahman [13] have given an expression for D which is exact within the LDA, but difficult to evaluate numerically.

2.4 Numerical Results for Fe and Ni and Their Alloys

In this section we summarize the results of Cooke et al. [18] and some of our own results [19,22]. Cooke et al. start with a reasonable tight-binding paramagnetic band structure and choose their two interaction parameters $U(t_{2g})$ and $U(e_g)$ to obtain the observed ground-state magnetization and the correct t_{2g} to e_g character of the moment as observed in neutron form-factor measurements. They obtain spin-wave dispersion curves which are generally in good agreement with experiment for both Fe and Ni. In Fe the dispersion is isotropic in $\underline{q}$ and above 100meV the spin wave peaks in the imaginary part of the transverse susceptibility broaden rapidly, with a consequent loss in intensity which agrees with the observed disappearance of spin-wave scattering. For Ni the situation is similar in the [111] direction, but in the [100] direction the calculated spin-wave dispersion curve turns over above 120 meV and exists right out to the zone boundary. The turn-over seems to be related to interaction with an "optic" mode. Experimental evidence for this interaction has been reported [23] although the continuation of the mode towards the boundary was not seen. The dispersion curves below 100 meV are largely determined by D which does not depend on the form of interaction W assumed. Higher energy details such as optic modes presumably do depend on W. In calculations for Ni Callaway et al. [21] use an empirical ferromagnetic band structure and, rather inconsistently, an interaction W given by (22) with $f(\underline{r})$ taken as a constant. They find a weak broad optic mode at about twice the energy of Cooke's and no turning-down of the main dispersion curve. Further experimental and theoretical work on this point might yield useful information about W, assuming that the justification for RPA, via SDFT, extends to frequencies above 100 meV.

We have used the formula (24) to carry out extensive calculations on D in Ni and its alloys [19] and in Fe and its alloys [22]. The alloys were treated in a virtual crystal approximation which is known [24] to give reasonable results for D, compared to CPA calculations, as long as the difference in atomic number Δz between the alloy components is not greater than 2. The ferromagnetic band structure for each system was obtained from a reasonable paramagnetic one by imposing an exchange splitting between the energy of $\uparrow$ and $\downarrow$ spin d orbitals so as to reproduce the observed magnetization. When hybridization with s and p orbitals is included the exchange splitting between bands is not rigid. In Ni the magnetization is not very sensitive to the d exchange splitting Δ, owing to the majority spin d band being full, so we allowed Δ to vary. We calculated D both with and without hybridization. The main results are as follows:

(i) In Ni D is very sensitive to changes in Δ but not much affected by turning off hybridization. The observed value of D is obtained with $\Delta = 0.48$ eV. This is larger than the quasi-particle exchange splitting observed in photoemission, as expected theoretically [25] (c.f. §4.1).

(ii) In Fe D is very sensitive to the precise band structure but using essentially the band structure of Cooke et al. [18] we obtain a good value for D, thus establishing consistency with their calculations. If hybridization is turned off D is reduced almost to zero, so that hybridization is essential for the stability of the ferromagnetic state in bcc Fe.

(iii) The calculations of D as a function of composition in alloys of Ni (with Cu, Co, Fe) and of Fe (with Ni, Co, Mn, Cr, V) agree rather well with experiment. The worst case is <u>Fe</u>V where $\Delta z = 3$ and the virtual crystal approximation is presumably breaking down.

2.5. The effect of spin-orbit coupling; application to actinide compounds

This section is a preliminary report on work in progress with J. Millar. This work was inspired by the inelastic neutron scattering measurements on uranium rock-salt compounds, praticularly ferromagnetic US and antiferromagnetic UN, which are described in the chapter by T.M. Holden. There is considerable evidence that in US and UN the 5f electrons must be regarded as itinerant and band calculations[26] yield 5f bandwidths of about eV, in reasonable agreement with photoemission measurements. The spin-orbit splitting between $j = 7/2$ and $j = 5/2$ orbital levels is about 0.7 eV, so that there is a tendency for the band to split into $j = 7/2$, $5/2$ subbands. In the self-consistent band calculations spin-orbit coupling is first omitted and the LDA is used. It is found that in UN both the ferromagnetic and antiferromagnetic Stoner criteria are satisfied. When the effect of the spin-orbit term $\lambda \underline{\ell} . \underline{s}$ is added to the spin polarized band structure a large orbital moment ($-1.83\mu_B$) is induced antiparallel to the spin moment ($+1.1\mu_B$), the numbers referring to the ferromagnetic state. Band calculations at reduced lattice parameters show that under pressure the magnitude of the orbital moment decreases more rapidly than that of the spin moment thus eventually a state of zero total moment is reached in which the spin moment is positive and the magnetization density is non-zero, although its integral over a unit cell vanishes. Experimentally [27] in UN both the sub-lattice moment and Néel temperature reduce with pressure and it seems more likely that a paramagnetic state with zero magnetization density would ultimately be reached. To discuss the dynamical magnetic response it is necessary to introduce electron interaction more symmetrically than first using SDFT and introducing spin-orbit coupling later. We consider a tight-binding f band with interaction energy $\frac{1}{2}Un_i(n_i-1)$, where n_i is the total number of electrons on site i. This commutes with all total atomic spin and orbital angular momentum operators S_i^a, L_i^a, as required of an effective Coulomb interaction. The Hamiltonian is thus

$$H = \sum_{\mu\nu\underline{k}} T_{\mu\nu}(\underline{k})\, c^\dagger_{\mu\underline{k}} c_{\nu\underline{k}} + U \sum_i \sum_{\mu<\nu}\sum n_{\mu i} n_{\nu i} \tag{27}$$

where the index μ now refers to an atomic orbital with quantum numbers (j,m) and $T_{\mu\mu}$ includes the spin-orbit splitting between j=7/2 and 5/2 levels. We follow the HF-RPA procedure of §2.3 and, in the notation of that section, the HF equations are

$$\sum_\mu \{T_{\lambda\mu}(\underline{k}) + Un\,\delta_{\lambda\mu} - U\langle n_{\lambda\mu}\rangle\}\, a_{n\mu}(\underline{k}) = E_{n\underline{k}}\, a_{n\lambda}(\underline{k}) \tag{28}$$

where the density matrix $\langle n_{\lambda\mu}\rangle = \langle c^\dagger_{\mu i} c_{\lambda i}\rangle$ is given by

$$\langle n_{\lambda\mu}\rangle = N_a^{-1} \sum_{\underline{k}\ell} a_{\ell\lambda}(\underline{k})\, a^*_{\ell\mu}(\underline{k})\, f_{\ell\underline{k}} \tag{29}$$

and n is the total number of electrons per atom. For U greater than a critical value the HF equations have a magnetic solution with $\langle n_{\lambda\lambda}\rangle \neq \langle n_{\bar{\lambda}\bar{\lambda}}\rangle$ where $\lambda = (j,m)$ and $\bar{\lambda} = (j,-m)$. To consider the crystal magnetized in different directions it is convenient to consider such magnetic solutions with different axes of quantization for (j,m). Clearly this will result in different magnetic moments, band structures and total energies for different directions of magnetization. This is the origin of large magnetic anisotropy and magnetostrictive effects, and we are currently investigating these theoretically.

Calculations of the dynamical susceptibility involve the solution of equations like (20) with a $14^2 \times 14^2$ matrix to invert. In preliminary calculations we are considering an fcc d band, instead of an f band, in the limit of large spin-orbit splitting where we need consider only the j = 3/2 sub-band. The matrices are then 16 x 16. At this stage we only wish to make one qualitative point which is in fact quite general. This point explains the observed absence [28a] of well-defined collective magnetic excitations (spin waves) in UN and US.

The inelastic neutron intensity is determined essentially by response functions $\langle\langle \frac{1}{2}L^a_{\underline{q}} + S^a_{\underline{q}};\ \frac{1}{2}L^b_{\underline{q}} + S^b_{\underline{q}}\rangle\rangle_\omega$. In the spin-only case, which is a good approximation for 3d metals with small spin-orbit coupling, we have for q = 0 in a ferromagnet a sharp spin-wave (Goldstone) mode at $\omega = 0$ because $S^a_0 = \Sigma_i S_i{}^a$ (a = x,y,z) is a constant of the motion. However with spin-orbit coupling included $\frac{1}{2}L^a_0 + S^a_0$ does not commute with the first term of (27). This is particularly easy to see in the case of strong spin-orbit coupling (single j sub-band considered) when L_0^a, S_0^a are both proportional to J_0^a. For example

$$J_0^z = \sum_{\underline{k}} \sum_{\mu=-(2j+1)}^{2j+1} \mu \, c_{\mu\underline{k}}^\dagger c_{\mu\underline{k}}$$

and

$$[H, J_0^z] = \sum_{\underline{k}} \sum_{\mu\nu} (\nu-\mu) \, T_{\mu\nu}(\underline{k}) \, c_{\mu\underline{k}}^\dagger c_{\nu\underline{k}} \ . \qquad (30)$$

Thus $[H, J_0^z] \neq 0$ because of off-diagonal terms $T_{\mu\nu}$ which allow electrons to change their m value as they hop between atoms. Clearly J_0^a has matrix elements between the ground state and a whole continuum of single-particle excitation states so that no sharp mode will exist even at $q = 0$. In other words the large anisotropy due to large spin-orbit coupling in the f band places the potential spin wave in the Stoner continuum. In the case of normal rare earth systems, or actinide systems with localized f electrons, J_0^a commutes with the exchange part of the Hamiltonian but single-site crystal anisotropy leads to an anisotropy gap in the spin wave spectrum. However, J_0^a has matrix elements between sharp crystal field levels, leading to well-defined crystal field excitations at $q = 0$, and there are no itinerant f electrons to give damping. It is the combination of strong spin-orbit coupling and itineracy which washes out spin waves in actinide compounds like ferromagnetic US and similarly in antiferromagnetic UN.

Since there are no well-defined collective modes to be thermally excited in these materials it is possible that the paramagnetic state above T_c bears some relation to a simple paramagnet without local moments. This would contrast with the case of Fe discussed in §3.1 where disordering occurs by transverse excitations which leave the atomic moments intact. We therefore propose to investigate whether the HF-RPA theory can explain the interesting observed anisotropy[28b] of the critical scattering in UN. In this connection we note that although the static susceptibility at $q = 0$ (in the ferromagnetic case) diverges at T_c, with associated critical scattering, there is also scattering at $\omega \neq 0$ since the order parameter $\frac{1}{2}L_0^z + S_0^z$ is not a constant of the motion. Similarly in a band paramagnet at $T = 0$ with strong spin-orbit coupling, such as $UA\ell_2$, standard spin fluctuation theory must be modified.

Finally, it is important to see how with increasing lattice constant, i.e. decreasing 5f band-width, the HF theory describes a continuous transition to more localized behaviour in USb and UTe. This is shown schematically in fig. 2 for the case of ferromagnetic ordering and considering just a half-filled (f^3) $j = 5/2$ band for simplicity. The lattice constant, and consequently the ordered magnetic moment, increases across the figure from left to right.

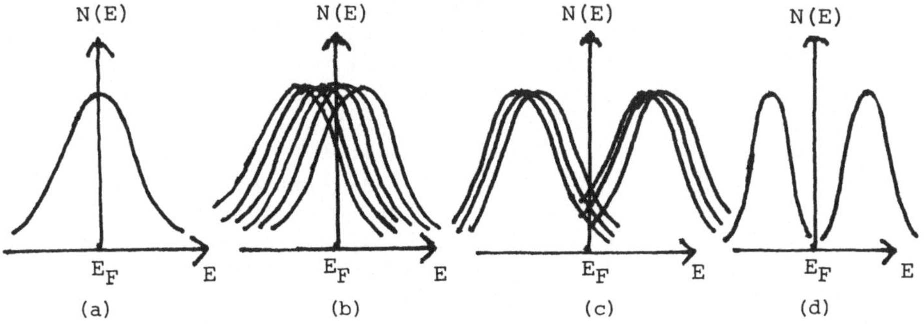

Fig. 2: Schematic $j = 5/2$ f bands in uranium compounds.

The six curves in figs. 2b and 2c represent schematically projected densities of states for $m = 5/2, 3/2, \ldots -5/2$ (larger m to the left). In the paramagnetic state of fig. 2a all these curves are drawn as degenerate although the true projected densities of states will of course have different shapes. The essential point is that states with $m > 0$ (contributing positively to the magnetization) move to lower energy, and states with $m < 0$ to higher energy, when magnetic moments appear in figs. 2b, 2c. In fig. 2d the states $m = 5/2, 3/2, 1/2$ (drawn as degenerate) are completely occupied and the states $m = -\frac{1}{2}, -3/2, -5/2$ are completely unoccupied. The splitting between the two sets of states in this limit is equal to the interaction energy U. Fig. 2d corresponds to the $J = 9/2$ ground state of the f^3 atomic configuration. Since we have neglected Hund's rule exchange in our model, this atomic limit corresponds to jj coupling with a moment gJ of $3.86\,\mu_B$ instead of the LS coupling result of $3.27\,\mu_B$. Since some intermediate coupling is appropriate for uranium this error is not serious, and in any case it is obviously better to switch to a localized point of view in the fig. 2d situation and include crystal field effects which will no longer be negligible. Fig. 2a corresponds to UN under very high pressure, sufficient to suppress the magnetism; fig. 2b to US with an ordered moment of $1.55\mu_B$ (the antiferromagnet UN with a smaller lattice constant corresponds to even weaker magnetism with an ordered sub-lattice moment of $0.75\mu_B$); fig. 2c to UTe with a larger moment of $1.91\,\mu_B$ (antiferromagnetic USb with a still larger lattice constant has a moment of $2.82\,\mu_B$); fig. 2d to an essentially localized situation as in UPd_3. Actually uranium in UPd_3 has an f^2 configuration[28c] so the HF solution of our model then corresponds to a modified fig. 2d with states $m = 5/2, 3/2$ occupied and the rest unoccupied. We remark that in general the quantum number m may be defined with respect to the direction of the <u>local</u> atomic moment so that the present description is also applicable in a magnetically disordered state (c.f. the alloy analogy of §3). In the more localized case itinerancy of the 5f electrons is partially (fig. 2c) or totally (fig. 2d) suppressed, so that one part of the deadly combination "s.o. coupling + itinerancy" becomes ineffective; hence well-defined magnetic excitations will begin to emerge, as observed in UTe and USb. The photoemission data on UPd_3[29] and UTe[30] can be interpreted as evidence for the magnetic splitting of the f band. The observation of this splitting above T_c in UTe indicates that in materials lying to the right of fig. 2 well-defined atomic moments persist in the paramagnetic state. This might not be the case in the weakly magnetic situation of UN, as suggested in the previous paragraph. A manifestation of the splitting of the f band at $T = 0$ is a reducing electronic specific heat coefficient γ as one goes from UN and US to UTe. The alloy system U_xY_{1-x} Sb is an interesting one [31]. In dilute alloys (x small) uranium appears to be non-magnetic and ferromagnetic order sets in at about $x = 0.15$ which is close to the percolation concentration at which an f band will form with nearest neighbour hopping. Antiferromagnetic order appears for $x > 0.45$. The lattice parameter is almost constant for $x < 0.15$ and for $x > 0.15$ it increases linearly with x, the slope indicating that the radius of additional U atoms is about 2% larger than for $x < 0.15$. It has been suggested [31] that these anomalies are associated with a valence change from U^{4+} to U^{3+} but an increase in atomic radius of about 13% would then be expected. The observed 2% increase seems more compatible with the appearance of moments of about $2.8\,\mu_B$ on each U atom, as indicated by magnetic measurements. We therefore suggest that the U atom remains approximately U^{3+} for all x and that the lattice parameter anomaly is of magnetic origin, just as in transition metal alloy systems [32] such as Fe_xV_{1-x}. This seems to be confirmed by optical absorption measurements [31] in which a peak associated with an $f \rightarrow d$ transition at 0.85 eV in USb shifts continuously to zero energy as x decreases to 0.15. The transition takes place between occupied f states and unoccupied d states near the Fermi level, so the observed shift may be understood in terms of a change from the f band of fig. 2c to that of fig. 2a as x decreases.

It is interesting to ask why we are able to describe within HF theory a situation ranging from a weakly correlated f band (fig. 2a) to one exhibiting Hubbard splitting (fig. 2d) (see §4.1 for a general discussion of Hubbard splitting). In the absence of other bands this would correspond to a metal-insulator (Mott) transition. The reason is that the present unrestricted HF theory, in which the potential of an electron at a site depends on the quantum number m, is more flexible than the usual unrestricted HF theory with spin ↑ and spin ↓ potentials. Rather similar ideas have been applied to transition metal oxides by Cyrot and Lyon-Caen [33], and particularly Brandow [34], but our use of jj coupling in this connection seems to be new, and suitable for materials with moderately narrow f bands. The present method is appropriate for systems with unquenched orbital moments; this is known to be the case in a system such as US from magnetic form factor measurements [35]. To obtain results analogous to the present HF theory using density functional theory it is clearly necessary to develop the relativistic form of the theory in which the exchange-correlation functional is a functional of density and current-density [12]. Presumably the dependence on current-density can lead to a non-local potential which depends on the angular momentum of the electron around a site, as in the present theory. The simple model of f bands considered here is more appropriate for the uranium chalcogenides (e.g. US) than the pnictides (e.g. UN) since the anion p bands lie well below the uranium f and d bands in the chalcogenides. Presumably superexchange involving the p bands leads to the antiferrmagnetism of the pnictides and this effect may also lead to the decreasing Curie temperature found in the chalcogenides as one moves to the right in fig. 2, ferromagnetic exchange associated with itinerant f electrons becoming weaker with increasing localization. In §4.2 we apply HF theory to cerium-based compounds with approximately an f^1 configuration.

3. FINITE TEMPERATURE THEORY

3.1. The Alloy Analogy

In view of the success at $T=0$ of SDFT, in the guise of HF-RPA, as described in §2.4, it is highly desirable to extend its application to $T\neq 0$. However the finite temperature form of SDFT involves an unknown temperature-dependent functional which must inevitably be non-local to take account of correlation in the presence of thermally excited spin fluctuations. A way round this difficulty is to note that these fluctuations are slow compared with electron hopping frequencies and may thus be treated reasonably within an adiabatic approximation. The trick is to break translational symmetry, bu considering "frozen" spin configurations, and to make them approximately self-consistent by applying SDFT, considering the many-electron system to be at $T=0$. Temperature enters the theory only in virtue of the entropy of the disordered spin configuration. The simplest case is a metal such as Fe where, as discussed in §2.2, totally disordered atomic moments may exist above T_C. Considering their directions as frozen we have a disordered 'alloy', and for simplicity we may consider the binary alloy case where the moments, Ising-like, are either up or down. Hasegawa [36] has shown that within the single-site CPA this involves no loss of generality in the paramagnetic state. Self-consistency is achieved either using the Hartree-Fock approximation in a one-band model [37] or, more realistically, using spin density functional theory applied to real d bands [38]. On the basis of this 'alloy' picture we may see how the simplest mean field theory of the itinerant electron system differs from that of a Heisenberg, or Ising, system. In the latter case the internal energy per atom takes the form $-\frac{1}{2}AM^2$ where M is the average magnetization per atom and A is proportional to the exchange parameter. On combining with the entropy of disordered moments one obtains mean field theory with a Curie-Weiss law for the susceptibility. In the itinerant 'alloy' picture the mean field state is not describable by the single parameter M. We must specify also the alloy composition by a parameter δc such that there are a fraction $0.5 + \delta c$ spin-up sites and $0.5-\delta c$ spin-down sites. In the simple Ising case M is related to δc by $M = 2\sigma_p\delta c$, where σ_p is the magnitude of the atomic moment. However in the itinerant case, given δc, M may be varied at will theoretically by applying a magnetic field. This is because the metallic 'alloy' responds to a field according to an enhanced Pauli susceptibility. Thus in the simplest mean field description we have two independent parameters M and δc. It is more convenient to introduce parameters M_l, M_i defined by $M_l = 2\sigma_p\delta c$, where σ_p is the magnitude of the local moment in zero field, and $M = M_l + M_i$. Thus M_l is associated with the reversal of local moments of fixed magnitude and M_i is an essentially itinerant contribution of a Pauli type. The energy per atom now becomes, for small M_i and M_l,

$$E = \tfrac{1}{2}\chi_i^{-1} M_i^2 - \gamma M_i M_l - \tfrac{1}{2}\beta M_l^2 \qquad (31)$$

and the coefficients, including the Pauli-like susceptibility χ_i, can be calculated within CPA. On combining with the entropy of disordered local moments one obtains a mean field theory [39,40] which leads to expressions for T_c and susceptibility in exact agreement with those of Hasegawa's saddle-point approximation [36]. First principles calculations along these lines using real bands lead to too large a value of T_c in Fe, but perhaps good agreement is too much to expect in view of the sensitivity of exchange interactions in Fe to the precise band structure (c.f. the discussion of D in §2.4).

A more general treatment of the alloy analogy due to Hubbard [41] and Hasegawa [42] employs the functional integral formalism beyond the saddle-point approximation. The physical principle is as follows. Suppose that the probability distribution of magnetic moment σ on any site is proportional to $\exp(-\psi(\sigma)/k_BT)$ where $\psi(\sigma)$ is an energy which must be calculated self-consistently as a function of σ. The exchange fields associated with the moments give rise to a disordered potential for the electrons as in a multi-component alloy. (In the actual calculations a distribution function of exchange fields is used, rather than of moments). The CPA leads to a coherent potential for the effective medium which is thus a functional of $\psi(\sigma)$. The latter function can be calculated self-consistently by identifying it with the free energy associated with a moment σ induced on the central site in the effective medium by means of a local field. The function $\psi(\sigma)$ may have a single minimum at $\sigma = 0$ (fig. 3a), which is the case of Ni above T_c, or two minima at $\pm\sigma_p$ (fig. 3b). In the latter case, as in Fe, the saddle-point approximation may be applied and the binary alloy picture emerges with a quasi-stable array of disordered moments $\pm\sigma_p$. A third possibility (fig. 3c) is discussed in §4.2. Hasegawa [43] has recently extended his general method to calculate the static wave-vector-dependent susceptibility $\chi(q)$ of Fe and Ni above T_c. The method, being based on the static approximation to the functional integral, cannot be used to calculate $\chi(\underline{q},\omega)$. Edwards [40] had previously discussed $\chi(q)$ using an effective Hamiltonian which yields the energy (31) in the molecular field approximation. Some results for $\chi(\underline{q},\omega)$ above T_c, have also been obtained [44] by this method. In the case of Fe the effective Hamiltonian is dominated by Heisenberg exchange so that calculations of $\chi(\underline{q},\omega)$ based on an effective Heisenberg model should be reasonable. In fact Shastry et al. [45] were able to explain many features of Lynn's neutron results [46] for the dynamical structure factor $S(\underline{q},\omega)$, apart from his observed peaks at finite ω in constant q plots ('spin waves above T_c') at quite small q ($q \gtrsim 0.5Å^{-1}$) Recent work at Brookhaven [47] using polarized neutrons does not show a peak for $q = 0.5Å^{-1}$ so it may be that an effective Heisenberg model gives a crude account of the spin dynamics of Fe in the paramagnetic state. Detailed comparison with the Brookhaven data is planned. We also plan to integrate the calculated $S(\underline{q},\omega)$ over the relevant energy window ($|\hbar\omega| < 60\,\text{meV}$) for comparison with the paramagnetic diffuse scattering date of Brown et al. [48]. Of course the effective Heisenberg model cannot be the whole story. Decay of precessing moments due to itinerant electrons is important at larger q even at $T = 0$, as discussed in §2.4. This means that whereas the magnetic response in an $S = 1$ Heisenberg model is confined to an energy range of 2 or 3 times kT_c, i.e. about 250 meV, it may spread considerably further at larger q due to moment decay. An estimate of this spread is $(kT_cW)^{\frac{1}{2}}$, where W is the d band width, and this is of order 1 eV. We hope to discuss this further elsewhere.

3.2. Other theories

The single-site CPA treatments of the alloy analogy neglect magnetic short-range order (SRO) and Hasegawa [49,43] has shown that for Fe and Ni this is a reasonably consistent approximation. The effective Heisenberg model of Fe also exhibits little SRO well above T_c. A quite different picture, involving a high degree of temperature-independent SRO, was introduced by Prange and Horenman[50], Capellmann[51] and Sokoloff [52] to explain Lynn's observation [46] of spin waves above T_c. The diffuse scattering data of Brown et al. [48] was interpreted as supporting this picture but a different analysis has been given by Edwards [53]. Prange and Korenman propose that above T_c the direction of the atomic magnetic moments varies rather slowly spatially with an angle of only about 36° between neighbouring spins. All the proponents of large SRO assume that it exists and investigate its consequences. No basis for large SRO has been given using statistical mechanics. Edwards, Shastry and Young [39, 45] argue that the observed specific heat and susceptibility of Fe are incompatible with large SRO. With regard to the susceptibility χ per atom we may consider the exact formula

$$k_BT\chi = \sum_j \langle m_i m_j \rangle \qquad (32)$$

where m_i is the z component of magnetic moment on atom i. The Prange-Korenman picture combines large atomic moments, essentially equal in magnitude to those at T = 0, with a slow spatial decay of the correlation function $\langle m_i m_j \rangle$. This leads to the right hand side of (32) being much larger than $\langle m_i{}^2 \rangle$ and to a value of χ very much larger than observed.

In the case of Ni the moments on single atoms may fluctuate rapidly and the validity of the single-site alloy analogy is questionable. One can estimate from the work of Terakura et al.[54] that a cluster of about six Ni atoms is required to support a self-consistent moment in the paramagnetic effective medium. Such long-lived moments may figure importantly in the thermodynamic properties and it is interesting that a simple model of $N_a/6$ disordered moments, each of $3\mu_B$ corresponding to $0.5\mu_B$/atom, yields the observed Curie constant in the susceptibility.

There is no space to discuss theories of very weak itinerant magnets such as $ZrZn_2$ and Ni_3Al, with low T_c and low saturation moment. In such systems the correlation length remains large over a wide range of temperature, down to T = 0 and up to many times T_c. Clearly in this case no single-site alloy analogy is possible.

4. STRONGLY CORRELATED SYSTEMS

In this section we consider the role of band theory in discussing strongly-correlated systems such as intermediate valence materials. We must distinguish clearly between the band structure which is used to calculate static properties ("bare bands") and the bands observed in a dynamic experiment such as photoemission, or inverse photoemission (BIS). The latter are related to "many-body" bands derived from the one-electron spectral function $-\pi^{-1}\,\mathrm{Im}\,G(\underline{k},\omega)$, where $G(\underline{k},\omega)$ is the one-electron Green's function. These bands may exhibit satellites due to Hubbard splitting and this already occurs in a 3d metal such as Ni. We discuss this, and the nature of Hubbard splitting in degenerate d or f bands, in §4.1. In §4.2 we apply unrestricted HF theory to Ce-based compounds. It is useful here to point out that Fermi liquid theory and SDFT (or a HF theory with effective interactions), are essentially 'orthogonal'. The spectral function $-\pi^{-1}\,\mathrm{Im}\,G(\underline{k},\omega)$ exhibits sharp quasi-particle peaks for $\underline{k}$ near the Fermi surface, as well as other structure corresponding to Hubbard splitting etc., and Fermi liquid theory deals with single-particle excitations involving these quasi-particles. SDFT does not purport to calculate such excitation properties. Both theories can in principle, be used to calculate a quantity such as the spin susceptibility, χ, of a paramagnet at T = 0, although only SDFT is a practical procedure for a real material. Although they both will yield, in principle, the same number for χ it will be achieved in different ways. Thus Fermi liquid theory will combine exchange enhancement (due to quasi-particle interaction) with a quasi-particle density of states which is enhanced compared with the bare band-structure. SDFT will combine a larger exchange enhancement (between "bare electrons") with the bare density of states. The shape of the Fermi surface in SDFT is probably, in principle, the same as in Fermi liquid theory although this has not been proved.

4.1. Hubbard Splitting

In Ni the d configurations appearing in the ground state are predominantly d^9 and d^{10}. d^8 is strongly suppressed although the interaction between two holes is not as strong as one might suppose, due to screening by s and p electrons, i.e. appearance of d^8s^2 configurations. Nevertheless this interaction is large enough to lead to a weak Hubbard splitting which manifests itself as the "two-hole" (d^8 final state) or "6eV" satellite occurring in photoemission at about 6eV below the Fermi level. To calculate "many-body bands" exhibiting this structure one requires a treatment which is correct in the atomic limit (bandwidth $\to$ 0) but which preserves a proper Fermi surface. This has been achieved in the ferromagnetic ground state by Edwards and Hertz [1,2] and, more recently, by Liebsch [3] and many others [4]. The general problem is that labelled 1(a) in §1. Some progress has recently been made using a "time-dependent alloy" analogy [55]. This corrects the major defect of the paper commonly known as Hubbard III, [56] namely the loss of a sharp Fermi surface.

In the ferromagnetic ground state of Ni the satellite in the many-body bands is most pronounced in the majority-spin band. Its appearance is accompanied by a reduction in the exchange splitting from its bare-band value. It is this reduced exchange splitting which is observed in angle-resolved photoemission. The exchange splitting in the bare bands of SDFT has significance as the one which appears in the calculation of a static quantity such as the spin-wave stiffness constant (c.f. §2.4).

In Co and Fe the d configurations appearing in the ground state are predominantly d^8, d^9 and d^7, d^8 respectively. The interaction which leads to suppression of d^{10} and d^9, respectively, cannot be screened by s and p electrons; on the other hand d^7 and d^6 configurations, in Co and Fe respectively, may not be so energetically unfavourable due to the formation of d^7s^2 and d^6s^2. Thus Hubbard splitting has been predicted [57] to lead to a satellite above the majority spin D band in Co and Fe and should be observable in BIS (d^{10} and d^9 final states in Co and Fe, respectively).

To see the effects of Hubbard splitting in a degenerate d or f band (N_f states per atom) we recall some results of Hubbard II [58]. This corresponds to the simplest decoupling of the equation of motion for $G(\underline{k},\omega)$ which gives the correct atomic limit. The many-body bands are given as solutions of the equation

$$F(E) = \varepsilon_{\underline{k}} \qquad (33)$$

where $F(E) = E - \Sigma(E)$, Σ being the self-energy, and $\varepsilon_{\underline{k}}$ is the bare-band energy. The function $F(E)$ is given by

$$\frac{1}{F(E)} = \frac{1}{N_f} \sum_{N=0}^{N_f-1} \frac{(N_f-N)\,\rho_N + (N+1)\,\rho_{N+1}}{E - UN} \qquad (34)$$

in the "zero configuration-width" case corresponding to an interaction energy $\frac{1}{2}Un(n-1)$ on an atom with n electrons. Here ρ_N is the probability of an atom having N electrons. Thus in the large U limit with n_f electrons/atom and $n_f < 1$ (e.g. Ce)

$$\rho_0 = 1 - n_f, \quad \rho_1 = n_f, \quad \rho_2 = \rho_3 \; \ldots \; = \rho_{N_f} = 0.$$

In this case

$$\frac{1}{F(E)} = \frac{1}{N_f}\left\{ \frac{N_f(1-n_f) + n_f}{E} + \frac{(N_f - 1)n_f}{E - U} \right\} . \qquad (35)$$

A graphical solution of (33) shows clearly that the bare band, with energies $\varepsilon_{\underline{k}}$ spread over a width W, splits into two bands. The lower band, centered around the original band-centre, contains $N_f(1 - n_f) + n_f$ states per atom; the upper band is at an energy U higher and contains $(N_f - 1)\,n_f$ states. The widths of these bands are proportional to the number of states they contain and are therefore $W(1 - n_f)$ and Wn_f respectively in the limit $N_f \gg 1$. The lower band, containing the Fermi level, corresponds to f^0 final states in photoemission (one electron removed) or f^1 final states in BIS (one electron added); the upper band corresponds to f^2 final states in BIS. Regarding Ce as a sort of inverted Ni, with less than one electron instead of less than one hole, the f^2 satellite corresponds to the d^8 (two-hole) satellite in Ni. The relative weights of $f^0 + f^1$ and the f^2 bands agree with more recent large U theories of the impurity problem [59].

4.2. A Hartree-Fock Theory of Ce-based compounds

In many materials such as αCe, CeN, $CeSn_3$, ... the calculated bare f bandwidth $W(=2\Delta)$ is in the range 1.5 - 2 eV. Generally the f bandwidth is expected to vary as R^{-7}, where R is the Ce-Ce interatomic distance, and R may vary from about 3Å in $CeRu_2$ to nearly 4.5Å in $CeAl_3$. Thus we may expect a wide range of f bandwidths, ranging from 2eV to 0.1eV. Clearly band theory is inappropriate when $2\Delta \sim 0.1$eV but it is reasonable to ask whether it makes sense for αCe with $2\Delta \sim 2$eV and $U \sim 6$eV (from photoemission and BIS data). Clearly

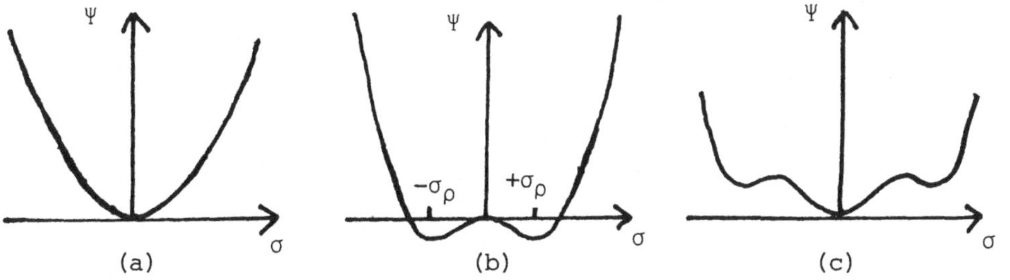

Fig. 3: Possible forms of the local free energy ψ as a function of local moment σ.

the f electrons in αCe are more highly correlated than in UN, for example, where $U \sim 2\Delta \sim 2$eV. Since in αCe U is considerably larger than the band-width we may ask why it does not order magnetically. In fact band calculations based on SDFT predict no magnetic order in Ce, the Stoner criterion not being satisfied. The reason is that in Ce, with an approximately f^1 configuration, the Fermi level E_F lies in the tail of the f band, where strong f - d hybridization occurs, on the low energy side. The density of states is comparatively low there so it is difficult to satisfy a Stoner criterion. In the case of uranium compounds, with an f^3 configuration say, E_F lies well into the f band and in fact corresponds roughly to a half-filled $f^{5/2}$ sub-band. The Stoner criterion is then easily satisfied for $U \sim 2\Delta$.

The HF theory of eqn. (28) may be applied to the Ce f band. To make some numerical estimates we consider a much-simplified model in which we take a $j = 5/2$ sub-band with $T_{\lambda\mu}(\underline{k}) = \varepsilon(\underline{k})\,\delta_{\lambda\mu}$ so that $\langle n_{\lambda\mu}\rangle = \langle n_\mu\rangle\,\delta_{\lambda\mu}$. Then, from (28), the energy bands are given by

$$E_{\mu\underline{k}} = \varepsilon(\underline{k}) + U(n_f - \langle n_\mu\rangle)\,. \qquad (36)$$

For simplicity we take the density of states N(E) corresponding to $\varepsilon(\underline{k})$ to be Lorentzian with fwhm 2Δ. The band label μ corresponds to the azimuthal quantum number m and in the absence of a magnetic field the six $j = 5/2$ bands are degenerate in the paramagnetic state. In the presence of a magnetic field this degeneracy is lifted and the resultant paramagnetic susceptibility is given by

$$\chi = (g\mu_B)^2\,\frac{j(j+1)}{3}\,\frac{N_{tot}(E_F)}{1 - UN(E_F)}\,, \qquad (37)$$

where $N_{tot}(E_F) = (2j+1)\,N(E_F)$ and in the present case $j = 5/2$, $g = 0.857$. This result is just that of the Newns-Hewson[60] Fermi liquid theory with the addition of exchange enhancement due to the interaction U. The criterion for a spontaneous ordered moment is $UN(E_F) > 1$ and in the present simple degenerate-resonance model we do not distinguish this from the criterion for an isolated local moment to appear on Ce. If we take $n_f = 0.8$, which determines the position of E_F in the band (each Lorentzian being a fraction 0.8/6 occupied), this criterion becomes $U > 19\Delta$. It is therefore not suprising that αCe, with $U \sim 6\Delta$, does not order magnetically. In fact for the present situation, with E_F lying in an upward-curving region of N(E), the HF total energy as a function of moment σ takes the form of fig. 3c, for fairly large U. For $U > 11.75\Delta$ the subsidiary minimum falls lower in energy than the $\sigma = 0$ minimum and when this occurs a sudden transition to a large moment takes place. The tendency of the interaction U is to put all the electrons into one sub-band, thus reducing the interaction term of (27) to zero, and in the large moment state only the $m = 5/2$ orbital is substantially occupied. (The quantization of m is with respect to the direction of the local moment; in a paramagnetic state these directions may be disordered as in the alloy analogy of §3.1). In the present example, the moment jumps to a fraction 0.4 of the full Ce f^1 moment $gj\mu_B$. For larger values of U/Δ the stable moment increases in magnitude. In these calculations we have kept n_f constant so that no valence change occurs.

The situation in Ce ($\sim f^1$) is quite different from that in U ($\sim f^3$). In the latter case the progress to large moments with increasing U/Δ occurs through small moments as in the series of rock-salt compounds discussed in §2.5. In Ce the transition to a large moment occurs well before the Stoner criterion is satisfied so that the exchange enhancement factor in (37) never exceeds 2.6. It is interesting that this is close to the value of 2.8 found by Podloucky and Glötzel [61] in their band calculation for αCe using the LDA to SDFT. The corresponding enhanced Pauli susceptibility $(2.8\mu_B^2 N_{tot}(E_F))$ accounts for only 55% of the measured susceptibility, although the calculated $N_{tot}(E_F)$ is in reasonable accord with the observed electronic specific heat. Equation (37) which includes the orbital contribution to χ, yields the observed χ if we take the calculated $N_{tot}(E_F)$ and an enhancement factor of 2.38.

The transition to magnetic moments occurs at $U \sim 12\Delta$ i.e. $\Delta \sim \frac{1}{2}$eV if U = 6 eV. Thus if the bare f bandwidth is greater than about 1 eV we expect that no local moments will have formed and a reasonable description of the system is a fairly strongly correlated paramagnetic metal. This critical bandwidth should correspond roughly to the critical Ce - Ce distance of the Hill plot [62]. The bare f bands of HF theory are shown schematically in the upper part of fig. 4. In the lower part of the figure we sketch the corresponding "many-body bands" Figures A correspond to the wide-band situation without local moments. Since U may already be considerably larger than 2Δ correlation effects are strong and in the "many-body bands" an f^2 satellite may appear together with a narrowing of the main band. (A2) corresponds to the smallest lattice constant, e.g. α´Ce or $CeRu_2$, with a weak satellite and not much narrowing. (A3) corresponds to the more strongly correlated situations of αCe, $CeSn_3$ and $CePd_3$, for example. On the right-

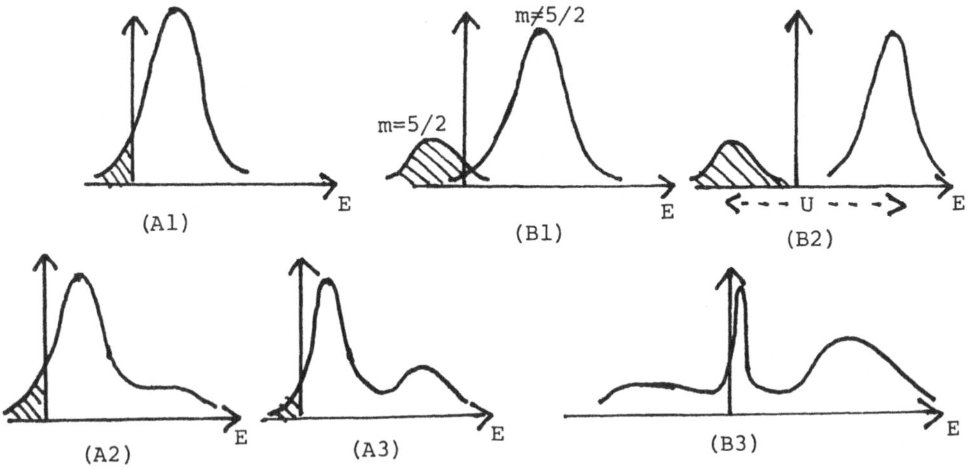

Fig. 4: Schematic 4f bands in Ce compounds. A1, B1 and B2 are bare bands and A2, A3 and B3 are "many-body bands".

hand side of fig. 4 moments have just formed (B1) or formed strongly (B2), only the $m = 5/2$ sub-band being substantially occupied. At high temperatures the alloy analogy may be used and the quantization of m refers to the direction of the local moment. At low temperatures the moments either order in some way or undergo a Kondo effect with a moderate (case B1) or low (case B2) Kondo temperature T_K. The many-body bands then exhibit a Kondo resonance at the Fermi level [59]. This Kondo-lattice behaviour probably applies to systems like $CeAl_3$ and $CeIn_3$. We distinguish "Kondo-lattice" behaviour (figs. B) sharply from true "mixed valence" behaviour (figs. A). In case A in our picture photoemission from f states well below E_F is associated with the far tail of the f band (hybridized with d) and not with a well-defined f level as in case B. We summarize some experimental support for our picture below together with some final comments.

1. A sharp transition from mixed valence to Kondo behaviour is found in many alloy systems as the lattice constant increases. In our picture this can occur with no significant change in the "real valence" (e.g. n_f can be close to 1 throughout). Examples of such alloy systems are $Ce(In_xSn_{1-x})_3$ [63], $Ce(Sn_xPb_{1-x})_3$ [64] and, in a dilute system, $Ce_{0.1}(La_xTh_{1-x})_{0.9}$ [65]. A striking feature of the transition, stressed by Crow and coworkers, is that in the mixed valence regime the resistivity shows a marked peak at a temperature T_{max} which tends to zero at the alloy composition corresponding to the "mixed valence-Kondo" boundary.

2. Close to the "mixed valence-Kondo" boundary on the "mixed valence" side, the energy of a Ce ion as a function of moment σ is as shown in fig. 3c, with the secondary minimum occurring at a small energy ΔE above the energy of the $\sigma = 0$ state. Thus in the "mixed valence" case

the magnetic state of the Ce ion can be induced thermally, n_f being approximately 1 in both the magnetic and non-magnetic states of the ion. It is clearly possible to construct a thermodynamic theory of the α-γ transition in Ce which involves the entropy of mixing of magnetic and non-magnetic ions, the spin-disorder entropy of the magnetic ions, and the lattice free energy. This is along the lines of Wohlleben[66] with the important difference that the two different types of ion in the present theory correspond to magnetic and non-magnetic ions with essentially the same valency, not to ions having different valence. The peak in the resistivity should occur where there is maximum disorder between the magnetic and non-magnetic ions; hence $T_{max} \to 0$ as $\Delta E \to 0$ at the "mixed valence - Kondo" boundary as observed (see point 1 above).

3. An important distinction between the "mixed valence" and "Kondo-lattice" situations is the width of the f^1 peak near the Fermi level. In our "mixed valence" case, with bare band-width greater than 1eV, the narrowed many-body band can hardly be less than about 0.2eV in width. The width of the Kondo resonance is $kT_K/N_f \sim 0.01$eV for $CeSn_3$, taking a Kondo picture with $T_K \sim 200$K. Unfortunately this difference cannot be resolved in BIS. The Kondo resonance corresponds to a very strong mass-enhancement of the quasi-particle bands near E_F. Thus the observation [67] of the Fermi surface in $CeSn_3$, with mass enhancements of only about 5 compared with the calculated bare bands, strongly suggests the "mixed valence" picture.

4. There is a close similarity between the thermally-induced moments we propose in a system such αCe and the temperature-induced moments occurring in some transition metal systems. We intend to return to this point, and to the whole subject matter of this section, elsewhere. The application of unrestricted Hartree-Fock theory to Ce in §4.2 is formally similar to that of Coqblin & Blandin[68]. However, they considered that $U/\delta \sim 150$ and achieved a transition from the magnetic (γ) state to non-magnetic (α) state by raising the f band relative to the Fermi level. This led to a first order change in n_f with low occupation in the α state. Their model of the γ-α transition is thus a 'promotional' one which is unlikely on several grounds[69]. The picture presented in §4.2 of a transition from a magnetic to a non-magnetic state at constant n_f is in line with the ideas of Hill & Kmetko[70].

REFERENCES

1. D. M. Edwards and J. A. Hertz, J. Phys. F 3 2191 (1973).
2. J. A. Hertz, Proc. Int. Conf. Magnetism (ICM-73) Vol. 3 p. 354 ("Nauka" Moscow 1974).
3. A. Liebsch, Phys. Rev. B 23 5203 (1981).
4. J. Igarashi, J. Phys. Soc. Japan 52 2827 (1983).
5. W. Kohn and L. J. Sham, Phys. Rev. 140 A1133 (1965).
6. U. von Barth and L. Hedin, J. Phys. C 5 1629 (1972).
7. M. Levy, Proc. Natl. Acad. Sci. USA 76 6062 (1979).
8. J. P. Perdew and A. Zunger, Phys. Rev. B 23 5048 (1981).
9. J. E. Harriman, Phys. Rev. A 24 680 (1981).
10. J. F. Janak, Phys. Rev. B 18 7165 (1978).
11. A. R. Williams and U von Barth, in "Theory of the Inhomogeneous Electron Gas" eds. S. Lundqvist and N. H. March (Plenum 1983).
12. J. Callaway and N. H. March, to be published.
13. D. M. Edwards and M. A. Rahman, J. Phys. F 8 1539 (1978).
14. S. Vosko and J. P. Perdew, Can. J. Phys. 53 1385 (1975).
15. J. F. Janak, Phys. Rev. B 16 225 (1977).
16. V. Heine, J. H. Samson and C. M. M. Nex, J. Phys. F 11 2645 (1981).
17. J. F. Cooke, Phys. Rev. B 7 1108 (1973).
18. J. F. Cooke, J. W. Lynn and H. L. Davis, Phys. Rev. B 21 4118 (1980).
19. D. M. Edwards and R. B. Muniz, to be published.
20. R. D. Lowde and C. G. Windsor, Adv. Phys. 19 813 (1970).
21. J. Callaway, A. K. Chatterjee, S. P. Singhal and A. Ziegler, to be published.
22. R. B. Muniz, J. F. Cooke and D. M. Edwards, to be published.
23. H. A. Mook and D. Tochetti, Phys. Rev. Lett. 43 2029 (1979).
24. D. M. Edwards and D. J. Hill, J. Phys. F 6 607 (1976).
25. D. M. Edwards, Proc. Int. Conf. Transition Metals 1977: Inst. Phys. Conf. Ser. No. 39 pp. 279-81.
26. P. Weinberger, C. P. Mallett, R. Podloucky and A. Neckel, J. Phys. C 13 173 (1980): M. S. S. Brooks and P. J. Kelly, to be published.
27. J. M. Fournier, J. Beille, A. Boeuf, A. Boeuf and A. Wedgwood, Physica 102B 282 (1980).

28a. W. J. L. Buyers, T. M. Holden, J. A. Jackman, A. F. Murray, P de V. DuPlessis and O. Vogt., J. Mag. Magn. Mater. 31 229 (1983);

28b. T. M. Holden, W. J. L. Buyers and E. C. Svensson, Phys. Rev. B 26 6227 (1982); see also chapter by T.M. Holden.

28c. A.F. Murray & W.J.L. Buyers in "Crystalline Electric Field and Structural Effects in F-Electron Systems," eds. J.E. Crow, R.P. Guertin & T.W. Mihalisin (Plenum 1980) p. 257.

29. Y. Baer, H. R. Ott and K. Andres, Solid State Commun., 36 387 (1980).
30. B. Reihl, N. Martensson, P. Hermann, D. E. Eastman and O. Vogt, Phys. Rev. Lett. 46 1480 (1981).
31. Seesee shapter by J. Schoenes.
32. M. Shiga, A.I.P. Conf. Proc. No. 18, 463 (1974).

33. M. Cyrot and C. Lyon-Caen, J. Phys. C 6 L247 (1973); J. Phys. Paris 36 253 (1975).
34. B. H. Brandow, Advances in Physics 26 651 (1977).
35. see chapter by G.H. Lander.
36. H. Hasegawa, Solid State Commun. 39 1229 (1981).
37. H. Hasegawa, J. Phys. Soc. Japan 46 1504 (1979).
38. T. Oguchi, K. Terakura and N. Hamada, J. Phys. F 13 145 (1983); B. L. Gyorffy, J. Kollar, A. J. Pindor, G. M. Stocks, J. Staunton and H. Winter, to be published.
39. D. M. Edwards, J. Mag. Magn. Mater. 15-18 262 (1980).
40. D. M. Edwards, J. Phys. F 12 1789 (1982).
41. J. Hubbard, Phys. Rev. B 20 4584 (1979); Phys. Rev. B 23 5974 (1981).
42. H. Hasegawa, J. Phys. Soc. Japan 49 963 (1980).
43. H. Hasegawa, to be published.
44. R. B. Muniz, Ph.D. Thesis (1983), University of London.
45. B. S. Shastry, D. M. Edwards and A. P. Young, J. Phys. C 14 L665 (1981).
46. J. W. Lynn, Phys. Rev. B 11 2624 (1975).
47. J. P. Wicksted, G. Shirane and O. Steinsvoll, Phys. Rev. B29, 488 (1984).
48. P. J. Brown, D. Deportes, D. Givord and K. R. Ziebeck, J. Appl. Phys. 53, 1973 (1982) and J. Mag. Magn. Mater. 31-34, 295 (1983).
49. H. Hasegawa, Solid State Commun. 38, 401 (1981).
50. R. E. Prange and V. Korenman, Phys. Rev. B 19 4691 (1978).
51. H. Capellmann, Z. Phys. B 34 29 (1979).
52. J. B. Sokoloff, J. Phys. F 5 1946 (1975).
53. D. M. Edwards, J. Magn. Mag. Mater. 36 213 (1983).
54. K. Terakura, N. Hamada, T. Oguchi and T. Asada, J. Phys. F 12 1661 (1982).
55. D. M. Edwards and J. A. Hertz, unpublished.
56. J. Hubbard, Proc. Roy. Soc. A 281 401 (1964).
57. D. M. Edwards, Phys. Lett. 33A 183 (1970).
58. J. Hubbard, Proc. Roy. Soc. A 277 237 (1964)
59. See Chapters by Y. Kuramoto and O. Gunnarsson.
60. D. M. Newns and A. C. Hewson, J. Phys. F 10 2429 (1980).
61. R. Podloucky and D. Glötzel, Phys. Rev. B 27 3390 (1983).
62. H. H. Hill, "Plutonium 1970 and Other Actinides" ed. W. N. Miner (New York: AIME 1970) pp. 2 - 19.
63. A. Maury, R. Freitag, J. E. Crow, T. Mihalisin and A. I. Abou-Aly, Phys. Lett. 92A 411 (1982).
64. J. Teter, R. Freitag, A. Maury, J. E. Crow and T. Mihalisin, J. Appl. Phys. 53 7910 (1982).
65. T. M. Holden, W. J. L. Buyers, P. Martel, M. B. Maple and M. Tovar, "Valence Instabilities" ed. P. Wachter and H. Boppart (North Holland 1982).
66. See Chapter by D. Wohlleben.

67. W. R. Johansson, G. W. Crabtree, A. S. Edelstein & U.D. McMasters, Phys. Rev. Lett. 46 504 (1981).
68. B. Coqblin and A. Blandin, Adv. Phys. 17 281 (1968).
69. B. Johannsson, Phil. Mag. 30 469 (1974).
70. H. H. Hill and E. A. Kmetko, J. Phys. F5 1119 (1975).

THEORY OF ELEMENTARY EXCITATIONS IN INTERMEDIATE VALENCE MATERIALS*

A. J. Fedro** and S. K. Sinha†

Materials Science and Technology Division

Argonne National Laboratory, Argonne, Illinois 60439

We review the formalism for calculating the properties of systems represented by an Anderson lattice Hamiltonian. First, a mean-field theory for the one electron Green's function is presented and then extended to include spin and charge fluctuations which lead to a many-body resonance near the Fermi level. The dynamical spin susceptibility and neutron scattering cross section are also evaluated. Finally, we present a calculation which indicates the possibility of a Cooper-pairing mechanism induced by electron correlations.

1. Introduction

In these lectures, we will present various levels of a theoretical framework to understand the thermodynamics and elementary excitations of a model system involving strong intrasite electron correlations, with a view to understanding the properties of the Anderson and Kondo lattices. First, we present the underlying operator formalism and indicate how it may be applied to obtain the simplest mean-field type theory for the Anderson lattice. Then, we show how the formalism may be extended to calculate the dynamical susceptibility, and also to the Kondo regime. Finally, we shall present some results relating to Cooper pairing of the electrons, driven by strong correlation effects, as a possible new mechanism for superconductivity in such materials.

Our starting point is the assumption that strong Coulomb correlations between f-electrons have the result that all f-shell configurations other than f^n and f^{n-1} are projected out. This is equivalent to stating that $U_{ff} \to \infty$. For simplicity, consider the case where n=1, and let $|i,o\rangle$ denote the state at site i where there are no f-electrons (or a full multiplet) and $|i,m\rangle$ denote the state corresponding to the mth state of the multiplet corresponding to one f-electron (or f-hole). (The states may equally well be crystal-field-split states.) We define operators (analogous to the standard basis operators[1]),

$$f^+_{im} = |i,m\rangle\langle i,o| \quad ; \quad f_{im} = |i,o\rangle\langle i,m| \; . \tag{1.1}$$

It is obvious from (1.1) that

$$f^+_{im}\, f^+_{im'} = 0 \tag{1.2}$$

for all m,m', and the completeness property requires that

$$|i,o\rangle\langle i,o| + \sum_m |i,m\rangle\langle i,m| = 1 \tag{1.3}$$

for each site i. Such operators, however, do not obey the usual Fermion anticommutation rules, but satisfy

$$[f^+_{im}, f^+_{jm'}]_+ = [f_{im}, f_{jm'}]_+ = 0 \quad ; \tag{1.4a}$$

$$[f_{im}, f^+_{jm'}]_+ = \delta_{ij}\, Q^i_{mm'} \; , \tag{1.4b}$$

where

$$Q^i_{mm'} \equiv \delta_{mm'}\,(1 - \sum_{m''} f^+_{im''} f_{im''}) + f^+_{im'} f_{im} \; . \tag{1.4c}$$

Note that the expectation value $\langle Q^i_{mm'}\rangle$ is of $O(1/N)$ where N is the number of states in the multiplet. We now write down the Hamiltonian for the Anderson lattice in the form

$$\begin{aligned} H = & \sum_{i,m} \varepsilon_{fm}\, f^+_{im} f_{im} + \sum_{k\sigma} \varepsilon_{k\sigma}\, C^+_{k\sigma} C_{k\sigma} \\ & + \sum_{i,m}\sum_{k\sigma} (N_\ell)^{-1/2} V^m_{k\sigma}\, [f^+_{im} C_{k\sigma}\, e^{-i\vec{k}\cdot\vec{r}_i} + \text{h.c.}] \\ & + \sum_{i,m} e_m\, f^+_{im} f_{im}\, \delta V \\ & + \frac{1}{2} B(V)\, \delta V^2 - g_J \mu_B H \sum_{i,m} \langle m|J_z|m\rangle\, f^+_{im} f_{im} \end{aligned} \tag{1.5}$$

where ε_{fm} is the energy of the multiplet state m, $\varepsilon_{k\sigma}$ that of an unperturbed conduction state of wavevector k and spin σ, N_ℓ the total number of lattice sites, $C_{k\sigma}$, $C^+_{k\sigma}$ the Fermion operators for conduction states, δV a volume deformation, B(V) the bulk modulus and e_m a deformation potential for state m, H an applied magnetic field and g_J the Lande factor for the multiplet.

2. Mean-Field Theory for the One-Electron Green's Functions

We are now in a position to use the Equation-of-Motion method[2] to calculate the single electron Green's Functions in lowest order. While this may appear to be a very crude solution to the problem (analogous to the "Hubbard I"-type approximations used for transition metals[3]), it has certain interesting features worth noting on the way to a better solution. It gives a natural generalization for the lattice case of the known lowest-order results for the single impurity problem, such as those of Varma and Yafet[4] or Ramakrishnan.[5] It exhibits a renormalization of the bare f-level similar to that obtained by Haldane[6] using rather general scaling arguments. One also obtains, almost trivially, results derived for the spin 1/2 f-level case using diagrammatic techniques[7], and is easily generalized to the full multiplet case. The basic idea is to decouple higher order Green's functions arising from the non-trivial anticommutation rules (Eqs. (1.4)) back down to lowest order. Physically, this corresponds to imbedding each f-electron shell in a mean lattice; thus, it is basically a mean-field type of theory. We start by defining operators for propagating excitations on the lattice, analogous to the $C_{k\sigma}$, $C^+_{k\sigma}$:

$$f^+_{km} = N_\ell^{-1/2} \sum_i f^+_{im} \, e^{-i\vec{k}\cdot\vec{r}_i}, \text{ etc.} \tag{2.1}$$

One may attempt to write down a ground-state Fermi liquid type wavefunction using these operators but it turns out to be very hard to evaluate because of the non-trivial anticommutation rules for the f-operators. The Green's function method, on the other hand, avoids the necessity of having to consider wavefunctions at all. Let us consider the usual (retarded) thermal Green's functions[8]

$$G^{AB}(\omega) \equiv \langle\langle A;B\rangle\rangle_\omega \tag{2.2}$$

and let us choose A, B to be f_{km} or $C_{k\sigma}$ and f^+_{km} or $C^+_{k\sigma}$, respectively. Their equations of motion are given by[8]

$$\omega\, G_k^{mm'}(\omega) = \langle[f_{km}, f^+_{km'}]_+\rangle + \langle\langle [f_{km},H] ; f^+_{km'}\rangle\rangle_\omega \; , \tag{2.3}$$

with similar equations for the other one-particle Green's functions. For simplicity, let us consider first the case where we have a simple spin-1/2 f-doublet so m can be replaced by a spin index σ.

By Eq'. (1.4), (1.5) we have

$$[f_{k\sigma},H] = (\varepsilon_f - g\mu_B H - e\langle n^f\rangle/B(V))\, f_{k\sigma}$$

$$+ N_\ell^{-1/2} \sum_{jk'\sigma'} e^{i(\vec{k}-\vec{k}')\cdot\vec{r}_j} V_{k'\sigma'}\, Q^j_{\sigma\sigma'}\, C_{k'\sigma'} \, , \qquad (2.4)$$

where ε_f is the unperturbed position of the f-level, and $e_\sigma \equiv e$. We have eliminated δV by minimizing the energy with respect to it. Using the representation (1.4c) for $Q^j_{\sigma\sigma'}$, and decoupling the last product in Eq. (2.4), we obtain

$$[f_{k\sigma},H] \simeq \varepsilon^*_{f\sigma} f_{k\sigma} + V_{k\sigma}\langle Q_\sigma\rangle\, C_{k\sigma} \, , \qquad (2.5)$$

where

$$\varepsilon^*_{f\sigma} = \varepsilon_f - g\mu_B H\sigma - e\langle n^f\rangle/B(V) + \sum_{k'} \{- V_{k'\sigma}\langle f^+_{k'\bar{\sigma}}\, C_{k'\bar{\sigma}}\rangle\}, \qquad (2.6)$$

where $\bar{\sigma}$ denotes the spin opposite to σ and

$$\langle Q_\sigma\rangle = 1 - \langle n^f_{\bar{\sigma}}\rangle \quad . \qquad (2.7)$$

The equation of motion for the conduction electron operator $C_{k\sigma}$ is rigorously

$$[C_{k\sigma},H] = \varepsilon_{k\sigma} C_{k\sigma} + V_{k\sigma} f_{k\sigma} \, . \qquad (2.8)$$

Let $\underset{\sim}{G}_{k\sigma}(\omega)$ stand for the matrix

$$\begin{pmatrix} G^{ff}_k(\omega) & G^{cf}_k(\omega) \\ G^{fc}_k(\omega) & G^{cc}_k(\omega) \end{pmatrix}$$

where f refers to the f-spin index and c to the conduction electron spin index, both being σ in this case. From Eqs. (2.3), (2.5) and (2.6), we obtain

$$\underset{\sim}{G}^{-1}_{k\sigma}(\omega) = \begin{pmatrix} \dfrac{\omega - \varepsilon^*_{f\sigma}}{\langle Q_\sigma\rangle} & V_{k\sigma} \\ V^*_{k\sigma} & \omega - \varepsilon_{k\sigma} \end{pmatrix} \quad . \qquad (2.9)$$

Thus the poles of the one-electron Green's function are at

$$E^{\pm}_{k\sigma} \equiv \frac{1}{2} \left(\varepsilon^{*}_{f\sigma} + \varepsilon_{k\sigma} \pm \Delta_{k\sigma} \right) ; \tag{2.10a}$$

$$\Delta_{k\sigma} \equiv \left[(\varepsilon^{*}_{f\sigma} - \varepsilon_{k\sigma})^2 + 4|V_{k\sigma}|^2 \langle Q_\sigma \rangle \right]^{1/2} \tag{2.10b}$$

which would correspond to one-electron hybridized "bands" for the non-interacting case (i.e., for $f_{k\sigma}$, $f^{+}_{k\sigma}$ obeying ordinary fermion statistics) if $\varepsilon^{*}_{f\sigma}$ was replaced by ε_f and $\langle Q_\sigma \rangle$ by 1. There is a "hybridization gap" of order $\pi|V|^2 \langle Q_\sigma \rangle / W$ (W is the conduction band width), which is, however, an indirect (zone-boundary) gap (see Fig. 1). From Eq. (2.9) it is obvious that the self-energy matrix $\underset{\sim}{\Sigma}_{k\sigma}(\omega)$ (defined as $\underset{\sim}{G}^{(0)-1}_{k\sigma}(\omega) - \underset{\sim}{G}^{-1}_{k\sigma}(\omega)$, where $\underset{\sim}{G}^{(0)}_{k\sigma}(\omega)$ is the Green's function matrix for the non-interacting case) is real for all ω, i.e., Fermi-liquid-like behavior has been "built in" the model from the outset by the very nature of the decoupling approximation. The thermodynamic averages of the f-electron occupation, etc. can be obtained from the general relations of the form[8]

$$\langle A\ B \rangle_T = \frac{1}{\pi} \int_{-\infty}^{+\infty} d\omega\, f(\omega - \mu) \lim_{\eta \to 0^+} \mathrm{Im} \langle\langle B; A \rangle\rangle_{\omega + i\eta} \tag{2.11}$$

where A, B are any one-electron operators, $f(\omega - \mu)$ is the Fermi function, and μ the chemical potential. The explicit expressions in the case of our model are given in Ref. (9). The chemical potential μ is determined by the condition

$$\sum_{\sigma} \{ \langle n^f_\sigma \rangle + \langle n^c_\sigma \rangle \} = n_e ; \tag{2.12}$$

n_e being the given number of electrons per unit cell. Eqs. (2.6), (2.9), (2.11) and (2.12) must be solved self-consistently at any temperature and applied field H for given ε_f, $V_{k\sigma}$, $\varepsilon_{k\sigma}$. The self-consistent determination of the chemical potential μ is, of course, a new feature not present in the impurity case.

From the results of such calculations we may obtain quantities such as the f-occupation, susceptibility, electronic specific heat and unit cell volume as functions of the position of the f-level, temperature, etc. Such numerical calculations were carried out for simple model systems and reported in Ref. 9 (although lattice effects were ignored there). We gain further insight into our model by neglecting the k-dependence of the hybridization matrix elements and assume the unperturbed conduction band is characterized by a constant density of states $\rho (= 1/W$ per spin).

Let us define an effective density of f-states $\tilde{n}^f_\sigma(\omega)$ such that

$$\langle n^f_\sigma \rangle = \int_{-\infty}^{+\infty} \tilde{n}^f_\sigma(\omega)\, f(\omega - \mu)\, d\omega \quad ; \tag{2.13}$$

and let $n^c_\sigma(\omega)$ be the conduction electron density of states (DOS).

From Eq. (2.11) it may be shown that

$$\begin{aligned} n^c_\sigma(\omega) = \rho; \tilde{n}^f_\sigma(\omega) &= V^2 \langle Q_\sigma \rangle^2 \rho/(\omega - \varepsilon^*_{f\sigma})^2 \quad \text{for } \omega \text{ outside the gap } \Delta \\ n^c_\sigma(\omega) = \tilde{n}^f_\sigma(\omega) &= 0 \quad \text{for } \omega \text{ inside the gap } \Delta. \end{aligned} \tag{2.14}$$

with

$$\Delta = 4V^2 \langle Q_\sigma \rangle \rho \quad , \tag{2.15}$$

and where

$$\langle Q_\sigma \rangle = 1 - \langle n^f_\sigma \rangle = 1 - \frac{1}{2} \langle n^f \rangle \tag{2.16}$$

for the paramagnetic case with zero applied field. The effective density of one-electron-like excitations is shown in Fig. 2. A peculiarity (compared to conventional one-electron DOS) is that the total number of available states is given by

$$\sum_\sigma \int_{-\infty}^{\infty} [\tilde{n}^f_\sigma(\omega) + n^c_\sigma(\omega)]\, d\omega = 2[1 + \langle Q_\sigma \rangle] = 4 - \langle n^f \rangle \tag{2.17}$$

so that the maximum filling of these 4 hybridized bands (2 per spin) is with 3 electrons, not 4, as required by the correlation restriction of only one f-electron per site. It may also be shown that the total number of states in the lower band K^- satisfies $1 < K^- < 2$. Thus, at T=0, for one electron per unit cell, μ must always lie in the lower band; whereas for two electrons per unit cell, μ must lie in the upper band. This result was also obtained by Roberts and Stevens [7]. For the case of 2 electrons/cell, in the non-interacting case, μ would always lie in the gap. Thus the present result implies a violation of Luttinger's Theorem[10] which states that the Fermi surface volume in the interacting case at T=0 should be identical to that in the non-interacting case, as has been emphasized by Martin[11]. In the present model it may be shown for the paramagnetic case that

$$V_{FS} = V_{FS}^{(0)} + \frac{2\pi^3}{\Omega_a} \frac{<n^f>^2}{[1 - 1/2 <n^f>]} , \tag{2.18}$$

where V_{FS} is the volume inside the Fermi surface, $V_{FS}^{(0)}$ that in the non-interacting case, Ω_a the unit cell volume and $<n^f>$ the average number of f-electrons per site. In the case where $<n^f> = 1$, this would yield a difference corresponding to one electron exactly between V_{FS} and $V_{FS}^{(0)}$. In the full multiplet case the discrepancy between V_{FS} and $V_{FS}^{(0)}$ vanishes as 1/N, however. Evaluation of $\varepsilon_{f\sigma}^*$ defined in Eq. (2.6), gives using Eq. (2.11)

$$\varepsilon_{f\sigma}^* = \varepsilon_f - g\mu_B H\sigma - \frac{e<n^f>}{B(V)} + V^2 \rho <Q_\sigma> \ln \left| \frac{W}{\varepsilon_{f\sigma}^* - \mu} \right| . \tag{2.19}$$

The second and third terms are merely the Zeeman and deformation potential terms respectively, while the last term yields a logarithmic upward renormalization of the f-level which has been found for the single-impurity problem previously[4,5] and is similar to the general scaling result of Haldane[6] for the single impurity.

The generalization of the above theory to the full N-fold multiplet case is quite straight forward, but will not be given here for reasons of space. The main point is that, as stated above, $<Q_{mm}>$ becomes $O(\frac{1}{N})$. The hybridization of the multiplet with a single conduction band produces (N+2) hybridized "bands", and the hybridization splitting of these bands is reduced by a factor of $O(\frac{1}{N})$. The f-levels thus can renormalize up towards the Fermi level or even above it.

While calculations for quantities such as the f-occupation, susceptibility, resistivity, etc. from the above model are qualitatively in agreement with experiment (at least in the strongly mixed-valence regime where $\varepsilon_f^* \to \mu$), a rigorous justification for the above simple mean-field decoupling theory (e.g., from scaling ideas or other methods) is lacking. It is interesting to note, however, that for large N the above results become almost identical to those derived by Read and Newns[12] by completely different methods. Our present feeling is that there is some merit to carrying out detailed numerical calculations for a real (as opposed to a model) system with the full multiplet structure included, in the case of large N and in the strongly mixed valence limit. This has not as yet been done. In the Kondo regime, however, the above type of theory is clearly not good enough, and we shall discuss improvements to it in Section 4.

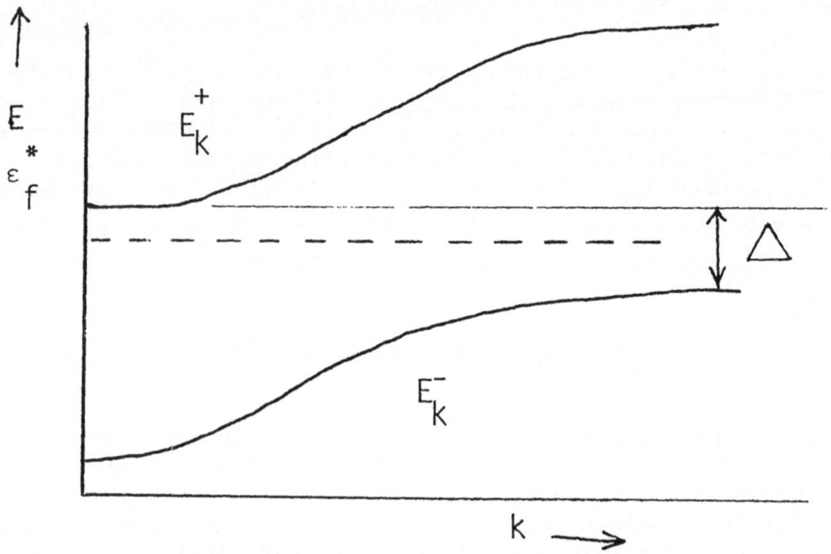

Fig. 1. Schematic of k-dependence of positions of poles of the one-electron Green's function from the mean-field theory.

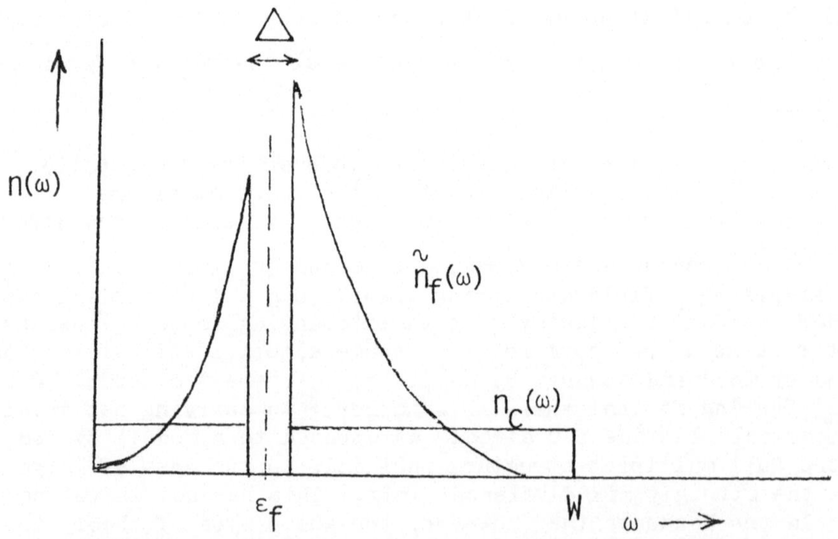

Fig. 2. Effective densities of "f-like" and conduction-like states for our simple model obtained from the mean-field theory.

3. The Two-Particle Green's Function and the Dynamical Susceptibility

As is well-known[13], the magnetic scattering of neutrons from a crystal is given by the scattering function

$$S(\vec{Q},\omega) = f^2(\vec{Q}) \left[1 - e^{-\beta\omega}\right]^{-1} \sum_{\alpha\beta} (\delta_{\alpha\beta} - \hat{Q}_\alpha \hat{Q}_\beta) \operatorname{Im} \chi^{\alpha\beta}(\vec{q},\omega) \tag{3.1}$$

where $\vec{Q}$ is the wavevector transfer of the neutron, $\hbar\omega$ its energy loss, $\hat{Q}$ is a unit vector along $\vec{Q}$, β is $1/kT$, $f(\vec{Q})$ is the magnetic form factor for the magnetic ions, and $\chi^{\alpha\beta}(\vec{q},\omega)$ is the dynamical susceptibility at the reduced wavevector $\vec{q}$ and frequency ω. Let us consider only scattering by the f-electrons for our simple model system of spin -1/2 f-doublets. In the paramagnetic case, we need only consider $\chi^{+-}(\vec{q},\omega)$, which may be related to the 2-particle Green's function by

$$\chi^{+-}(\vec{q},\omega) = (g\mu_B)^2 \frac{1}{N_\ell} \sum_{k_1 k_2 \sigma} \langle f^+_{k_1+q\bar{\sigma}} f_{k_1\sigma} ; f^+_{k_2\sigma} f_{k_2+q\bar{\sigma}} \rangle_\omega . \tag{3.2}$$

We evaluate the Green's functions by the Equation-of-Motion Method. Let us consider the operators

$$N^+_1(k\sigma,q) \equiv f^+_{k+q\bar{\sigma}} \cdot f_{k\sigma} \quad ; \quad N^+_2(k\sigma,q) \equiv C^+_{k+q\bar{\sigma}} \cdot f_{k\sigma}$$

$$N^+_3(k\sigma,q) \equiv f^+_{k+q\bar{\sigma}} \cdot C_{k\sigma} \quad ; \quad N^+_4(k\sigma,q) \equiv C^+_{k+q\bar{\sigma}} \cdot C_{k\sigma} . \tag{3.3}$$

Define the general Green's function

$$G^{ij}_{kk'\sigma\sigma'q}(\omega) \equiv \langle N^+_i(k\sigma,q) \; ; \; N_j(k'\sigma',q) \rangle_\omega . \tag{3.4}$$

In evaluating $[N^+_i(k\sigma,q), H]$ we may use the definitions (1.1) to treat products of many f operators on a single site <u>exactly</u>, but on different sites, we decouple four-operator products back down to two in all possible ways. We use the commutation relations given by Eqs. (1.4), and obtain

$$[N^+_1(k\sigma,q),H] \simeq (\varepsilon_{f\sigma} - \varepsilon_{f\bar{\sigma}})\, N^+_1(k\sigma,q) + V_{k+q}\langle Q_\sigma \rangle N^+_2(k\sigma,q)$$

$$- V_k \langle Q_\sigma \rangle N^+_3(k\sigma,q) + \langle f^+_{k\sigma} f_{k\sigma} \rangle \frac{1}{N_\ell} \sum_K V_{K+q} N^+_2(K\sigma,q)$$

$$- \langle f^{+}_{k+q\bar{\sigma}} f_{k+q\bar{\sigma}} \rangle \frac{1}{N_\ell} \sum_K V_K N_3^{+} (K\sigma, q)$$

$$- \{V_k \langle f^{+}_{k\sigma} C_{k\sigma} \rangle - V_{k+q} \langle f^{+}_{k+q\bar{\sigma}} C_{k\bar{\sigma}} \rangle\} \frac{1}{N_\ell} \sum_K N_1^{+}(K\sigma, q) \tag{3.5}$$

with similar equations for the commutators of H with N_2^{+}, N_3^{+}, N_4^{+}. The last three terms in Eq. (3.5) arise from the kinematic restrictions to single occupation of the f-sites built into the f-operators, which we may term a "blocking effect". Substitution of Eq. (3.5) into the equations of motion leads to a set of four coupled integral equations for the $G^{ij}_{kk'\sigma\sigma' q}(\omega)$, which can, however, be solved, since by Eq. (3.2) we need only $\sum_{kk'\sigma} G^{11}_{kk'\sigma\sigma q}(\omega)$. The solutions for $\chi^{+-}(\vec{q}, \omega)$ have the following structure

$$\chi^{+-}(\vec{q}, \omega) = [\{\underset{\sim}{1} + \underset{\sim}{\pi}(\vec{q}, \omega)\}^{-1} \cdot \underset{\sim}{Z}(\vec{q}, \omega)]_1 \tag{3.6}$$

where $\underset{\sim}{\pi}(\vec{q}, \omega)$ is a 4x4 matrix and $\underset{\sim}{Z}(\vec{q}, \omega)$ is a four-vector. These are functions of $\langle n^{f}_{k\sigma} \rangle$, $\langle n^{f}_{k+q\sigma} \rangle$, $\langle n^{c}_{k\sigma} \rangle$, $[\omega + E^{\pm}_{k\sigma} - E^{\pm}_{k+q\bar{\sigma}}]^{-1}$, etc.

$Z_1(\vec{q}, \omega)$ is analogous to a single non-interacting band susceptibility for this two-band model, so that $\underset{\sim}{\pi}$ gives an effective enhancement. The numerical calculations for $S(\tilde{Q}, \omega)$ are shown in Fig. 3, for a model fcc lattice of ions containing both f and conduction orbitals (with only nearest neighbor hopping integrals). The occupation numbers and bands were calculated from the self-consistent one-electron theory described in the last section. Two points should be noted. First, the enhancement effect due to the f-electron correlations is appreciable only in the dynamic regime $\omega \lesssim \Delta$, the hybridization gap ($\Delta \sim \pi V^2 \rho$). Secondly, in the strongly mixed-valence regime, $S(Q, \omega)$ exhibits a central peak and a peak at $\omega \sim \Delta$. The latter peak is a maximum when q is at the zone-boundary and decreases to zero intensity as $q \to 0$, although it is essentially dispersionless. This is consistent with the gap being an indirect one, as may be seen from Fig. 1. This peak also broadens and decreases in intensity as (kT/Δ) increases. It is interesting to note that such qualitative behavior was found for the peak observed at 10 meV energy transfer in the inelastic neutron scattering from single-crystal paramagnetic TmSe by Grier and Shapiro[14], although the simple two-band model presented here is certainly a very crude representation of the more complex "band structure" in a real multiplet system. Finally, the central peak around $\omega = 0$ has a width which is fairly temperature insensitive, in agreement with neutron scattering experiments on mixed-valent systems[15].

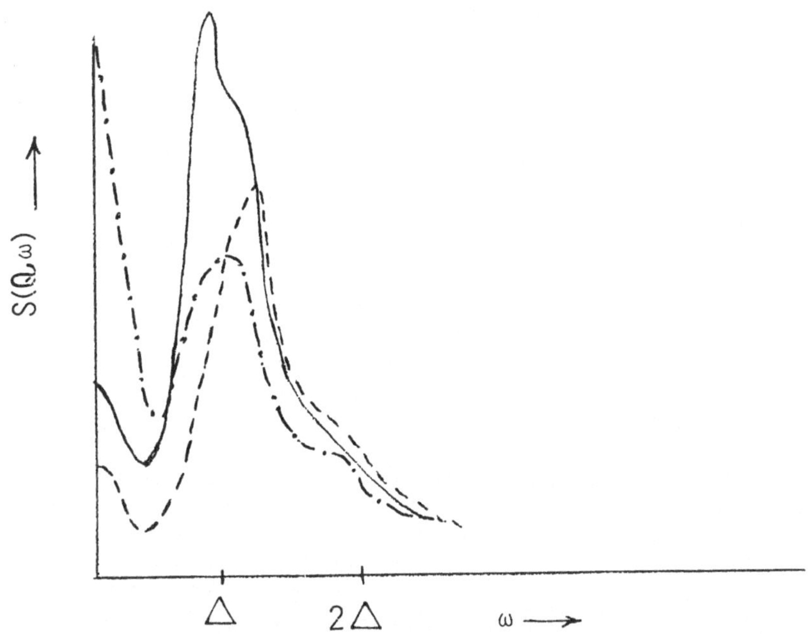

Fig. 3. S(Q, ω) for Q at the point 2π/a(1,0,0) for the model fcc Anderson lattice discussed in the text. The full curve denotes the calculation with the enhancement factor and the dashed curve the unenhanced S(Q, ω) for kT/Δ = 0.18. The dot-dash curve shows the enhanced S(Q, ω) for kT/Δ = 1.76.

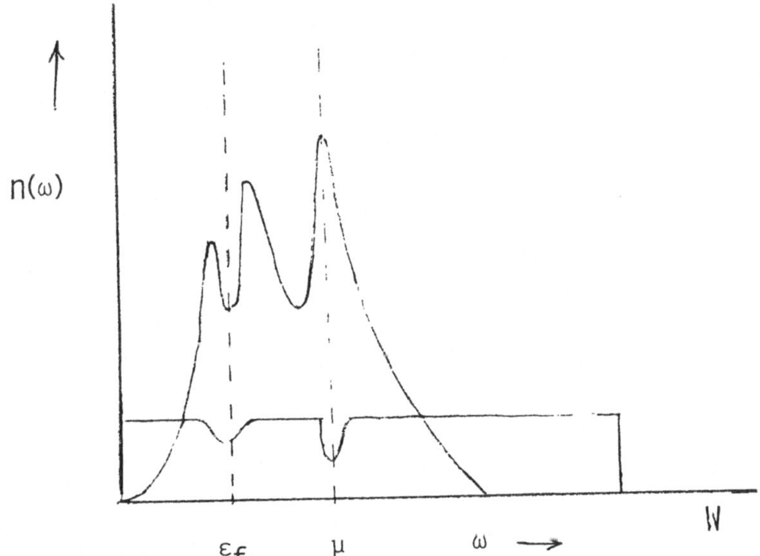

Fig. 4. Effective densities of "f-like" and conduction-like states for an Anderson lattice with ε_f well below the Fermi-level from the one-electron Green's function including spin and charge fluctuation effects.

4. Incorporation of Fluctuation Effects in the One-Electron Green's Function

We may extend the theory given in Section 2 by using a higher-order decoupling procedure. Using Eq. (1.4c), we may write Eq. (2.4) as

$$[f_{k\sigma},H]=\overset{*}{\varepsilon}_{f\sigma}f_{k\sigma}+V_kC_{k\sigma}-\frac{1}{N_\ell}\sum_{j,k_1}V_{k_1}e^{-i(\vec{k}-\vec{k}_1)\cdot\vec{r}_j}[Z^\sigma_{jk_1}-S^\sigma_{jk_1}], \quad (4.1)$$

where we have assumed that $V_{k\sigma}$ is independent of σ. The operators Z and S are defined as

$$Z^\sigma_{jk_1}\equiv f^+_{j\bar{\sigma}}f_{j\bar{\sigma}}C_{k_1\sigma} \ ; \ S^\sigma_{jk_1}\equiv f^+_{j\bar{\sigma}}f_{j\sigma}C_{k_1\bar{\sigma}} \ . \quad (4.2)$$

The equations of motion which result from Eqs. (2.8) and (4.1) yield Green's functions involving Z and S. The procedure is to take next the equations of motion of these Green's functions and decouple operators which involve f-operators or c-operators on different sites, except those that occur in the combinations given in Eq. (4.2). In this way, we keep the effects of spin and charge fluctuations on an individual f-shell explicitly in the problem. The method is similar to that used by Nagaoka[16] in his treatment of the single Kondo impurity, although, in that case, charge fluctuations (given by the Z-operators cannot occur). The next result of this procedure is that Eq. (2.9) in the mean-field theory is replaced by

$$\underset{\sim}{G}^{-1}_{k\sigma}(\omega)=\begin{pmatrix}\dfrac{\omega-\overset{*}{\varepsilon}_{f\sigma}(\omega)}{\langle Q_\sigma(\omega)\rangle} & V_k \\ V^*_k & \omega-\varepsilon_{k_\sigma}\end{pmatrix} \quad (4.3)$$

i.e., the quantities $\overset{*}{\varepsilon}_{f\sigma}$ and $\langle Q_\sigma\rangle$ in the mean-field theory now become complex and frequency-dependent. The explicit expressions for $\overset{*}{\varepsilon}_f(\omega)$ and $\langle Q(\omega)\rangle$ are[17]

$$\overset{*}{\varepsilon}_{f\sigma}(\omega)=\varepsilon_{f\sigma}+\frac{1}{N_\ell}\lim_{\eta\to o^+}\sum_k|V_k|^2[\langle n^c_{k\bar{\sigma}}\rangle-\langle n^f_{k\bar{\sigma}}\rangle+\langle n^f_{\bar{\sigma}}\rangle]/[\omega+i\eta-\varepsilon_k]; \quad (4.4a)$$

$$\langle Q_\sigma(\omega)\rangle=1-\langle n^f_{\bar{\sigma}}\rangle-\frac{1}{N_\ell}\lim_{\eta\to o^+}\sum_k V^*_k\langle f^+_{k\bar{\sigma}}C_{k\bar{\sigma}}\rangle/[\omega+i\eta-\varepsilon_k]. \quad (4.4b)$$

The expectation values $\langle n^f_{k\sigma}\rangle$, etc. are still given by Eqs. (2.11), e.g.,

$$\langle n^{f}_{k\sigma} \rangle = \frac{1}{\pi} \int_{-\infty}^{+\infty} d\omega\, f(\omega-\mu)\ \mathrm{Im}\ G^{ff}_{k\sigma}(\omega) \qquad (4.5)$$

etc., and the self-consistent solutions of Eqs. (4.3) to (4.5), together with the condition (2.12) for the chemical potential leads to a set of coupled integral equations for $\varepsilon^{*}_{f\sigma}(\omega)$ and $\langle Q_{\sigma}(\omega)\rangle$.

These integral equations are very hard to solve analytically, since they are highly singular as $\omega \to 0$ and $T \to 0$. We may, however, solve for $\varepsilon^{*}_{f}(\omega)$ in the limit where $|W_1|$, $|W_2| \gg k\,T \gg V$, where W_1 and W_2 are respectively the lower and upper edges of the conduction band, obtaining

$$\varepsilon^{*}_{f}(\omega) = \varepsilon_{f} + V^{2}\rho[\phi_{1}(\omega) - \gamma] + i\pi f(\omega) + V^{2}\pi\ \psi(\omega, W_{1}, kT), \qquad (4.6)$$

where γ is a frequency-independent constant, which must be determined self-consistently from Eqs. (4.4), and

$$\psi(\omega,\ W_{1},\ kT) \equiv \int_{W_{1}}^{\infty} dx\, \frac{f(x)}{x-\omega}\ . \qquad (4.7)$$

Here $f(x)$ is the Fermi function $[e^{x/kT} + 1]^{-1}$.

In the limit $\omega\to 0$, the behavior of this solution is

$$\varepsilon^{*}_{f}(0) \sim \varepsilon_{f} + |\varepsilon_{f}|\ \ln\,(T/T_{k}) \quad , \qquad (4.8)$$

where

$$T_{k} = T_{o}\ e^{-|\varepsilon_{f}|/(V^{2}\rho)} \quad , \qquad (4.9)$$

with $T_{o} \sim W$.

This shows that even if the bare ε_f is well below the Fermi level, its renormalization at $\omega = 0$ (i.e., at the Fermi level) can bring the effective ε^{*}_{f} back up to the Fermi level, producing a sharp temperature-dependent resonance at the Fermi level. It can then be shown that the f DOS, $\tilde{n}_{f}(\omega)$, will have peaks both at $\omega \sim \varepsilon_{f}$ and in the vicinity of the Fermi level (see Fig. 4). This is the lattice analogue of the so-called "many-body resonance" or Abrikosov-Suhl resonance in the case of the single impurity[18]. Similar results for the single impurity case have been derived by Lacroix and Cyrot[19] and for the lattice case by Lacroix[20] and by Baumgartel and Muller-

Hartmann.[21] In order to see what happens in the more interesting low temperature region, the integral equations will have to be solved (probably numerically). This has not yet been done. Equations (4.4) also have to be generalized to the full multiplet case, where in analogy with the single impurity case, we expect the amplitude of the resonance at the Fermi level to be increased by a factor of N. The existence of this resonance for $T \lesssim T_k$ would be consistent with several experimental results for "dense Kondo" systems which is the limit of our Anderson lattice as ε_f sinks well below the Fermi level.

5. Correlation-Induced Pairing Instability

We may also use the Equation-of-Motion Method for the two-particle Green's functions to investigate whether the Anderson Lattice Hamiltonian (1.5) has a Cooper-pairing type instability, which is not driven by the usual phonon-mediated attraction. In order to show this, we study the dynamical response functions for pair creation. These are related to the two-particle Green's functions of the form $\langle\langle \tilde{N}_i^+ (k, q) ; \tilde{N}_j(k', q)\rangle\rangle$, where

$$\tilde{N}_1^+ (k, q) \equiv f^+_{k+q\uparrow}f^+_{-k\downarrow} \quad ; \quad \tilde{N}_2^+ (k, q) \equiv c^+_{k+q\uparrow}f^+_{-k\downarrow}$$

$$\tilde{N}_3^+ (k, q) \equiv f^+_{k+q\uparrow}c^+_{-k\downarrow} \quad ; \quad \tilde{N}_4^+ (k, q) \equiv c^+_{k+q\uparrow}c^+_{-k\downarrow} \quad . \qquad (5.1)$$

If these response functions diverge for $q, \omega \to 0$ at some $T = T_c$, it means that there is a spontaneous formation of Cooper pairs, i.e., a pairing instability. Using a decoupling procedure similar to that described in Section 3 for the calculation of the spin response function (where products of several f-operators on the same site were treated exactly but different sites decoupled) we obtain equations of motion of the following form:

$$[\tilde{N}_1^+(k, q), H]=(\varepsilon_{f\sigma} + \varepsilon_{f\bar{\sigma}})\, \tilde{N}_1^+ (k, q) + V_{k+q} \langle Q_{\bar{\sigma}}\rangle \tilde{N}_2^+(k, q)$$

$$+ V_k \langle Q_\sigma\rangle \tilde{N}_3^+(k, q) - \sum_\sigma (\varepsilon^*_{f\sigma} - \varepsilon_{f\sigma}) \frac{1}{N_\ell} \sum_{k_1} \tilde{N}_1^+ (k_1, q)$$

$$+[\langle n^f_{k+q\sigma}\rangle+\langle n^f_{f\bar{\sigma}}\rangle-1 - \langle f^+_{j\sigma}f_{j\sigma}\rangle] \frac{1}{N_\ell} \sum_{k_1} V_{k_{1+q}}(k_1,q)\, \tilde{N}_2^+(k,q)$$

$$+ [\langle n^f_{k+q\sigma}\rangle+\langle n^f_{k\bar{\sigma}}\rangle-1 - \langle f^+_{j\bar{\sigma}}f_{j\bar{\sigma}}\rangle] \frac{1}{N_\ell} \sum_{k_1} V_{k_1} \tilde{N}_3^+(k_1,q) \quad . \qquad (5.2)$$

By summing this equation over k, we may show that $\langle \frac{1}{N} \sum_{\ell} \sum_{k_1} \tilde{N}_1^{+} (k_1, q) \rangle$ is always identically zero as it must be since it corresponds to $\langle f^{+}_{i\uparrow} f^{+}_{i\downarrow} \rangle$. Thus, the decoupling scheme does not violate the restriction that two f-electrons cannot be on the same site. Hence the fourth term on the right of Eq. (5.2) vanishes. Similar equations may be written down for the other operators in Eq. (5.1), and the solution of the resulting equations for the Green's function yields a vanishing denominator if the following equation is satisfied for $T = T_c$

$$\frac{\langle Q \rangle}{N} \sum_{\ell k} \frac{2V_k^2 \, \varepsilon_k}{A_k \Delta_k} \left[\frac{1}{2E_k^+} \mathrm{Tanh} \left(\frac{\beta_c E_k^+}{2}\right) - \frac{1}{2E_k^-} \mathrm{Tanh} \left(\frac{\beta_c E_k^-}{2}\right) \right]$$

$$+ \left[\langle Q \rangle - 1\right] \frac{1}{N} \sum_{\ell k} \frac{8 \, V_k^2 \, \varepsilon_k}{A_k (A_k^2 - \Delta_k^2)} = 1 \quad , \qquad (5.3)$$

where we have dropped the spin index, $\beta_c \equiv 1/k_B T_c$,

$$A_k = \varepsilon_f^* + \varepsilon_k \quad , \qquad (5.4)$$

and $E_k^{\pm}$, Δ_k are defined in Eqs. (2.10).

An examination of the solutions of this equation reveals the following features: (1) No solution for T_c exists if the Fermi Level μ lies in the lower band E_k^- (i.e., if ε_f^* lies above μ). This would be the case for a system with one electron per unit cell. (2) T_c decreases rapidly as ε_f^* moves down below μ. Thus it is only in the strongly mixed-valent regime that T_c is largest. Insertion of typical values of V, conduction bandwidth W and ($\varepsilon_f^* - \mu$) yield T_c of the order of 1 K or less.

The primary instability is in the response of $(C^{+}_{k\uparrow} f^{+}_{-k\downarrow} - f^{+}_{k\uparrow} C^{+}_{-k\downarrow})$, i.e., singlet pairing between f-electrons and conduction electrons. We ascribe this to an attractive interaction between these elcotrons of a form similar to the Schrieffer-Wolff[22] $- J \vec{\sigma} \cdot \vec{S}$ interaction in the Kondo regime and arising from the same physical origin. In the mixed-valence regime, however, there are f-electrons at the Fermi surface, and thus the attractive interaction leads to a pairing instability. The tendency towards instability with regard to singlet f-d pairing in mixed valence systems has also been discussed by DeMenezes,[23] who, however, considers the usual phonon mechanism as the source of the attractive interaction. Our theory has not considered possible pair-breaking mechanisms such as the real scattering of f-electrons by spin or charge fluctuations. One can only appeal to the experimental observation that Fermi-liquid behavior is estimated in these systems at low temperatures. The fluctuation theory of Section 4 can in principle be incorporated into our formalism but this has as yet not been carried out. One might expect that, even in the Kondo regime, a sharp resonance of f-like states at the Fermi level might lead once again to significant T_c's due to the above mechanism. Further investigation of this pairing mechanism is necessary before one can attempt to identify it with what is observed in the recently discovered "heavy-fermion" superconductors.[24,25]

REFERENCES

* Work supported by the U.S. Department of Energy.
** On leave from Northern Illinois University, DeKalb, IL 60115.
† Present address: Exxon Research and Engineering Company, Corporate Research Labs. Annandale, NJ 08801.

1. Hubbard, J., Proc. Roy. Soc. A285:542 (1965).
2. Tiablikov, S. V., "Methods in the Quantum Theory of Magnetism" Plenum Press, New York (1967).
3. Hubbard, J., Proc. Roy. Soc. A276:238 (1963).
4. Varma, C. M. and Yafet, Y., Phys. Rev. B 13:2950 (1976).
5. Ramakrishnan, T. V., in: "Valence Fluctuations in Solids", L. M. Falicov, W. Hanke and M. B. Maple, ed., North-Holland, Amsterdam (1981).
6. Haldane, F. D. M., Phys. Rev. Lett. 40:416 (1978).
7. Roberts, M. and Stevens, K. W. H., J. Phys. C 13:5941 (1980).
8. Zubarev, D. N., USP. Fiz. Nauk. 71:71 (1960).
9. Fedro, A. J. and Sinha, S. K., in: "Valence Fluctuations in Solids", L. M. Falicov, W. Hanke and M. B. Maple, eds., North-Holland, Amsterdam (1981) p. 329.
10. Luttinger, J. M., Phys. Rev. 119:1153 (1960).
11. Martin, R. M., Phys. Rev. Lett. 48:362 (1982).
12. See chapter by Read, N. and Newns, D.M.
13. Izuyama, T., Kim, D. and Kubo, R., J. Phys. Soc. Japan 18:1025 (1963).
14. Grier, B. H. and Shapiro, S. M., bid Ref. 11, p. 325; Phys. Rev. B 25:1457 (1982).
15. Loewenhaupt, M. and Holland-Moritz, E., J. Appl. Phys. 50:7546 (1979).

16. Nagaoka, Y., Phys. Rev A. 138:1112 (1965).
17. Fedro, A. J. and Sinha, S. K., Proc. Int. Conf. on Valence Instabilities, P. Wachter and H. Boppart, eds., Zurich, North-Holland, Amsterdam (1982) p. 371.
18. Zlatic, V., Gruner, G. and Rivier, N., Solid State Comm. 14:139 (1974).
19. Lacroix, C. and Cyrot, M., Phys. Rev. B 20:1969 (1979).
20. Lacroix, C., J. Appl. Phys. 53(3):2131 (1979).
21. Baumgartel, H. and Muller-Hartmann, E., Proc. Int. Conf. on Valence Instabilities, P. Wachter and H. Boppart, eds., Zurich, North-Holland, Amsterdam (1982) p. 57.
22. Schrieffer, J. R. and Wolff, P. A., Phys. Rev. 149:491 (1966).
23. De Menezes, O. L. T., ibid, p. 53.
24. Steglich, F., Aarts, J., Bredl, C. D., Lieke, W., Meschede, D., Franz, W. and Schäfer, H., Phys. Rev. Lett. 43:1892 (1979).
25. Ott, H. R., Rudigier, H., Fisk, Z. and Smith, J. L., Phys. Rev. Lett. 50:1595 (1983).

ASPECTS OF THE DYNAMICS OF ELECTRONS IN MIXED VALENCE SYSTEMS

Erwin Müller-Hartmann

Institut für Theoretische Physik
Universität Köln
Köln, West Germany

INTRODUCTION

These lectures will be concerned with dynamical properties of electrons in rare earth mixed valence systems. Emphasis will be put on properties which are specific to the mixed valence state. We will be primarily aiming at discovering features of mixed valence systems which are basically different from what we know of conventional solids.

One message of the lectures will be that the electron dynamics at high temperatures is already interesting enough to be addressed here exclusively. We will confine ourselves to the dynamics at high temperatures in a sense to be specified later. This restriction has the benefit that we are able to use rather elementary methods for deriving our results. Due to a suitable choice of methods all calculations boil down to lowest order perturbation expansions. Y. Kuramoto is going to show in his seminar how one does go beyond the high temperature theory.

We will specifically address ourselves to two types of dynamical properties of electrons in mixed valence systems. The first type is the dynamics of the 4f-shells as represented in terms of dynamical susceptibilities. The method we will choose to approach this type of problem is the method of memory functions. We will see that the dynamics of mixed valence 4f-shells is quite different from the dynamics of stable 4f-shells. The second type is the dynamics of single particle excitation in mixed valence compounds described by fermion propagators. We will get some insight into how conduction electrons are affected by valence fluctuations and into what kind of itineracy 4f-electrons acquire in mixed valence compounds. The method we will use to investigate this aspect is the decoupling of equations of motion.

TWO TYPES OF ELECTRONS

What kind of model do we have in mind if we want to study the electron dynamics of mixed valence systems? The feature we want to emphasize particularly by our model is that two very different types of electronic states are involved: the highly localized 4f-orbitals and the delocalized Bloch states of the conduction bands. Electrons occupying a 4f-shell are strongly correlated with one another, whereas electrons in the conduction bands can be regarded as dynamically mutually independent. The dynamics of electrons in mixed valence systems receives a high degree of complexity through the fact that a hybridization couples 4f-orbitals and conduction band states.

The model describing this kind of situation is commonly called the Anderson model (of magnetic impurities)[1]. The circumstances under which the model will be used here are, however, somewhat different than in ref.[1] : We are aiming at describing a truly mixed valence situation where two configurations $4f^{n}$ and $4f^{n+1}$ of the 4f-shell are virtually degenerate such that both of these configurations are realized in the system with a finite probability. The model to be used here will also include the sevenfold orbital degeneracy of 4f-shells and the intraionic correlations of the 4f-shells will be described realistically. A model with this emphasis is often called a Hirst model[2]. It is a popular model of the electrons in mixed valence systems and we expect it to have some relevance for the dynamics we want to study.

Our model is explicitly given by the (grand canonical) Hamiltonian

$$H = H_{4f}+H_{cb}+H_{hybr}, \tag{1}$$

$$H_{4f} = \sum_{l\alpha\gamma} (E_{\gamma}^{(\alpha)}-\alpha\mu)P_{l\gamma}^{(\alpha)} \tag{2}$$

$$H_{cb} = \sum_{\underline{k}\nu\sigma} (\epsilon_{\underline{k}\nu}-\mu)c_{\underline{k}\nu\sigma}^{+}c_{\underline{k}\nu\sigma} \tag{3}$$

$$H_{hybr} = \frac{1}{\sqrt{N}} \sum_{lm\underline{k}\nu\sigma} [V_{\underline{k}\nu m}e^{i\underline{k}\cdot\underline{R}_l} f_{lm\sigma}^{+}c_{\underline{k}\nu\sigma}+V_{\underline{k}\nu m}^{*}e^{-i\underline{k}\cdot\underline{R}_l} c_{\underline{k}\nu\sigma}^{+}f_{lm\sigma}] . \tag{4}$$

The Hamiltonian H_{4f} of the 4f-shells includes all intraionic exchange and spin-orbit interactions and is expressed in terms of the projection operators $P_{l\gamma}^{(\alpha)}$ onto the 4f-level γ of the configuration $4f^{\alpha}$ at the rare earth ion l with position $\underline{R}_l$. The ground state energies $E_0^{(n)}-n\mu$ and $E_0^{(n+1)}-(n+1)\mu$ of the configurations $4f^n$ and $4f^{n+1}$ are assumed to be almost degenerate, whereas the energies of all other configurations ($\alpha \neq n$, n+1) are assumed to be infinite. Most of the excited levels ($\gamma>0$) of the configurations $4f^n$ and $4f^{n+1}$ have such a large excitation energy that their energy will be assumed infinite, too. Thus, for systems with Ce, Tm and Yb only the Hund's rule ground states ($\gamma=0$) of the configurations $4f^n$ and $4f^{n+1}$ will appear explicitly (n=0 for Ce, n=12 for Tm and n=13 for Yb). For Sm- and Eu-systems, however, it is often important to include a finite occupation of the first excited level of the configuration $4f^6$ which is only about 400 K (Sm) and 500 K (Eu) above the ground state.

The valence of the system is related to the probability

$$\delta = <\sum_{\gamma} P_{\gamma}^{(n+1)}> \tag{5}$$

of having the configuration $4f^{n+1}$. We will often use this parameter δ to characterize a mixed valence system rather than the excitation energy

$$E_f = E_0^{(n+1)}-E_0^{(n)}-\mu \tag{6}$$

which according to (2) is the model parameter primarily controlling the valence.

At some points we will include a crystal field Hamiltonian of the 4f-orbitals into H_{4f}.

The conduction band states in (3) are Bloch states orthogonal to the atomic 4f-orbitals. They are labeled by their wave vector $\underline{k}$, their band index ν (in case there are several conduction bands) and their spin direction σ.

The hybridization (4) is assumed to be a one-body force. It describes the transfer of an electron from a Bloch state into a 4f-orbital with magnetic quantum number m (m=-3,..,3) and vice versa. This is a purely orbital process which does not depend on spin. Two-body hybridizing forces do certainly exist, but are usually considered negligibly small. We will come back to this point later, because some of our results will be sensitive to the assumption of a one-body hybridization. Some dynamical properties will turn out to be even sensitive to the variation of the matrix element $V_{\underline{k}\nu m}$ with the magnetic quantum number m. In order to simplify matters we will often assume that the 4f-orbitals hybridize with a single spherically symmetric band. With this assumption we still have

$$V_{\underline{k}m}^{*} = \sqrt{4\pi}\, V_k \cdot Y_{3m}(\hat{k}), \tag{7}$$

where the non-trivial dependence on m and on the direction of $\underline{k}$ is given by a spherical harmonic. We want to stress here that replacing $V_{k\nu m}$ by a constant ignores the f-like symmetry of the conduction electron that hybridized at the 4f site.

The hybridization matrix elements V can be assumed to be small compared to most intraionic excitation energies and to the width of the conduction bands. In a true mixed valence case they are, however, not small compared to the excitation energy (6). This is what makes mixed valence systems difficult to understand fully, since in general hybridization may not be treated as a perturbation. In these lectures, we will nevertheless concentrate on features which are obtainable from low order perturbation calculations. This will limit the validity of our results to high temperatures.

DYNAMICAL SUSCEPTIBILITIES OF MIXED VALENT 4f-SHELLS

Basic Notions

The first topic we want to turn to now is the dynamics of the 4f-shells. In particular, we will be interested in the dynamics of the 4f-magnetization which is susceptible to direct experimental inspection with neutron scattering. We will also include into the discussion the dynamics of other 4f-shell properties like the 4f-occupation (4f-charge) or the electric quadrupolar moment. In the model we have in mind for mixed valence systems all these quantities would be conserved in the absence of hybridization. Our aim will therefore be to understand what kind of dynamics the hybridization introduces into the 4f-shells in a mixed valence situation.

The standard quantity* containing all information about the dynamics of a physical observable A (the magnetization, e.g.) in the neighbourhood of thermal equilibrium is the dynamical susceptibility

$$\chi(\omega) = i\int_0^\infty < [A(t),A]> e^{i\omega t} dt. \tag{8}$$

It describes the response of the expectation value of A to an external field of frequency ω. If A is such that neutrons couple to it, the neutron scattering cross section will be proportional to the dynamical structure function

$$S(\omega) = \frac{\omega}{1-e^{-\beta\omega}} \cdot \frac{\mathrm{Im}\chi(\omega)}{\omega}, \tag{9}$$

where ω denotes the energy loss of the neutron ($\hbar=1$).

---

* For a detailed presentation of the quantities introduced in the following as well as of the memory function method see, e.g. ref.[3].

Another useful quantity is the relaxation function

$$C(t) = \int_0^\beta \langle A(t-i\tau)A\rangle d\tau \tag{10}$$

which describes the way in which the expectation value of A relaxes to its equilibrium value $\langle A\rangle=0$, after a stationary external field is suddenly switched off at time t=0. There is an intimate relation between the relaxation function and the dynamical susceptibility. This relation is expressed by the identity

$$i\omega\cdot c(\omega) = \chi(\omega) - \chi(o), \tag{11}$$

where

$$c(\omega) = \int_0^\infty C(t)e^{i\omega t}dt \tag{12}$$

is the Laplace transform of the relaxation function.

The simplest type of relaxation behavior is the exponential decay of an oscillatory motion given for not too short times by

$$C(t) = \chi(o)e^{-\Gamma t}\cos\Delta t\ . \tag{13}$$

The dynamical susceptibility corresponding to this relaxation is easily derived from (12) and (11). It implies a structure function (9) with

$$\mathrm{Im}\chi(\omega)/\omega=\chi(o)\left[\frac{\Gamma/2}{\Gamma^2+(\omega-\Delta)^2}+\frac{\Gamma/2}{\Gamma^2+(\omega+\Delta)^2}\right]. \tag{14}$$

In the special case of a purely exponential decay, $\Delta=0$, this corresponds to the observation of a "quasi-elastic line", a Lorentzian centered at $\omega=0$ the half-width of which is given by the relaxation rate Γ. For $\Delta > 0$ two inelastic lines of Lorentzian shape centered at $\omega=\pm\Delta$ will be observed. This indicates the existence of an internal excitation of energy Δ and of half-width Γ which the neutron is able to excite (energy loss $\omega\approx\Delta$) and to de-excite (energy gain $-\omega\approx\Delta$).

The dynamical susceptibilities of the mixed valence systems we want to study may contain several quasi-elastic or inelastic lines of the type (14). This indicates that the magnetization (or whatever the meaning of A is) is composed of several physical ob-observables which are independent of each other and each of which has the simple dynamical behavior given by (13). In this case, instead of looking at the susceptibility (8) only one should better study the complete dynamics of an appropriate set of obervables A_k which is described by the susceptibility matrix

$$\chi_{kl}(z) = i\int_0^\infty \langle[A_k(t),A_l]\rangle e^{izt}dt \tag{15}$$

or by the corresponding matrix of relaxation functions $c_{kl}(z)$.

For technical reasons we replace the frequency ω with the complex frequency z (Im z>0) from here on. By diagonalizing these matrices one may find those observables in the subspace spanned by the A_k which have a simple dynamics. At the same time, the observable of original interest A which is contained in the subspace is decomposed into those observables.

Essentials of the Memory Function Method

We will now discuss the memory function approach[4,5,3] to dynamics. It is a method by which dynamical susceptibilities of the functional form (14) are derived most elegantly. The method makes use of a geometrical interpretation of dynamics in the vector space of physical observables. To this aim the Liouville operator

$$\mathcal{L}A = [H,A] \tag{16}$$

is introduced as a linear operator in this vector space and time evolution of observables in the Heisenberg picture is described as a linear map

$$A(t) \equiv e^{iHt}Ae^{-iHt} = e^{i\mathcal{L}t}A. \tag{17}$$

The crucial feature of the method is that a scalar product is introduced in the vector space of observables which serves as a means for judging the dynamical independence of observables (via orthogonality). The natural choice for a scalar product between two observables A and B of a quantum system is

$$\langle A|B\rangle = \int_0^\beta \langle A^+(-i\tau)B\rangle d\tau. \tag{18}$$

In terms of this scalar product the Liouville operator $\mathcal{L}$ is a hermitean operator and the time evolution (17) is a unitary map.

The susceptibilities (15) and their relaxation functions assume a particularly compact form in terms of our new notation. We obtain

$$\chi_{kl}(z) = \langle A_k|\frac{\mathcal{L}}{\mathcal{L}-z}A_l\rangle, \quad \chi_{kl}(o) = \langle A_k|A_l\rangle \tag{19}$$

and

$$c_{kl}(z) = \langle A_k|\frac{i}{z-\mathcal{L}}A_l\rangle. \tag{20}$$

The memory function method is based on the assumption that a set of observables can be sorted out which move on a much slower timescale than the remaining variables of the system. The slow variables relax due to their coupling to the remaining variables which serve as a thermal bath for the slow variables and the dynamics of which need not be treated in much detail. The slow variables are sorted out with the help of the operator

$$\mathcal{P} = \sum_{k,l} |A_k>(\chi(0)^{-1})_{kl}<A_l| \tag{21}$$

which projects ($\mathcal{P}^2=\mathcal{P}^+=\mathcal{P}$) the space of observables onto the subspace spanned by the slow variables; $\chi(0)^{-1}$ means the inverse of the matrix $\chi_{kl}(0)$. We will denote the complementary operator by

$$\mathcal{Q} = 1 - \mathcal{P}. \tag{22}$$

With a few simple algebraic manipulations the useful identity

$$c(z) = i\chi(0)\left[z\chi(0)-\eta+i\sigma(z)\right]^{-1}\chi(0). \tag{23}$$

is derived for the matrix (20) of relaxation functions. In this identity the relaxation functions are expressed in terms of the static susceptibility matrix (19), the matrix

$$\eta_{kl} = <A_k|\mathcal{L}A_l> \tag{24}$$

and the matrix of memory functions

$$\sigma_{kl}(z) = <\mathcal{Q}\dot{A}_k| \frac{i}{z-\mathcal{Q}\mathcal{L}\mathcal{Q}} (\mathcal{Q}\dot{A}_l)>. \tag{25}$$

The memory functions are themselves relaxation functions. They are the relaxation functions of the operators $Q\dot{A}_k$ under the action of the reduced Liouville operator $\mathcal{Q}\mathcal{L}\mathcal{Q}$. A simple physical interpretation can be given to the operators $\mathcal{Q}\dot{A}_k$: they are those components of the velocities $\dot{A}_k$ which are orthogonal to (i.e. dynamically independent of) the variables A_k. Altogether, the memory functions are relaxation functions of a reduced dynamics from which the slow variables are completely removed. This implies that $\sigma(z)$ can be expected to be a regular, slowly varying function of z in the slow frequency range.

A slight change of notation in (23) leads to

$$c(z)= i\chi(0)\left[z-\Delta+i\Gamma(z)\right]^{-1} \tag{26}$$

with

$$\Delta = \chi(0)^{-1}\cdot\eta \tag{27}$$

and

$$\Gamma(z) = \chi(0)^{-1}\cdot\sigma(z). \tag{28}$$

We have now cast the relaxation function into a form from which we can read off the physical meaning of Δ and Γ (see (13)). The eigenvalues of the relaxation matrix $\Gamma(z)$ are the relaxation rates of the subspace of slow variables A_k and the eigenvalues of Δ are the excitation energies. We will often quote as relaxation rates Γ

those eigenvalues of $\Gamma(z)$ where we have inserted the associated excitation energy Δ for z.

In applications of the memory function method the static susceptibility $\chi(0)$ is considered an input. The philosophy of the method is based on the expectation that $\sigma(z)$ and Δ are much less sensitive to approximations than $c(z)$ or $\chi(z)$ are. This expectation is justified physically through the separation of time-scales discussed above.

Mixed Valent 4f-Shells

For mixed valence systems there is a clear-cut separation between the fast time-scale of the broad conduction bands and the much slower time-scale of the 4f-shells due to their weak hybridization with the conduction bands. Our natural choice of slow variables is therefore the set of observables

$$X^{(n)}_{\gamma M\gamma' M'} = |J_\gamma M><J_{\gamma'} M'|, \quad X^{(n+1)}_{\gamma\bar{M}\gamma'\bar{M}'} = |\bar{J}_\gamma \bar{M}><\bar{J}_{\gamma'} \bar{M}'| \tag{29}$$

for each 4f-shell, where J_γ and $\bar{J}_\gamma$ denote the total angular momenta of those levels of the configurations $4f^n$ and $4f^{n+1}$, respectively, which we decide to include, and where M and $\bar{M}$ denote the eigenvalues of J^z_γ and $\bar{J}^z_\gamma$. As already mentioned before (5), for Ce, Yb and Tm the first excited levels of both configurations are so high that we are satisfied with including the ground states, $\gamma = \gamma' = 0$, only. On the other hand, for Sm and Eu it is necessary to include the first excited level of the configuration $4f^6$.

We will now discuss the calculation of the memory functions (25) associated with the set of observables (29). Since these operators are eigenoperators of the ionic Hamiltonian (2), it is obvious that $Q[H_{4f}, A_k]=0$ and thus

$$\mathcal{Q}\dot{A}_k = \mathcal{Q} i\left[H_{hybr}, A_k\right]. \tag{30}$$

The memory functions (25) are therefore explicitly proportional to the square of the hybridization. For a lowest order evaluation of the memory function we then evaluate the relaxation functions (25) with the unperturbed Hamiltonian $H_o = H_{4f} + H_{cb}$ replacing $\mathcal{L}$ by $\mathcal{L}_o$ and all thermal averages <...> by $<...>_o$. Note that this also affects the scalar product defined in (18) and the projection operator $\mathcal{Q}$ defined in (22,21). It turns out that now the operator $\mathcal{Q}_o$ can be omitted completely from the expression for $\sigma(z)$ because it is ineffective due to H_o conserving the 4f-occupation. We end up with second order memory functions of the form

$$\sigma^{(2)}_{kl}(z) = <i\left[H_{hybr}, A_k\right] \left| \frac{i}{z-\mathcal{L}_o} i\left[H_{hybr}, A_l\right] \right>_o \tag{31}$$

which are quite easily evaluated. The only technicality in the evaluation is that matrix elements $<\bar{J}\bar{M}|f^{+}_{m\sigma}|JM>$ between Hund's rule states of different configurations are involved. These matrix elements can be related to a few reduced matrix elements with standard methods of group theory (Wigner-Eckart theorem). They are related to the coefficients of fractional parentage of atomic physics.

It has to be mentioned that in all practical applications of the memory function method it is desirable to get rid of the explicit appearance of the operators $\mathcal{Q}$ in the denominator of (25), since these are quite difficult to handle.

The second order relaxation matrix $\Gamma^{(2)}(z)$ is obtained from $\sigma^{(2)}(z)$ via (28) where it is sufficient to insert the unperturbed static susceptibility $\chi^{(0)}(0)$. The excitation matrix Δ in (27) should be calculated by a straightforward second order perturbation expansion in order to be consistent with $\Gamma^{(2)}$. There is no coupling between different 4f-shells in this lowest order treatment. All results derived in this section apply therefore to mixed valence compounds, alloys as well as impurities.

The hybridization appears in the relaxation matrix $\Gamma^{(2)}$ in terms of hybridization integrals

$$v_{mm'} = \frac{1}{N} \sum_{\underline{k}\nu} V^*_{\underline{k}\nu m} V_{\underline{k}\nu m'} \, \delta(\epsilon_{\underline{k}\nu} - \mu). \tag{32}$$

They are energies of the order of $|V|^2 \rho_F$, where $|V|$ is an average hybridization matrix element and ρ_F the conduction band density of states at the Fermi-surface. They set the scale for the relaxation rates and also for the range of validity of the lowest order approximation discussed here: it is valid for temperatures larger than the energies (32).

Up to this point the relaxation functions (26) and the associated dynamical susceptibilities (15,11) form matrices of rather large dimensions. The number of operators we have started from in (29) is in fact equal to $(2J_0+1)^2+(2\bar{J}_0+1)^2$ if we include ground states only which makes a minimum matrix dimension of 37 for Ce up to a huge dimension of 233 for Tm. It would be disastrous if we had to cope with such large matrices. We do not have to because we can use symmetry considerations to cut these matrices down to much smaller sizes.

The highest symmetry we can assume is complete rotational symmetry. This assumption implies the neglect of crystal fields of the 4f-shells and a hybridization as in (7) with a spherical conduction band. We will discuss this case in detail.

The first simplification we obtain due to rotational invariance is a reduction of the 49 hybridization integrals (32) to just one such integral:

$$v_{mm'} = W_0 \cdot \delta_{mm'} \, . \tag{33}$$

The main simplification is however that the matrix of the dynamical susceptibility can be almost completely diagonalized by methods of group theory. To achieve the diagonalization we have to perform a unitary transformation of the set of variables (29) which transforms them into irreducible tensor operators. Using Clebsch-Gordan coefficients we write the tensor operators of the configuration $4f^n$ as

$$T^{(n)}_{\gamma\gamma'\Lambda\lambda} = \sum_{MM'} (-1)^{M'} \langle J_\gamma M J_{\gamma'}, -M' | \Lambda\lambda \rangle X^{(n)}_{\gamma M \gamma' M'} \tag{34}$$

and those of the configuration $4f^{n+1}$ correspondingly. The tensor indices $\Lambda\lambda$ are angular momenta assuming integer values according to the familiar rules

$$|J_\gamma - J_{\gamma'}| \leq \Lambda \leq J_\gamma + J_{\gamma'}, \qquad -\Lambda \leq \lambda \leq \Lambda . \tag{35}$$

The power 2^Λ is usually called the multipolarity of the tensor operator. It is a standard result of group theory that the susceptibility matrix (15) formed with tensor operators as (34) is diagonal in the index pairs Λ_k, Λ_l and λ_k, λ_l and its diagonal elements are independent of the index $\lambda_k (=\lambda_l)$. The same statement applies to the memory function matrix and all other matrices which appeared above. This leads in fact to an enormous reduction of the matrix dimensions. For each multipolarity 2^Λ we are left with one matrix the dimension of which is equal to the number of times Λ fulfills the rule (35) and the corresponding rule for $4f^{n+1}$.

Let us first look at the magnetic susceptibility in which we are primarily interested. Since the magnetization is a vector, the operator of the 4f-magnetization has to be a linear combination of the dipolar tensor operators ($\Lambda=1$) introduced above. The magnetic susceptibility is therefore obtained from the ($\Lambda=1$)-submatrix.

For the Anderson model ignoring orbital degeneracy the configurations $4f^0$ with J=0 (non-magnetic) and $4f^1$ with $\bar{J}=1/2$ (magnetic) are mixed. There is only one single dipolar operator among the 4f-shell operators in this case, the spin operator of $4f^1$, and the ($\Lambda=1$)-submatrix is one-dimensional. The dynamical susceptibility has a single quasi-elastic line with a half-width (relaxation rate)[6,7,8] *

$$\Gamma_{AM} = 2\pi W_o \frac{2(1-\delta)}{2-\delta} . \tag{36}$$

In an alternative version of the Anderson model the configuration $4f^0$ is replaced by $4f^2$ such that J=1/2 and $\bar{J}=0$. The relaxation rate for this model is obtained from (36) by replacing δ with $1-\delta$ (particle-hole transformation). Both of these relaxation rates of the Anderson model are plotted in Fig. 1 as dashed lines.

---

* My criticism of ref.[6] in ref,[9] was incorrect.

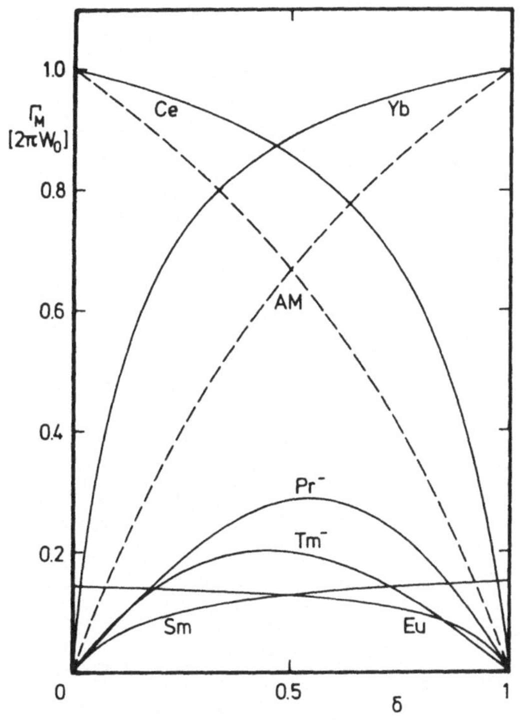

Fig. 1 Magnetic relaxation rates of various mixed valent ions as given by equations (36-39) and (43).

For mixed valence Ce ions the configurations $4f^{0}$ with J=0 and $4f^{1}$ with $\bar{J}_o$=5/2 are mixed (;the excited level at 3000 K with $\bar{J}_1$=7/2 can be ignored). Thus, Ce is represented qualitatively rather well by the spin-only Anderson model, there is just a larger value of $\bar{J}_o$ due to the orbital angular momentum. The relaxation rate[9,10]

$$\Gamma_{Ce} = 2\pi W_o \frac{6(1-\delta)}{6-5\delta} \tag{37}$$

differs from (36) only due to the larger degeneracy of the $4f^{1}$ ground state.

The case of Yb ions is closely related to the Ce case via a particle-hole transformation of the 4f-shell which implies a change of sign of the spin-orbit coupling. The relaxation rate of the magnetization of the J_o=7/2 ground state of the configuration $4f^{13}$ ($\bar{J}$=0 for $4f^{14}$) is given by[9,10]

$$\Gamma_{Yb} = 2\pi W_o \frac{8\delta}{1+7\delta} \ . \tag{38}$$

Before turning to the more intricate cases of the ions Tm, Sm and Eu we wish to insert a few general comments. The relaxation of mixed valence 4f-shells we have derived here is quite different from the familiar Korringa relaxation of stable 4f-shells in metals: the rates are not proportional to, but independent of temperature, and they are much larger than typical Korringa rates even at room temperature. The measured relaxation rates of mixed valence Ce and Yb systems [11,12] do in fact display these two features. Although our relaxation rates were derived in the high temperature regime they happen to be of larger significance because the rates measured turn out to be rather constant down to low temperatures.

A second comment has to be added to qualify the vanishing of the above relaxation rates in the limit of stable magnetic ions ($\delta\to 1$ in (26,37) and $\delta\to 0$ in (38)). This feature is physically not relevant, because the temperature above which the lowest order rates apply approaches infinity in this limit. At any fixed temperature a cross-over from a mixed valence to a "Kondo" relaxation takes place where higher order processes dominate the relaxation rate. A simple estimate shows that the cross-over temperature is of order $W_o/(\delta\ln^2\delta)$ if $\delta\to 0$ is the magnetic limit.

Mixed valent Tm ions are special because for them both configurations $4f^{12}$ with $J_o=6$ and $4f^{13}$ with $\bar{J}_o=7/2$ have a magnetic ground state. There is a dipolar operator ($\Lambda=1$), the operator of the total angular momentum, in both configurations. The ($\Lambda=1$)-submatrix of the susceptibility is two-dimensional, i.e. the dynamics of the 4f-magnetization follows from the coupled dynamics of both angular momenta. Diagonalizing the relaxation matrix (28) we obtain two relaxation rates [9,10]

$$\Gamma^{\pm}_{Tm} = 2\pi W_o \frac{169}{14(8+5\delta)} \left[1\pm\sqrt{1-\frac{5770}{4563}\delta\,(1-\delta)}\,\right] . \tag{39}$$

The two observables which are dynamically independent and which relax with these two rates are different linear combinations of the angular momenta of $4f^{12}$ and $4f^{13}$ and they depend on the valence. Their contributions $I^{\pm}$ ($I^{+}+I^{-}=1$) to the susceptibility of the 4f-magnetization also depend on the valence and are given by

$$I^{+} = (\Gamma^{o}-\Gamma^{-})/(\Gamma^{+}-\Gamma^{-}) , \tag{40}$$

where

$$\Gamma^{o}_{Tm} = \frac{149201\cdot\delta(1-\delta)}{7(2401-1537\delta)(8+5\delta)} . \tag{41}$$

In terms of the rates (39) and the weight factors (40) the dynamical suscpetibility of the Tm 4f-magnetization is

$$\chi(\omega) = \chi(o)\left[I^{+}\frac{\Gamma^{+}}{\Gamma^{+}-i\omega} + I^{-}\frac{\Gamma^{-}}{\Gamma^{-}-i\omega}\right] . \tag{42}$$

Theory thus predicts a superposition of two quasi-elastic lines for mixed valent Tm-systems. The half-width Γ^- of the narrow line is shown in Fig. 1, the half-width Γ^+ is always more than 10 times larger than Γ^-. This makes the experimental detection of the Γ^+-line very difficult. It is in fact virtually impossible to detect it at high temperatures, because I^+ is rather small [9,10].

Regardless of the difficulties of experimental verification it is interesting that detailed predictions as (39 - 42) can be made, esentially without adjustable parameters; the numbers in (39) and (41) are all given unambiguously through Hund's rule matrix elements. One might ask, however, how sensitively these predictions depend on certain features of the model used. We have pursued this question in a certain direction by studying the influence of two-body hybridizing interactions. Hybridizing interactions due to Coulomb matrix elements between three 4f-orbitals on the same atom and one conduction band state are not approximated well by the one-body interaction (4) because of the strong 4f-correlation. We have found [13] that these two-body hybridizing interactions tend to reduce the splitting between Γ^+ and Γ^- and to enhance I^+. Estimating the size of these effects with a one-OPW conduction band model we found that they are not negligibly small.

The 4f-dynamics of TmSe at low temperatures is quite different from that of mixed valent Ce- and Yb-systems [14]. The high temperature theory discussed here does not give a clue to this difference. For a discussion of this topic the reader is referred to the lectures by Y. Kuramoto, P. Wachter and M. Loewenhaupt.

The dynamics of Sm- and Eu-systems looks rather similar to that of Ce- and Yb-systems, if only Hund's rule ground states were relevant: there would be a single quasi-elastic line of half-width

$$\Gamma_{Sm} = 2\pi W_o \cdot \frac{22}{147}\,\frac{6\delta}{1+5\delta}\;, \quad \Gamma_{Eu} = 2\pi W_o \cdot \frac{1}{7}\,\frac{8(1-\delta)}{8-7\delta}\;. \tag{43}$$

The reduction by factors of 22/147 and of 1/7 as compared to (37, 38) is due to Hund's rule exchange correlations which reduce the effectiveness of the hybridization (4). This kind of effect is of course also present in (39).

The rates (43) are shown in Fig. 1. They are, however, not of much relevance, since the temperature window in which they apply is too narrow: in the high temperature region where our lowest order approximation is reliable the first excited state (J=1) of the configuration $4f^6$ has a non-negligable thermal population. If it is incorporated the number of tensor operators (34) gets larger. Instead of just one we obtain four dipolar operators, namely in addition to the angular momentum of the magnetic ground state (of $4f^5$ and $4f^7$, respectively) the angular momentum of the $4f^6$ excited state, and the transition operator from the $4f^6$ ground state to the excited state and its hermitean conjugate. These operators introduce two quasi-elastic and two inelastic lines into the magnetic susceptibility[15]. Detailed predictions without adjustable parameters can be made about the half-widths and weights of these lines. The half-widths and the weights vary conspicuously with increasing temperature due to the increasing thermal population of the excited state. An example of the quasi-elastic half-widths is shown in Fig. 2.

Experimental verification of the predictions for Sm- and Eu-systems would provide a sensitive test of the standard model of mixed valence systems. It is therefore worth to be tried despite the obvious experimental difficulties.

Besides the magnetic susceptibility discussed above the susceptibility of the 4f-charge is of interest[9,10]. It can be important for phonon self-energies, because the atomic radius of rare earth ions varies much with the 4f-occupation. In particular, if the relaxation of the 4f-charge is not much faster than the frequency of a longitudinal phonon, the phonon is expected to have

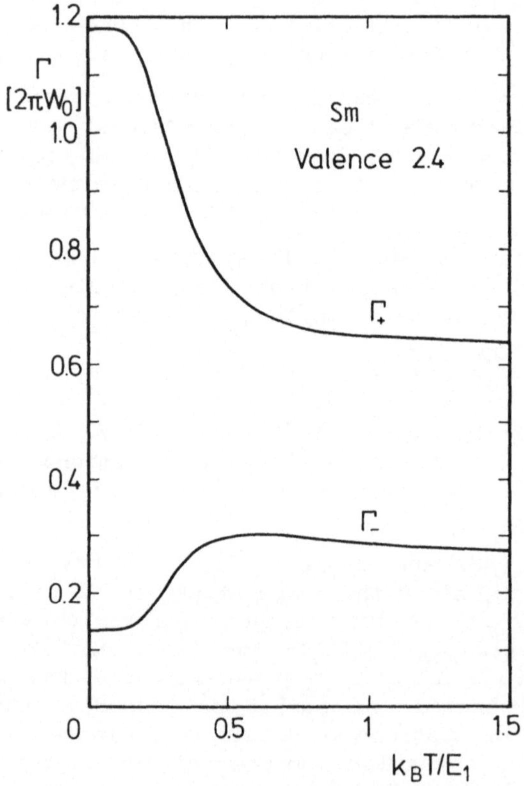

Fig. 2 The two quasi-elastic magnetic relaxation rates of Sm-ions at an arbitrarily chosen valence of 2.4; E_1 denotes the spin-orbit splitting of about 400 K.

an anomalously small life-time due to its non-adiabatic coupling to the valence fluctuation. It is therefore of interest to know the 4f-charge relaxation rates.

Since the 4f-charge is a scalar, the dynamical charge susceptibility follows from the dynamics of the monopolar operators, $\Lambda=0$. These monopolar operators are the projection operators $P_\gamma^{(n)}$ and $P_\gamma^{(n+1)}$ onto the levels of the 4f-shell. Thus, the ($\Lambda=0$)-submatrix of the susceptibility has always a dimensions larger than 1. There is, however, one particular operator in the subspace spanned by the $P_\gamma^{(\alpha)}$ which is strictly conserved by the Hamiltonian (1): $\sum_{\gamma\alpha} P_\gamma^{(\alpha)}$. Since this operator does not relax, the number of relaxation rates of charge dynamics is one less than the number of ($\Lambda=0$)-observables. It is also obvious from the features just explained that charge dynamics is different from the dynamics of the magnetization. One cannot be surprised if charge relaxation rates will not be the same as magnetic relaxation rates.

If excited states are disregarded, the variables $P_0^{(n)}$ and $P_0^{(n+1)}$ will relax with a single rate according to what was said before. These charge relaxation rates are given by the formulas ($\Gamma=2\pi W_0\cdot\gamma$)

$$\gamma_{Ce} = \frac{6}{6-5\delta}\ ,\ \gamma_{Yb} = \frac{8}{1+7\delta}\ ,\ \gamma_{Tm} = \frac{169}{7(8+5\delta)} \tag{44}$$

$$\gamma_{Sm} = \frac{22}{147}\,\frac{6}{1+5\delta}\ ,\ \gamma_{Eu} = \frac{1}{7}\,\frac{8}{8-7\delta}\ .$$

They always exceed the magnetic relaxation rates.

Including the $4f^6$ excited level for Sm and Eu adds the projection operator onto this level to the theory. It provides the charge susceptibility with two relaxation modes and leads to an increase with increasing temperature of the effective charge relaxation rates above the values shown in (44)[15].

Summarizing this section one can say that 4f-shell dynamics of mixed valence systems seems to imply a number of interesting and peculiar features which are not found in systems of stable valence. Because of space limitations we leave out effects of crystalline anisotropy on the dynamics[9,15].

DYNAMICS OF SINGLE PARTICLE EXCITATIONS IN COMPOUNDS

In this section we have a look at the dynamics of the fermion operators $c_{\underline{k}\nu\sigma}$, $c^+_{\underline{k}\nu\sigma}$ and $f_{lm\sigma}$, $f^+_{lm\sigma}$ in mixed valence compounds. The questions to be answered are, e.g., in what way the motion of conduction electrons is influenced by the hybridization with the localized 4f-electrons and in which sense the localized 4f-electrons acquire itineracy due to their hybridization with the conduction electrons.Because of space limitation we will only briefly describe some recent results[16] which we think are of basic interest.

It is convenient to use atomic transition operators[17] $X_{lM\bar{M}} = |lJ_oM\rangle\langle l\bar{J}_o\bar{M}|$ instead of the field operators $f_{lm\sigma}$, because the former are eigenoperators of H_{4f}. To make maximum use of the conservation of lattice momentum $\underline{k}$ in the compound we go over to the fourier transformed valence fluctuation operators

$$X_{\underline{k}M\bar{M}} = \frac{1}{\sqrt{N}} \sum_l e^{-i\underline{kR}_l} X_{lM\bar{M}} . \tag{45}$$

The method we employ is decoupling equations of motion of the propagators. In order to achieve a compact notation we introduce matrices. The conduction band propagators $G_{\underline{k}\nu\sigma,\nu'\sigma'}(z) = \langle\langle c_{\underline{k}\nu\sigma} | c^+_{\underline{k}\nu'\sigma'} \rangle\rangle$ form the matrix $G_{\underline{k}}(z)$, the 4f transition propagators $F_{\underline{k}M\bar{M},M'\bar{M}'}(z) = \langle\langle X_{\underline{k}M\bar{M}} | X^+_{\underline{k}M'\bar{M}'} \rangle\rangle$ form the matrix $F_{\underline{k}}(z)$, and the hybridization matrix elements $V_{\underline{k}M\bar{M},\nu\sigma} = \sum_m V_{\underline{k}\nu m} \langle \bar{J}_o\bar{M} | f^+_{m\sigma} | J_oM \rangle$ form the matrix $V_{\underline{k}}$. First of all, there is an exact relation,

$$G_{\underline{k}}(z) = G^o_{\underline{k}}(z) + G^o_{\underline{k}}(z) V^+_{\underline{k}} F_{\underline{k}}(z) V_{\underline{k}} G^o_{\underline{k}}(z) , \tag{46}$$

expressing the conduction band propagator in terms of the 4f transition propagator. The equations of motion of $F_{\underline{k}}(z)$ up to second order in the hybridization produce higher propagators. If these are decoupled, we obtain

$$F_{\underline{k}}(z) = \left[z - M(z) - N(z) V_{\underline{k}} G^o_{\underline{k}}(z) V^+_{\underline{k}}\right]^{-1} N(z) , \tag{47}$$

where M(z) and N(z) are determined self-consistently from integral equations. It follows from (47) and (46) that the self-energy $\Sigma_{\underline{k}}(z) = (G^o_{\underline{k}}(z))^{-1} - (G_{\underline{k}}(z))^{-1}$ of the conduction bands is given by the simple expression

$$\Sigma_{\underline{k}}(z) = V^+_{\underline{k}} \left[z - M(z)\right]^{-1} N(z) V_{\underline{k}} . \tag{48}$$

The first conclusion to be drawn from (46-48) concerns the high temperature limit of the self-energy $\Sigma_{\underline{k}}(z)$. In this limit M(z) and N(z) can be calculated perturbatively and we are confident that they are obtained correctly to second order in the hybridization from the decoupling approximation. Since $N(z) = (1-\delta)/(2J+1) + \delta/(2\bar{J}+1) + O(V^2)$ and since $\mathrm{Im}\, M(z) = O(V^2 \rho_F)$, we conclude from (48) that "within the 4f-band", $z \approx \mathrm{Re}\, M(z)$, the strength V of the hybridization cancels out of (48) and $\mathrm{Im}\, \Sigma_{\underline{k}}(z) = O(\rho_F^{-1})$. Thus the scattering rate of conduction electrons due to valence fluctuations is independent of the strength of their coupling[18].

A second conclusion follows from a closer inspection of the lowest order mass operator M(z). Its imaginary part has an energy dependence of the form Im $M(\omega) \approx V^2 \rho_F$ $(a+b \cdot \mathrm{tgh}\ \beta\omega/2)$. This leads to a strong intrinsic particle-hole asymmetry of the self-energy $\Sigma_{\underline{k}}(z)$. It has to be associated with a conspicuously large thermopower of predictable sign and size.

The valence fluctuation propagator (47) contains interesting information too. Its dependence on lattice momentum signals that 4f-electrons can travel through the compound. It is possible to make detailed statements about the nature of itineracy, because the dependence on $\underline{k}$ in (47) is explicit. Let us assume for simplicity that the conduction band is spherically symmetric with a hybridization given by (7). In this case of highest symmetry possible the propagator (47) is still a matrix. It is diagonalized by turning the angular momenta M and $\overline{M}$ into the direction of $\underline{k}$ and by coupling the operators (45) into tensor operators as in (34). The propagator (47) then describes the dynamics of valence transitions of definite angular momentum Λ with a component λ in the direction of $\underline{k}$. We obtain a strong intrinsic anisotropy of propagation, since only valence transitions with $|\lambda|=1/2$ can travel. This can be expected to lead to particular effects of anisotropy in magneto-transport.

REFERENCES

1. P. W. Anderson, Localized Magnetic States in Metals, Phys. Rev. 124: 41 (1961).
2. L. L. Hirst, Theory of Magnetic Impurities in Metals, Phys. Kond. Mat. 11:255 (1970).
3. D. Forster,"Hydrodynamic Fluctuations,Broken Symmetry, and Correlation Function", Benjamin, London and Amsterdam (1975).
4. R. Zwanzig, Statistical Mechanics of Irreversibility, in: "Lectures in Theoretical Physics", Vol. III, W.E. Brittin, B.W. Downs and J. Downs, eds., Interscience Publishers, New York (1961).
5. H. Mori, A Continued-Fraction Representation of the Time-Correlation Functions, Progr. Theoret. Phys. 34:399 (1965).
6. C. A. Balseiro and A. López, Dynamical Susceptibility of Intermediate Valence Systems, Solid State Commun. 17:1241 (1975).
7. M. E. Foglio, Theory of the Inelastic Magnetic Scattering in Mixed Valence Compounds: $CePd_3$, J. Phys. C11:4171 (1978).
8. Y. Kuramoto, Dynamical Susceptibilities of Intermediate Valence Compounds, Z. Physik B 37:299 (1980).

9. E. Müller-Hartmann, Dynamics of 4f-Electrons in Mixed Valence Compounds, in: "Electron Correlation and Magne-Magnetism in Narrow-Band Systems", T. Moriya, ed., Springer, Berlin Heidelberg New York (1981).
10. Y. Kuramoto and E. Müller-Hartmann, Dynamics of 4f-Electrons in Intermediate Valence Compounds, in: "Valence Fluctuations in Solids", L.M. Falicov, W. Hanke and M.B. Maple, eds., North-Holland, Amsterdam (1981).
11. E. Holland-Moritz, M. Loewenhaupt. W. Schmatz,and D. Wohlleben, Spontaneous Relaxation of the Local 4f Magnetization in $CePd_3$, Phys. Rev. Lett. 38:983 (1977).
12. S. M. Shapiro, J.D. Axe, R.J. Birgeneau, J.M. Lawrence, and R.D. Parks, Spin Dynamics in the Mixed Valence Alloy $Ce_{1-x}Th_x$, Phys. Rev. B 16:2225 (1977).
13. S. Schmitz, diploma thesis, Universität Köln, May 1983, unpublished.
14. M. Loewenhaupt and E. Holland-Moritz, Spin Dynamics of TmSe Studied by Neutron Scattering, J. Magn. Mag. Mat. 9:50 (1978).
15. H. J. Schmidt, diploma thesis, Universität Köln, July 1983, unpublished.
16. H. G. Baumgärtel and E. Müller-Hartmann, to be published.
17. J. Hubbard, Electron Correlations in Narrow Energy Bands, Proc. Roy. Soc. A 277:237 (1964).
18. D. Müller, E. Cattaneo, H. Schneider, W. Schlabitz, and D. Wohlleben, The High Temperature Resistivity of Some Ce and Yb Compounds with Unstable Valence, in: "Valence Instabilities", P. Wachter and H. Boppart, eds., North-Holland, Amsterdam (1982).

STATISTICAL THERMODYNAMICS OF MIXED VALENCE AT HIGH TEMPERATURES

Dieter Wohlleben

II. Physikalisches Institut
Universität zu Köln
5000 Köln 41, FRG

INTRODUCTION

In these lectures we present an expression for the free enthalpy of Rare Earth (RE) mixed valence compounds for thermal energies large compared to the energy gain associated with quantum mixing between the two integral valence states "$4f^{n+1}$" and "$4f^{n}$" involved in the mixture. This expression uses only well known or at least measurable properties of the two integral valence states, such as Hund's rule ground states and excited spin orbit (SO) multiplets and crystal field (CF) splittings of the 4f shell with their degeneracies, and volumes, thermal expansion coefficients and bulk moduli of the unit cell of the crystal in $4f^{n+1}$ and $4f^{n}$. The only bona fide mixed valence parameter entering the formalism is the temperature and pressure dependent fractional valence or the interconfigurational excitation energy, E_X, associated with it. We show that the experimental data available at high temperatures, such as the susceptibility, lattice constant anomaly, thermal expansion anomaly, L_{III} valence etc., can be explained on the basis of such a simple, semi-classical Ansatz. The high temperature behavior of mixed valence compounds turns out to be a good starting point to understand the more complicated anomalies which one observes at thermal energies comparable to or smaller than the energy gain due to quantum mixing, as we try to show in another set of lectures [1].

We first derive an expression for the elastic energy of mixed valence compounds as function of volume and valence from the observation that the volume of homogeneously (dynamically) mixed valence compounds and of static mixtures of RE compounds with different integral valence is a non-linear function of valence and concentration respectively (breakdown of Végards "law"). We then write down the free enthalpy, G, involving this elastic energy, F_e, a chemical energy, F_c, and an entropic part, F_s, which takes into account the SO and CF splittings and degeneracies of $4f^{n+1}$ and $4f^n$. We discuss the valence and temperature dependence of F_s for Ce, Pr, Sm, Eu, Tm and Yb with emphasis on the entropy limit of the valence. We then define the forces which determine the valence at high temperatures and show that the equilibrium valence of Ce and Yb compounds with small T_f is determined by the balance of the entropy- and the segregation forces: it saturates at a valence within 10% of valence three at very high temperatures and moves deeper into the mixed valence state with decreasing temperature due to CF splittings, by about 5-10% at 150 K in both cases.

In the second set of lecture notes [1] we introduce the valence fluctuation temperature T_f into our expression for the free enthalpy. We show that this quantity is a function of temperature, valence, CF and SO splittings and of an effective mixing matrix element, which turns out to be the only additional parameter necessary for a consistent and fairly quantitative description of the data available on a large number of RE mixed valence systems down to T=0.

2. THE NONLINEAR VOLUME-VALENCE RELATIONSHIP

One of the most obvious indications of mixed valence in a RE compound is the lattice constant anomaly, which one detects when plotting the lattice constants of the full series of isostructural RE compounds against the so-called ionic radius. A classical example, the $REAl_2$ series of compounds, is shown in Fig. 1[2]. The lattice constant of $YbAl_2$ is not on the straight line formed by the majority, which are trivalent, according to their susceptibility, nor does it even lie close to the hypothetical lattice constant of divalent $YbAl_2$, which one obtains via the dashed line from the ionic radius of divalent Yb, or by accepting $CaAl_2$ as a good "double" for divalent $YbAl_2$. It has become common practice to obtain a measure of the fractional valence v from such a lattice constant anomaly by linear interpolation from the lattice constants or volumes of the integral valence reference compounds:

$$v_V(\nu) = q_{n+1} + \nu_V \equiv q_{n+1} + \frac{a - a_n}{a_{n+1} - a_n} \simeq q_{n+1} + \frac{V - V_n}{V_{n+1} - V_n} . \qquad (1)$$

Here q_{n+1} is the lower integral valence and ν_V the fractional occupation of the higher valence state. We add the index V to the valence obtained in this way because eq. (1) implicitly assumes the validity of Végards law, which is observed in metallic alloys. The procedure has obvious deficiencies with regard to the choice of

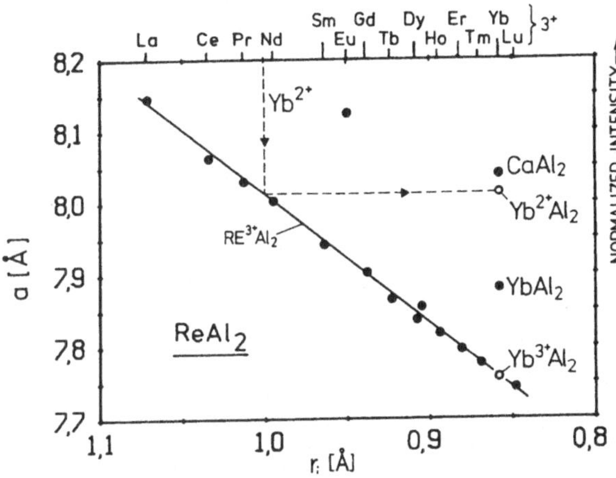

Figure 1
Lattice constants a of the $REAl_2$ intermetallic compounds[2] ms. the ionic radius r_i. Linear inter-oikatuib betweeh $Yb^{2+}Al_2$ and $Yb^{3+}Al_2$ yields the Végard valence v_V = 2.53 for $YbAl_2$.

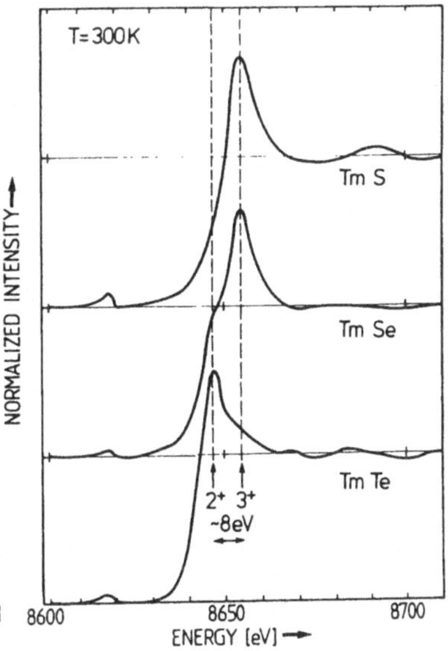

Figure 2
L_{III} X-ray absorption edges of three Tm monochalcogenides[6]. The intensity ratio of the two peaks underlying the TmSe spectrum yields the L_{III} valence v_L = 216.

the divalent reference volume as exemplified by Fig. 1. This is a general problem and especially serious for reference volumes of tetravalent Ce [3]. Nevertheless most numbers for the fractional valence in the literature have been obtained in this way because the lattice constants are always available, while other methods to determine the valence cannot be applied nearly as generally or have other drawbacks [3].

This was true until a few years ago, when the L_{III} x-ray absorption method to determine the valence came into widespread use[4-8]. This method is applicable to all RE species which exhibit mixed valence, at all temperatures and pressures, even down to fairly low concentrations of a few percent. The L_{III} absorption edge of RE atoms shows a characteristic peak of a few eV width at the edge, which is due to the large number of empty final 5d states for the 2p→5d L_{III} transitions in these atoms. The peak position is shifted by 7 to 10 eV to higher energies from a given valence state q_{n+1} to the next higher valence state q_n. Fig. 2 shows the edges for nearly divalent TmTe, for nearly trivalent TmS and for strongly mixed valent TmSe [6]. The valence is extracted from such absorption spectra by fits involving the superposition of two shifted absorption edges with the appropriate relative weight. Bauchspiess[8] gives an absolute

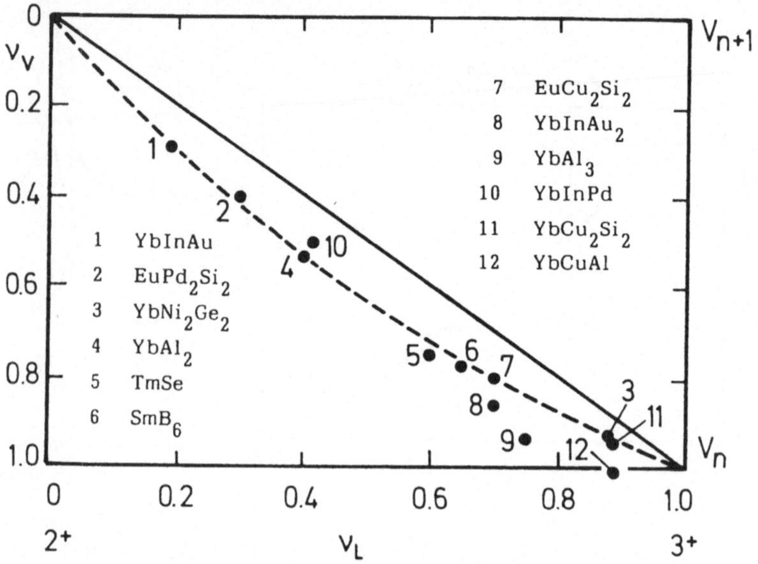

Figure 3
Végard valence ν_V against L_{III} valence ν_L for 12 homogeneously mixed valence compounds.

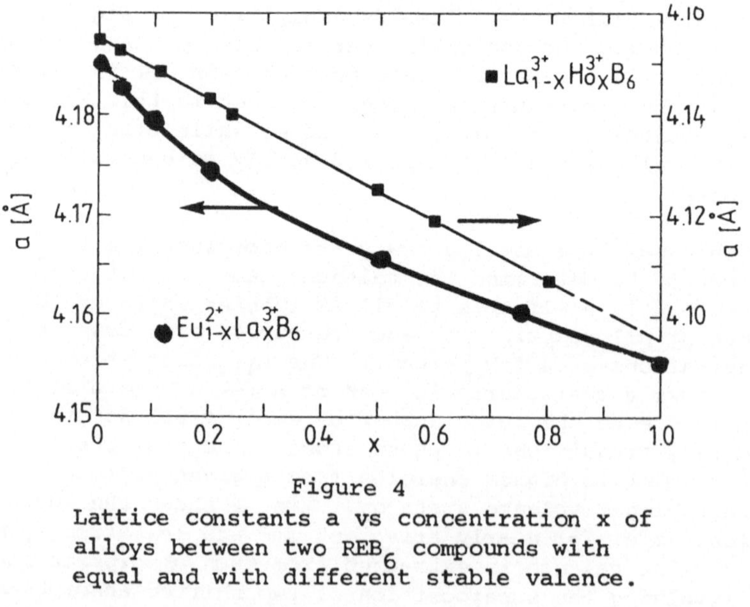

Figure 4
Lattice constants a vs concentration x of alloys between two REB_6 compounds with equal and with different stable valence.

error of $\nu_L {}^{+5}_{-2}$ % for this method; the relative error upon changes of temperature or pressure is of course smaller. Fig. 3 shows a plot of the Végard valence against the L_{III} valence, ν_V vs ν_L for 12 mixed valence compounds [9]. There is a clear discrepancy; ν_L is from 10 to 20% smaller than ν_V. Significantly, wherever available, valencies extracted from Mössbauer isomershifts, Curie constants and photoemission agree much better[9] with ν_L than with ν_V (for TmSe, the best documented case, see e.g.[10]). Therefore rather than discussing the relative merits and deficiencies of the Végard- and L_{III}-methods we look at the volume as function of concentration in mixtures of LaB_6 with stable valent HoB_6[11] and EuB_6[12] in Fig. 4. While the mixture of two trivalent compounds nicely follows Végard's law, the mixture of divalent EuB_6 with trivalent LaB_6 does not. The same type of deviation from linearity was also observed in mixtures of EuB_6 with GdB_6[13]. The deviations of the lattice constant from a linear concentration dependence are very similar to the deviations of the L_{III} valence from the linear volume valence relationship implied in the Végard valence (see right hand side of Fig. 3). Apparently the lattice constant is neither linear in concentration nor in valence.

It is well known that Végard's "law" must break down whenever the bulk moduli of the two components of an alloy are sufficiently different[14]. In RE mixed valence compounds, the components are the unit cells of the compounds with the RE atom with q_{n+1} and q_n valence electrons. The bulk moduli of trivalent RE compounds are indeed larger than for isostructural divalent compounds, by a factor between 1.3 and 2. Qualitatively the deviation from Végard's law observed in Fig.'s 3 and 4 may be explained as follows. Imagine a mixture of two types of atoms, in a ratio of $\nu/(1-\nu)$, or $x/(1-x)$, one hard and small, the other soft and large, held together by a common pressure from the chemical binding. Let this pressure be linearly related to the average number of valence electrons $p(\nu) \sim q_{n+1} + \nu$. Then the average volume $V(\nu)$ of the mixture will be smaller than the geometric mean of the separate volumes V_{n+1} and V_n under separate pressures $p_{n+1} \sim q_{n+1}$ and $p_n \sim q_{n+1} + 1$, because the soft spheres are compressed more by the increased pressure $p(\nu)$ they feel in the mixture than the hard spheres are expanded by the relative decrease of pressure they feel at $p(\nu)$.

Fig. 5 shows the bulk moduli of some $REAl_2$[15], RESe[16], REB_6[17] and $RECu_2Si_2$[18] compounds, plotted against the valence q divided by the volume V. One observes a very remarkable linear relationship between bulk modulus and valence electron density q/V for integral valence

$$B_i = \alpha\, q_i/V_i + B_o \tag{2}$$

with slope α and intersection at the ordinate B_o characterizing each

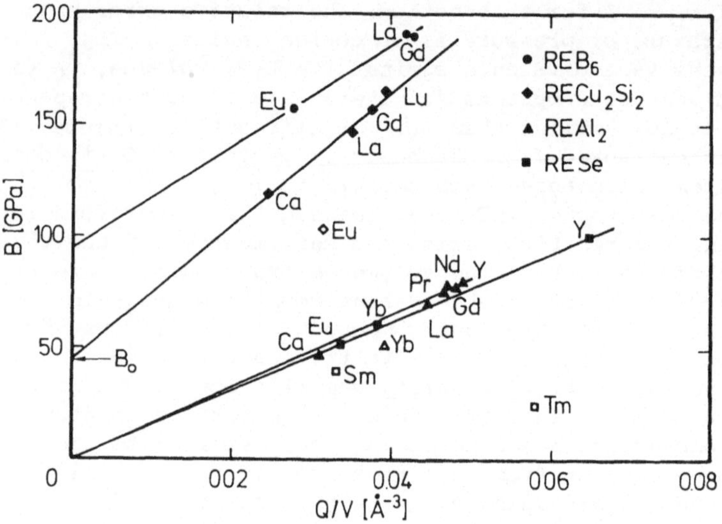

Figure 5
Bulk modulus B vs number of RE valence electrons Q divided by the unit cell volume V. Closed symbols refer to integral valence, open symbols to mixed valence. Q is denoted by q in the text.

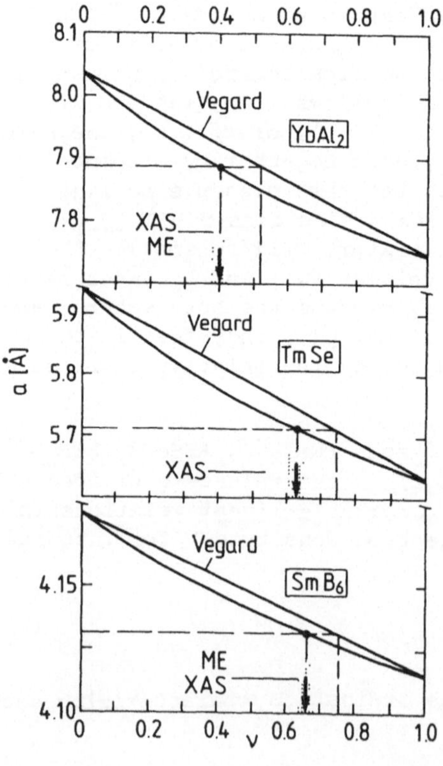

Figure 6
Comparison of valence from nonlinear $V(\nu)$ of eq. (5) with Végard valence and with valence from other methods.

series of isostructural compounds. Mixed valence compounds have bulk moduli smaller than given by ew. (2); they can therefore be recognized by their anomaly in such plots of B vs q/V (Fig. 5) just as clearly as in plots of volume against ionic radius (Fig. 1). The bulk modulus may be regarded as the pressure of the atom which compensates the pressure from the chemical binding forces $p(\nu)$ in equilibrium. We therefore make the Ansatz [9]

$$B(\nu) = (1-\nu)\ B_{n+1} + \nu\ B_n \qquad (3)$$

for the isovalent bulk modulus of a mixed valence compound, i.e. for a fictitious bulk modulus which one would measure if the valence would not change with a pressure applied from the outside (this could be realized in the static mixtures of Fig. 4). If one then generalizes eq. (2) to

$$B(\nu) = \alpha(q_{n+1}+\nu)/V(\nu) + B_o \qquad (4)$$

and equates (3) and (4), one obtains for the volume $V(\nu)$ [9]

$$V(\nu) = \frac{(q_{n+1} + \nu)}{(q_{n+1}/V_{n+1})(1-\nu)+(q_n/V_n)\nu} \ . \qquad (5)$$

Fig. 6 shows the application of this expression to measurements of the valence from the lattice constants of $YbAl_2$, TmSe and SmB_6 [9]. While the Végard valence is clearly higher than the valence determined by other methods, the valence extracted from the nonlinear $V(\nu)$ falls into the other values within the error bars. The relative difference of volumes between di- and trivalent compounds is small, $((V^{2+}- V^{3+})/V^{3+} \simeq 5 - 15\%)$. For mixtures of such compounds eq. (5) may therefore be approximated by

$$V(\nu) \simeq V^{2+} - (V^{2+}- V^{3+})\cdot 3\nu/(2+\nu) \ . \qquad (6)$$

The dashed line in Fig. 3 comes from this equation. It is a clear improvement for the volume-valence relationship over the linear Végard Ansatz $V_V(\nu) = V_{n+1} - (V_{n+1}- V_n)\cdot\nu$. Discrepancies remain and will be discussed later.

The apparent success of our model for $V(\nu)$ tells us something very important. The nonlinearity of $V(\nu)$ is caused by the large difference of valence electron densities (e.g. 2/V and 3/V); as is most clearly seen in eq. (6). Therefore it seems that the two types of valence states preserve different valence electron densities in the solid, i.e. the valence electrons are not smeared out evenly over the entire solid. Note that we are talking here almost about a ground state property. It would be a mistake to think of the density of electrons in the unit cell of a homogeneously mixed valent compound as a single number in the regime of real valence fluctuations, i.e. at thermal energies or at excitation energies larger than the mixing energy kT_f or larger than a few K to a few hundred K or 0.1 to 10 meV! Therefore the question arises: how does one construct bands of translationally invariant Bloch states in this situation at energies a few meV above and below ε_F at T=0 or anywhere at $T>T_f$? We shall come back to this point when discussing resistivity [1].

3. THE ELASTIC ENERGY OF MIXED VALENCE COMPOUNDS

In order to carry through the program outlined in the introduction we need an expression for the elastic energy of mixed valence compounds as function of valence and volume, $F_e(\nu,V)$. Since valence and volume are independent thermodynamic parameters (see later), the equilibrium volume $V(\nu)$, at constant temperature and pressure, and at the equilibrium valence ν, follows from

$$\partial(F_e(\nu,V)+pV)/\partial V\Big|_{T,p,\nu} = 0 \tag{7}$$

$V(\nu)$ should look like eq. (5) at p = 0. We make the Ansatz

$$\begin{aligned} F_e(\nu,V) = (1-\nu)\ \alpha\ q_{n+1}\left[\ln\frac{V_{n+1}}{V} + \frac{V-V_{n+1}}{V_{n+1}}\right] \\ + \nu\ \alpha\ q_n\left[\ln\frac{V_n}{V} + \frac{V-V_n}{V_n}\right] \\ + B_o\ (V-V_V)^2/2V\ . \end{aligned} \tag{8}$$

Here $V_V \equiv (1-\nu)V_{n+1} + \nu V_n$ is the Végard volume and α and B_o are the quantities appearing in eq. (2). Applying eq. (7) to eq. (8) yields

$$V(\nu,p) = \frac{(q_{n+1}+\nu) + B_o V_V/\alpha}{(1-\nu)(q_{n+1}/V_{n+1}) + \nu(q_n/V_n) + (B_o+p)/\alpha}\ . \tag{9}$$

Here a term $B_o(1-V_V/V)^2/2$ was neglected. For the fictitious bulk modulus one finds

$$B_V \equiv V\ \frac{\partial^2 F_e}{\partial V^2}\Big|_{\nu,T} = \alpha(q_{n+1}+\nu)/V(\nu,p) + B_o V_V/V(\nu,p) \tag{10}$$

quite close to the Ansatz of eq. (4), since $V_V/V(\nu,p) \simeq 1$. $V(\nu,p)$ is the equilibrium volume in eq. (9). Note that for p=0 and B_o=0, $V(\nu)$ in eq. (9) is identical with that in eq. (5). On the other hand, for p=0 but $B_{n+1} = B_n = B_o$ (equal bulk moduli of the two components, $\alpha \to 0$)

V(ν) goes to the Végard volume, as it should. Such a smooth transition to the Végard volume with increasing B_o cannot be achieved with eq. (5), since B_o and α do not appear in that expression. No expression for $F_e(\nu,V)$ containing B_o and α could be found which yields eq. (5) for $B_o \neq 0$. The difference between ew. (5) and eq. (9) is irrelevant for the $REAl_2$ and RESe (B_o=0) but becomes important for the $RECu_2Si_2$ and REB_6 (Fig. 5) because there $B_o \neq 0$. Apparently, for B_o= 0 the RE valence electron density q_i/V_i dominates the bulk modulus of the entire unit cell, while for $B_o \neq 0$ the bulk modulus of the other atoms in the cell (the hard boron and silicon cages) is felt and eventually becomes dominant.

According to eq. (9) V(ν,p) can now be traced as function of valence and pressure. How does it behave as function of tempearture? Let us assume that α is independent of temperature. Then eq. (2) yields

$$\partial \ln(B_i - B_o)/\partial T = -\partial \ln V_i/\partial T \equiv \beta_i \ . \qquad (11)$$

This result is exactly what one also gets from the Grüeisen constant $\gamma_G = \beta BV/C$ if one assumes $\partial\gamma_G/\partial T = \partial\beta/\partial T = \partial C/\partial T \approx 0$, conditions which apply well to a solid above the Debye temperature (C is the specific heat). Therefore we take account of the temperature dependence of $F_e(\nu,p)$ and V(ν,p) by simply using the temperature dependent volumes $V_{n+1}(T)$ and $V_n(T)$ of the reference compounds in eqs. (8) and (9).

Throughout the present section we have worked with a simple classical elastic model which completely ignores quantum mixing of the two configurations. If this quantum mixing depends on volume, or if there is volume dependence in the intraionic spectra, other terms must be added to the right-hand side of eq. (7) and the equilibrium volume will then no longer be described by eq. (9). This is a possible reason for the deviations of some of the data in Fig. 3 from the dashed line (eq. (6)).

4. STATISTICAL THERMODYNAMICS OF MIXED VALENCE AT HIGH TEMPERATURES

We assume that at high temperatures a mixed valence compound can be treated like a gas to two types of noninteracting "atoms" ("unit cell" would probably be a better name) and shall justify this assumption later. The two "atoms" are characterized by their different electronic configurations $(Xe)4f^n(5d6s)^{m+1}$ and $(Xe)4f^{n+1}(5d6s)^m$ (henceforth "$4f^n$" and "$4f^{n+1}$") with different ground state energies, E_n and E_{n+1}. These energies refer to the <u>entire</u> atom, including the valence electrons. (Note the change of notation: $q_n \rightarrow m+1$, $q_{n+1} \rightarrow m$). The total number N>>1 of atoms is fixed, but the valence fluctuation can be regarded as a chemical reaction in which the ratio of the two types of atom can shift with temperature and pressure. We define the valence by the fractional occupation

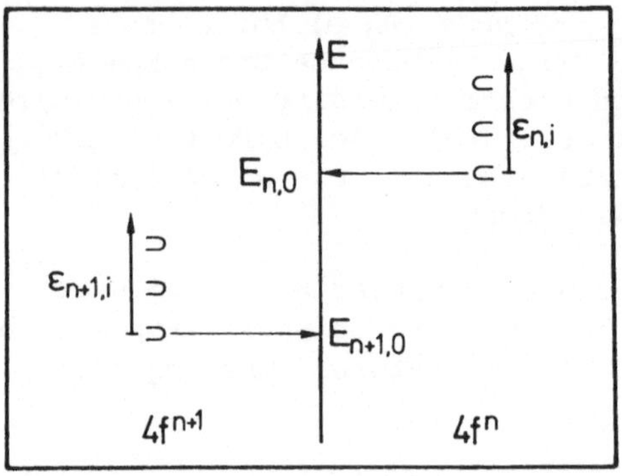

Figure 7
Schematic energy spectrum in a mixed valence situation.

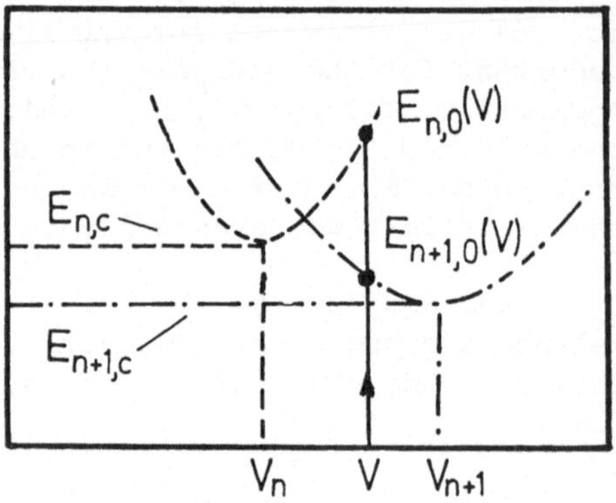

Figure 8
Volume dependence of the ground state energies of the two valence states in a mixed valence situation.

Table I The valence of RE atoms with unstable valence at the high and low temperature entropy limits. T_l and T_u are the approximate lower and upper ends of the relevant temperature range.

RE	n	g_n	g_{n+1}	ν	v	T_l(K)	T_u(K)
Ce	0	1	6	.14	3.14	≤400	~800
		1	14	.067	3.067	3000	∞
Pr	1	6	9	.4	3.4	≤400	~800
Tm	12	13	8	.619	2.619	≤400	∞
Yb	13	8	1	.89	2.89	≤400	∞
Eu	6	1	8	.11	2.11	0	100
Sm	5	2	1	.67	2.67	0	100
Ce	0	1	2	.33	3.33	0	100
Yb	13	2	1	.67	2.67	0	100

ν of $4f^n$, as before

$$N = N_{n+1} + N_n = N(1-\nu) + N\nu = \text{const.} \tag{12}$$

The grand canonical free energy of this system is

$$\mathcal{F} = -kT\,(N_{n+1}\ln(\zeta_{n+1}/N_{n+1}) + N_n\ln(\zeta_n/N_n) + N)$$
$$F \equiv \mathcal{F}/N \simeq -kT\,((1-\nu)\ln(\zeta_{n+1}/(1-\nu)) + \nu\ln(\zeta_n/\nu) + 1) \tag{13}$$

The partition functions ζ_{n+1} and ζ_n have their energies counted on a common scale in eq. (13). If one refers the energies in each type of atom to the respective atomic ground state (Fig. 7) one obtains

$$F = (1-\nu)E_{n+1} + \nu E_n \tag{14}$$
$$- kT\,((1-\nu)(\ln \zeta'_{n+1}/(1-\nu)) + \nu\ln(\zeta'_n/\nu) + 1)\ .$$

with the intra-atomic partition functions

$$\zeta'_{n+1} = \sum_i g_{n+1,i}\, e^{-\varepsilon_{n+1,i}/kT}\ ;\ \zeta'_n = \sum_i g_{n,i}\, e^{-\varepsilon_{n,i}/kT}\ . \tag{15}$$

Here the ε's refer to intra-atomic CF and SO splittings and the g's to the corresponding degeneracies. We assume that the CF and SO splittings do not depend on volume or temperature, while E_{n+1} and E_n do. Based on our expression for the elastic energy, (eq. (8)) we take the following view for E_{n+1} and E_n (Fig. 8). In the lattice in question, at a reference temperature T, at reference pressure p=0, at the corresponding volume V_{n+1} and at $\nu=0$, the ground state of $4f^{n+1}$ is at $E_{n+1,c}$ and in the same lattice at temperature T, at p=0, at volume V_n and $\nu=1$, the groundstate of $4f^n$ is at $E_{n,c}$. The letter c refers to "crystal-chemical" or "chemical" for short. Both energies are independent of temperature, pressure, volume and valence by the nature of their definition; they do, however, change with crystal structure. For the time being we interpret $F_e(\nu,V)$ of eq. (8) as follows. In the mixed valence state both $4f^n$ and $4f^{n+1}$ have the same volume V. The first term on the RHS of eq. (8) then measures the energy of compression of $4f^{n+1}$ from V_{n+1} to V, and the second the energy of dilatation of $4f^n$ from V_n to V. Only the third term measures an elastic interaction between $4f^{n+1}$ and $4f^n$, it can be disregarded for $B_0 = 0$ or added to eq. (14) as a separate term for $B_0 \neq 0$. Thus, at volume V, E_{n+1} and E_n are found as indicated in Fig. 8. The full high temperature ($T \gg T_f$) expression for the free enthalpy then reads

$$G(\nu,V,p,T) = F_c + F_e + F_s + pV \tag{16}$$

$$= (1-\nu)E_{n+1,c} + \nu\, E_{n,c} (1-\nu)\alpha\, q_{n+1}\left[\ln \frac{V_{n+1}(T)}{V} + \frac{V-V_{n+1}(T)}{V_{n+1}(T)}\right]$$

$$+ \nu\alpha q_n\left[\ln \frac{V_n(T)}{V} + \frac{V-V_n(T)}{V_n(T)}\right] + B_o (V-V_V(T))^2/2V$$

$$- kT\left[(1-\nu)\ \ln(\zeta'_{n+1}(T)/(1-\nu)) + \nu\ \ln\ (\zeta'_n(T)/\nu)\right] + pV\ .$$

F_c, F_e and F_s ($\equiv$-TS) are the chemical, elastic and entropic parts of the free energy. We have suppressed the term -kT of eq (14), since it appears implicitly in the elastic energy (phonons) .

The equilibrium values of valence and volume, ν_e and V_e, follow at constant T and p from

$$\left.\frac{\partial G}{\partial \nu}\right|_{T,p,V_e} = 0 \qquad \left.\frac{\partial G}{\partial V}\right._{T,p,V_e} = 0 \tag{17a,b}$$

After application of eqs. (17) to (16), the only unknown parameter left in G is $E_{n+1,c} - E_{n,c}$ = const. All other quantities can be measured, α and B_o by high pressure X-ray diffractometry or by ultrasound, $V_{n+1}(T)$ and $V_n(T)$ by X-ray diffractometry or capacitive thermal expansion on reference compounds and the CF splittings by neutron inelastic scattering on the mixed valence compound itself (if these plittings are larger than T_f). The SO splittings are known from atomic spectroscopy.

We now discuss the assumptions leading to eq. (16). In the literature, one often finds the symbol $4f^{n+1} \leftrightarrow 4f^n + e$ characterizing the valence fluctuation, and E_{n+1} and $E_n + \varepsilon_F$ for the two basic energies[3]. Where is the conduction electron in eqs. (12) to (16)? Is it reasonable to absorb ε_F in E_n? We think so. Consider the elastic energy. The bulk moduli B_{n+1} and B_n absorbed in F_e measure the volume dependence of the energy of the m and m+1 outer valence electrons per atom regardless of whether these electrons are conducting or insulating. After the transition $4f^{n+1} \rightarrow 4f^n + e$ the electron e remains on average on its atom; that was just the point in our discussion of the nonlinear volume-valence relationship. ε_F is meant to be a kinetic energy, but it is not necessarily a translational kinetic energy. It is to a large extent kinetic energy in the local 5d wave function occupied by e, and certainly only a part of the entire energy of rearrangement of the $(5d6s)^m$ to the $(5d6s)^{m+1}$ configuration. In short, it seems much more reasonable to absorb "ε_F" in the composite change of E_{n+1} to E_n associated with

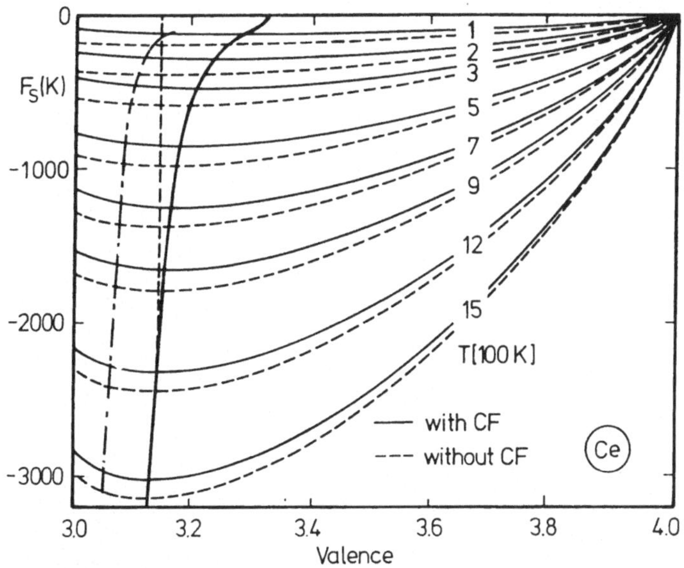

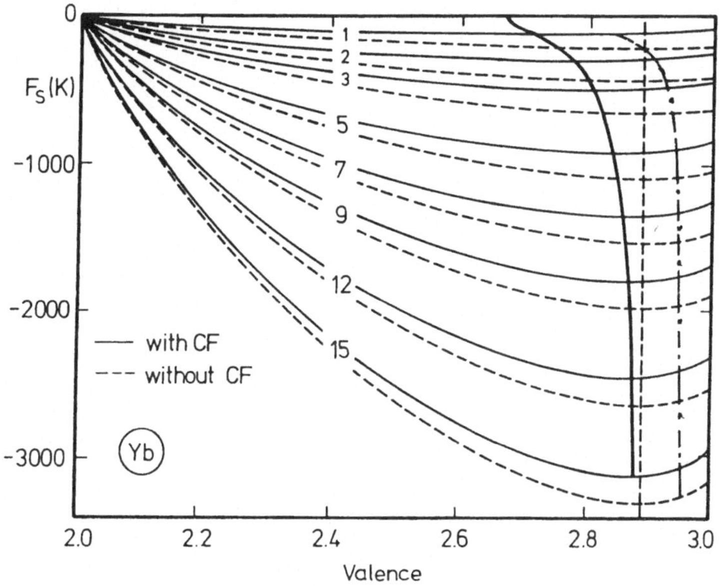

Figure 9a

The entropic part (without quantum mixing) of the free energy, $F_S = -TS\ (T,\nu)$ for Ce and Yb, without --- and with —— crystal field splittings. Without CF splittings the valence at the minimum of F_S is stationary (---) at the entropy limit $\nu_o = .14$ (Ce) and $\nu_o = .89$ (Yb). With finite CF splittings it moves away from this limit at lower temperatures (——). The valence resulting from equilibrium between the entropy and the segregation forces is also indicated for $dE_{xse}/dT = 1\ k_B$ (—— • —— • ——).

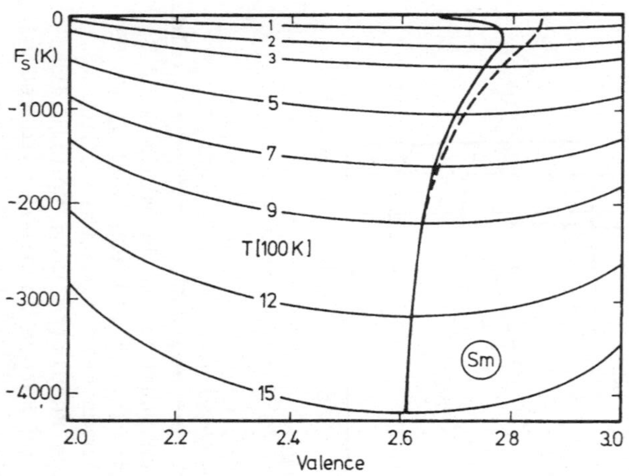

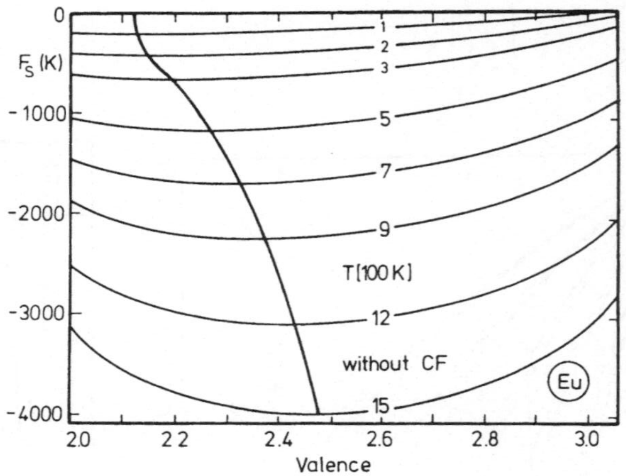

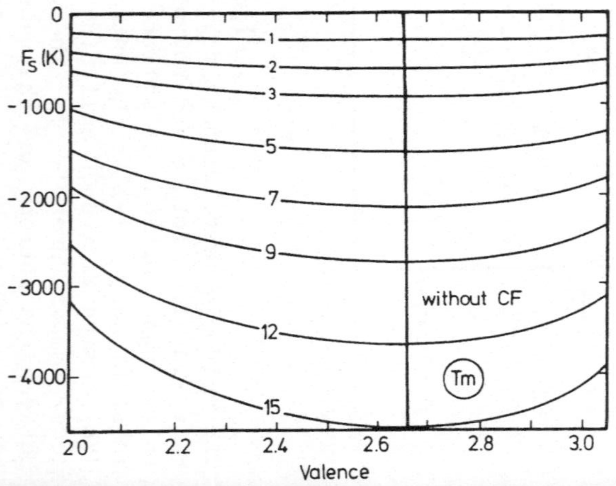

Figure 9b
The entropic part of the free energy $FS = - TS\ (T,\nu)$ for Sm, Eu and Tm. Even without CF splittings the valence at the minimum moves strongly with temperature for Sm and Eu because of the Van Vleck anomalies of $4f^5$ and $4f^6$.

the changes of local valence electron density rather than to write it as a separate energy with the wrong connotation ("free" electron) which also immediately brings up the problem of having to sort out the changes in screening, correlation etc. energies as separate and unmeasurable quantities. It turns out that $E_{n+1} - E_n$ is measurable (see later). Of course the mixed valence compounds are metals. Therefore what about the electronic specific heat? Clearly, there is no place for conduction electron excitations nor for phonons in eq. (16). At high temperatures we can take care of both by subtracting the specific heat of appropriate reference compounds, if necessary. On the other hand we have shown earlier[19] that the large low temperature electronic specific heat coefficient of YbCuAl is the low temperature part of a quasi-Schottky anomaly of $4f^{13}$, which is considered in eq. (16) via $\zeta'_n = \zeta'_{13}$ (see later).

We shall now discuss some consequences of eq. (16) at atmospheric pressure. In Fig. 9 we show $F_s(T,\nu)$, the entropic part of G, for Ce, Sm, Eu, Tm and Yb, first without any CF splittings and then also with CF splittings as detected by neutron scattering for $CeCu_2Si_2$[20] and $YbCu_2Si_2$[21] and a standard cubic CF splitting of 100 K in $4f^5$ between Γ_7 and Γ_8 for Sm. At high temperatures F_s decreases linearly with T at all ν's and one observes minima near $\nu_o = .14$, .62 and .89 for Ce, Tm and Yb (Table I). These high temperature minima correspond to absolute maxima of the atomic entropy, which are at $S = k \ln(g_{n+1} + g_n) = k \ln 7$, $k \ln 21$ and $k \ln 9$ for Ce, Tm and Yb. Thus the depth of the minima of F_s increases from Ce to Yb to Tm. For Sm and Eu the minimum moves strongly (even without CF splitting) with increasing T due to the SO splittings (Van Vleck anomalies of $4f^5$ and $4f^6$). For sufficiently low temperature (but still at $T \gg T_f$) the excited CF states are frozen out and therefore the CF splittings shift the minima at lower temperatures towards higher valence for Ce and towards lower valence for Yb. The atomic entropy is $k \ln 3$ for Ce, Sm and Yb at the low temperature minima, because $4f^1$, $4f^5$ and $4f^{13}$ are Kramers ions with a doublet left in the CF ground state ($\zeta'_1 \to 2$ for Ce, $\zeta'_5 \to 2$ for Sm and $\zeta'_{13} \to 2$ for Yb). The valence at the minima is then 3.33 for Ce and 2.67 for Sm and Yb (bottom two lines of Table I).

Let us apply eq. (17a) to (16):

$$\frac{\partial G}{\partial \nu}\Big|_{T,p,V} = \frac{\partial}{\partial \nu} (F_c + F_e + pV) + kT \ln(\zeta'_{n+1}\nu/\zeta'_n(1-\nu)) = 0. \quad (18)$$

The second term on the RHS is $\partial F_s/\partial \nu$. This term turns out to be the difference between the ground-state energies of $4f^{n+1}$ and $4f^n$. Rewriting it yields

$$(1-\nu)/\nu = (\zeta'_{n+1}/\zeta'_n) e^{-(\partial F_s/\partial \nu)/kT} . \quad (19)$$

Since $(1-\nu)/\nu$ is the ratio of the occupation probabilities of $4f^{n+1}$

and $4f^n$, the RHS of eq. (19) must be the ratio of their partition functions ζ_{n+1}/ζ_n. From the definitions of ζ'_{n+1} and ζ'_n (eq. (15)) it follows that

$$\partial F_s/\partial\nu = E_{n+1} - E_n \equiv E_x \quad . \qquad (20)$$

We call E_x the interconfigurational excitation energy[22]. Clearly we know this quantity if we know the CF and SO splittings, the temperature and the valence, all of which are measurable.

For $\partial F_s/\partial\nu = 0$, i.e. at the minimum of $F_s(\nu,T)$, we have $E_x=0$ (eq. (20)), or exact degeneracy of the configurational groundstates E_{n+1} and E_n. This situation is known as the "configurational crossover"[22].

Let us assume for a moment that $F_c + F_e + pV \equiv 0$. From the equilibrium condition eq. (18), we see that this means $\partial F_s/\partial\nu = 0$, i.e. in this situation the equilibrium valence is at the minimum of $F_s(\nu,T)$. Therefore, the entropy term F_s alone would drive the valence along the dashed or full lines in Fig. 9 as function of temperature always at exact configurational corssover! We call the valence at the high temperature end of these lines the entropy limit.

In the literature one often finds the term "strongly mixed valent" used in the sense of $\nu\simeq.5$. The term "configurational crossover" is also in use for this situation. We learn from our discussion of F_s that at high temperatures ($kT>\Delta_{CF}$, kT_f) "strongly mixed valence" and "configurational crossover" are only synonymous for $g_n \simeq g_{n+1}$, e.g. for Pr (ν=.4 at $E_x = 0$) or Tm (ν=.62 at E_x= 0, see Table I). On the other hand, "configurational crossover" occurs for Ce and Yb at $3.067<v<3.14$ and at $v = 2.89$, only 7-14% and 11% away from the trivalent state! Since it is experimentally difficult to distinguish these values of the valence from valence three, the existence of configurational crossover (or $E_x<kT$) for nearly all Ce and Yb compounds above a few 100 K has not been fully appreciated in the past, and this has led to a lot of confusion, especially in the case of Ce.

5. SUSCEPTIBILITY AND VALENCE AT HIGH TEMPERATURES

We shall now take a look at experimental data, with special attention to temperatures high compared to CF splittings and T_f. Since the CF splittings are a few 100 K the data should come from above room temperature. Such data are unfortunately rare. To our knowledge, only magnetic susceptibility, resistivity and a few lattice constant measurements exist for this region. About those data one can say generally that while below room temperature susceptibility and resistivity are complicated, above they are quite simple. Fig. 10 shows the inverse susceptibility of YbCuAl and

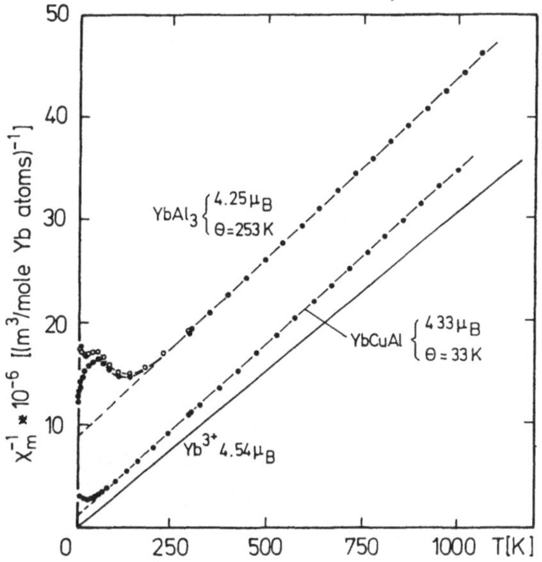

Figure 10
Inverse susceptibility χ^{-1} of YbCuAl and $YbAl_3$ vs temperature T[23].

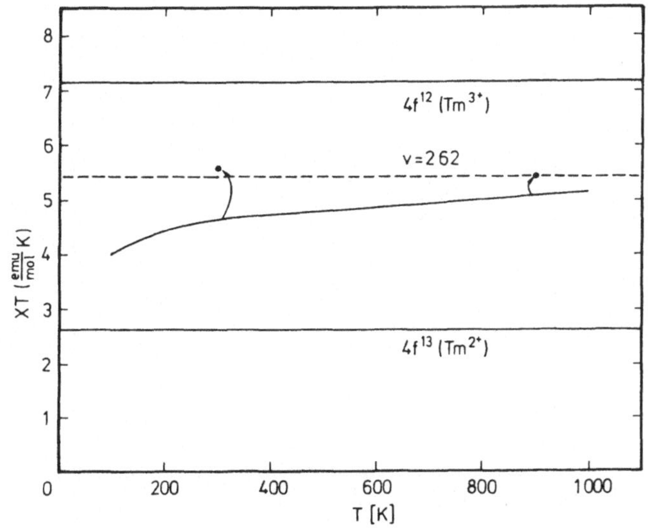

Figure 11
The product χT for TmSe vs temperature[24]. Dashed line is χT at entropy limit, with T_f=0. Arrows show corrections for T_f=80K.

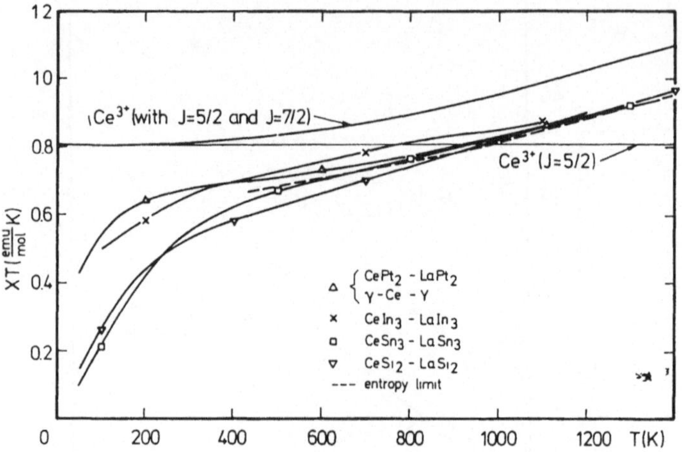

Figure 12

The product χT for 4 Ce metals. The dashed line at high temperatures indicates χT for the valence at the minimum of the entropic part of the free energy (Fig. 9).

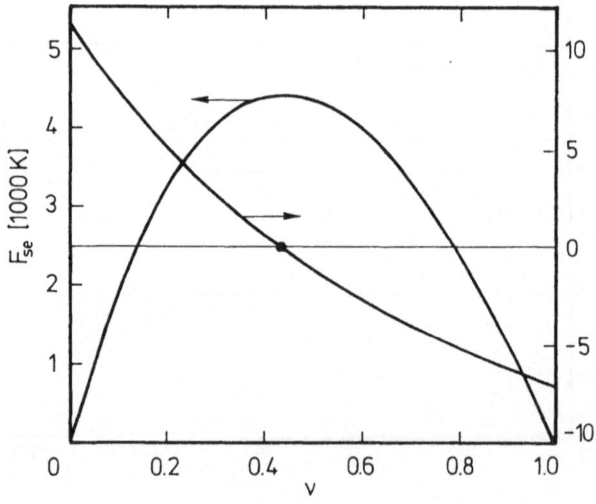

Figure 13

The segregational term of the elastic free energy and the resulting segregation force as function of valence for $YbCu_2Si_2$ (α = 1.9 eV, B_o = 0, V_{n+1} = 85 Å^3, V_n = 78 Å^3).

$YbAl_3$[23]. Sufficiently far above the temperature of the maxima of χ one observes good Curie Weiss behavior with Curie constants which are 88 and 91% of those of $4f^{13}$ for $YbAl_3$ and YbCuAl. These numbers are very close to the $4f^{13}$ occupation in the entropy limit ν_0=.89 (Table I). It is important to realize that the valence stays at these values from a few 100 out to 1000 K, the highest temperatures of the measurement.

Fig. 11 shows χT for TmSe [24]. If there are kinks or slight nonlinearities in χ^{-1} vs T, as there are in TmSe and in Ce compounds, such plots give a safer measurement of the fractional valence via the susceptibilities. One observes that χT approaches the entropy limit from below, as it should for finite fluctuation temperature. Correction by a factor $(1+T_f/T)$, with T_f= 80 K, makes T coincide with the entropy limit within a few % at all temperatures above 300 K. Fig. 12 shows χT vs T for γCe and for a number of Ce compounds[25]. The upper line is χT calculated for $4f^1$ with Curie and Van Vleck contribution from the Hund's rule ground state (J=5/2) and from the SO multiplet (J=7/2) 3000 K above it, which causes the rise of χT at high temperature. Clearly J=7/2 must be taken into account, since the measured susceptibility is above χT for J=5/2 alone (χT at T=0) at high temperatures. The dashed line is the locus of χT at the valence at the minimum ν_m of F_s for Ce in Fig. 9 ($\chi T = (1-\nu_m)\chi(4f^1)\cdot T$). All available data coincide at 1000 K with $\nu=\nu_m$ within the experimental scatter. This is truly surprising for $CePt_2$ and $CeIn_3$, which order magnetically at low temperatures and should therefore be classified as magnetically stable (trivalent). It is mildly surprising for γCe. For $CeSn_3$ and $CeSi_2$, which are thought to be mixed valent, the coincidence may be deceptive; a correction by $(1+T_f/T)$, with the $T_f \simeq 200$ K measured by neutrons at 300 K[21], puts $CeSn_3$ on the trivalent line at 1000 K, which is again surprising if one believes in mixed valence in $CeSn_3$. However, T_f decreases with increasing temperature for $CeSn_3$ at 300 K[21] and may therefore be considerably smaller than 200 K at 1000 K, which would make the correction by $(1+T_f/T)$ smaller, thus leaving room for mixed valence. Altogether it is safe to say that the data put the valence of these Ce systems at values between the entropy limit and valence three, and that the valence, if fractional, is nearly independent of temperature at these high temperatures.

6. THE FORCES DRIVING THE VALENCE AT HIGH TEMPERATURES

We have seen that all data of the susceptibilities available near 1000 K, for Ce, Yb and Tm compounds with low temperature 4f instabilities but with $T_f <$ 300 K, show the valence to be within a few percent of the entropy limit and to be temperature independent (i.e. to change by at most a few percent over a temperature range ΔT of at least 500 K). This suggests that the entropy term F_s of the free enthalpy dominates in determining the valence there. To answer the question why the other terms are so insignificant we go back to eq. (18) where we see that we must study the behavior of $\partial(F_c + F_e + pV)/\partial\nu \neq 0$. We have in equilibrium

$$\partial(F_c + F_e + pV)/\partial\nu\big|_{T,p,V} = E_{nc} - E_{n+1,c} \tag{21}$$

$$- \alpha\left\{q_{n+1}\left(\ln\frac{V_{n+1}}{V} + \frac{V-V_{n+1}}{V_{n+1}}\right) - q_n\left(\ln\frac{V_n}{V} + \frac{V-V_n}{V_n}\right)\right\}$$

$$- B_o(V-V_V)\cdot(V_{n+1} - V_n)/V = -\,\partial F_s/\partial\nu = -\,E_x$$

In the sense of a generalized thermodynamical force $F_x \equiv -\partial G/\partial x$, where x is any thermodynamical variable, eq. (21) is an equation between several forces driving the valence ν. We give these forces the following names

$(E_{n+1,c} - E_{n,c}) \equiv E_{xc}$	chemical force
$\alpha\{\quad\} \equiv E_{xse}$	segregation force
$B_o(V-V_V)(V_{n+1} - V_n)/V \equiv E_{xL}$	lattice force
E_x	entropy force .

Segregation force is the only name which is not immediately obvious. In Fig. 13 we show $F_e(\nu)$ and $\partial F_e/\partial\nu$ calculated for $B_o = 0$. $F_e(\nu)$ is zero at ν=0 and ν=1 and positive in between. Consequently $\partial F_e/\partial\nu$ is positive near ν=o, then decreases, crosses zero and becomes large and negative at ν≙1. It clearly tends to drive the valence towards integral values, i.e. to "segregate" the mixture into integral valent components in order to avoid the strain energies necessary to compress $4f^{n+1}$ from V_{n+1} to V and to expand $4f^n$ from V_n to V[26]. The chemical force is independent of ν,T,p and V and can have either sign. The lattice force tends to decrease the valence ($V<V_V$, $V_n<V_{n+1}$ always). The entropy force $F_s = kT\ln(\zeta'_{n+1}\nu/\zeta'_n(1-\nu))$ goes through zero at the minima of F_s, e.g. at the high temperature entropy limit, but diverges at ν=0 and ν=1 for T>0, with the opposite tendency of the segregation force, i.e. driving the valence towards the minimum of F_s away from integral valence (Fig. 9). It is therefore the only force so far in G to keep the mixed valence state stable. Because of its divergence at ν=0 and ν=1, it will always succeed in pulling the valence at least slightly away from integral values, the more so, the higher T.

According to their L_{III} and lattice constant valences all compounds of Figs. 10-12 are already within ± 10% of the valence demanded by the minimum of F_s at 300 K, which means that E_x, the sum of the chemical and elastic forces is already fairly small there. We shall now investigate why these other forces apparently succumb to the entropy force at higher temperatures or why they at least seem to be unable to drive the valence further away from the entropy limit with increasing temperature. The chemical force is independent of T, while the entropy force grows linearly with T at constant ν for temperatures large compared to the CF splittings ($\zeta'_{n+1} \to g_{n+1}$, $\zeta'_n \to g_n$), so the chemical forces will never be able to pull the valence at any value away from a minimum of F_s with rising T. The segregation and lattice forces are proportional to strains of the type $(V-V_i)/V \ll 1$ (eq). (21)). Depending on whether $(V-V_i) \lessgtr 0$ they will therefore either increase or decrease linearly with temperature assuming linear thermal expansion of V and of the volumes V_i of the reference compounds. Therefore at agiven constant ν, the elastic forces are linear in T (above the Debye temperature) just as the entropy force (at $kT > \Delta_{CF} \simeq k\Theta_D$) and the valence will therefore be determined between the two at some constant value.

Consider Ce and Yb compounds. In both the volume is quite close to that of the trivalent references, or $|V-V_{n+1}| \ll |V-V_n|$ for Ce, and the reverse for Yb. Moreover $\alpha \simeq B_i V_i \gg B_0(V_{n+1}-V_n)$. From these inequalities it follows that for Ce the second term and for Yb the first term of the segregation force dominates the other and the lattice force. Using the Grüneisen relationship,

$$\gamma_G = V\ \beta\ B/C \simeq 2.3\ , \tag{22}$$

the temperature coefficient of the dominant term of the segregation force becomes, after a little algebra,

$$\frac{dE_{xse}}{dT} = \begin{cases} +(V_1 - V_0)\beta_1\alpha/V_0 & \text{(Ce)} \\ -(V_{14} - V_{13})\beta_{13}\alpha/V_{14} & \text{(Yb)} \end{cases} \tag{23}$$

$$= \gamma_G 3k\ n_c(V_{n+1} - V_n)/V_n \quad (T>\theta_D)\ .$$

Since $(V_{n+1} - V_n)/V \simeq$ 5-15% for ternary and binary RE compounds with n_c=2-5 atoms per unit cell, dE_{xse}/dT turns out to have magnitude quite close to one or somewhat less (in units of the Boltzmann constant k) ($dE_{xse}/dT \simeq (0.2-1)K/K$).

Fig. 14 shows a construction to find the high temperature valence to Ce and Yb compounds at the entropy limit by the intersection of the temperature coefficient of the entropy force ($dE_x/dT = k\ \ln(g_n(1-\nu)/(g_{n+1}\nu))$) with that of the segregation force (eq. (23)). We see that this construction puts the high temperature valence even closer to valence three than the entropy limits but the valence definitely remains fractional by at least a few percent and temperature independent (if $\partial\beta/\partial T = 0$).

At lower temperatures the finite CF splittings shift the minima of Fs away from the high temperature entropy limit and away from integral valence from both Ce and Yb. This shift amount to about 10% at 150K

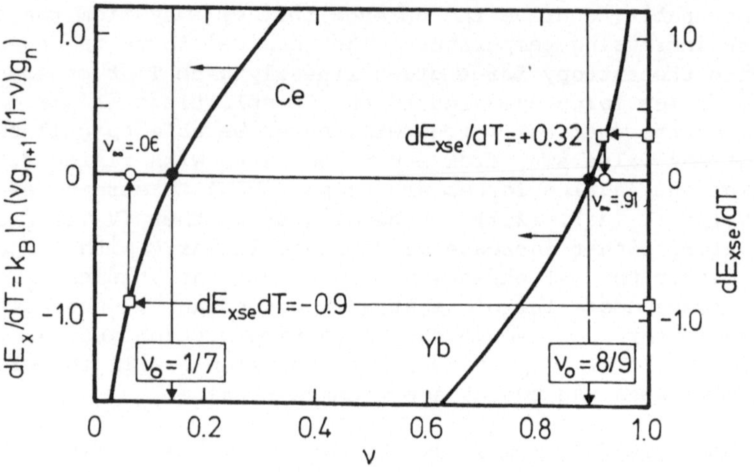

Figure 14
Equilibrium between entropy force and segregation force for $CeCu_2Si_2$ and $YbCu_2Si_2$ at very high temperatures.

with the exemplary CF splittings of $CeCu_2Si_2$ and $YbCu_2Si_2$ (Fig. 9). On the other hand, the thermal expansion coefficient β of the reference compounds remains in general constant, i.e. dE_{xse}/dT = const, down to ∿150 K. Therefore, the equilibrium valence shifts essentially in parallel with the minimum of F_s, as long as entropy force and segregation force remain dominant. We have indicated this behavior by the dot dashed lines for $|dE_{xse}/dT| = 1$ K/K for Ce and Yb in Fig. 9. The behavior of the valence indicated by these lines is semiquantitatively observed without exception by thermal expansion and x-ray diffractometry for all Ce and Yb compounds studied so far, e.g. $CeCu_2Si_2$[27], $CeAl_3$[28], $CeSn_3$[29], $CeBe_{13}$[30], $CePd_3$[5], $CeNi_2Ge_2$[29], YbCuAl[19], $YbCu_2Si_2$[29,3], $YbNi_2Ge_2$[29], etc. Between 100 and 400 K, the valence of Ce compounds shifts from 4 to 10% towards Ce^{3+}, and that of Yb compounds by a few percent towards Yb^{3+}. The shift is the weaker, the higher the temperature.

CONCLUSION

Our expression for the free enthalpy at high temperatures seems to describe the few data available on mixed valence compounds with T_f< 300 K at temperatures above 300 K quite well. It shows that the Hund's rule degeneracies and the excited spin-orbit multiplets are decisive in determining the valence above a few 100 K for compounds with sufficiently small T_f, i.e. except for the so-called "tetravalent" Ce compounds for which the fluctuation temperature can be many 1000 K [31,32]. Each type of RE species capable of 4f instabilities has therefore its very own "handwriting", which extends also to lower temperatures as exemplified by the shift of the valence due to finite crystal field splittings for Ce and Yb. Elastic forces are unable to shift the valence by more than a few percent away from the entropy limit above a few 100 K. The characteristic behavior of the valence of a given RE species determines not only the thermal expansion anomaly but also the characteristic differences of the transport properties, i.e. "Kondo" anomalies for Ce, Pr and Tm, and the absence thereof for Eu and Yb, as shall be shown elsewhere [33]. Quite clearly, eq. (16) provides a useful frame for thinking about mixed valence, and the proper understanding of high temperature properties will be a key to understanding the more complex phenomena when $T \lesssim T_f$. For a generalization of eq. (16) into that temperature range see reference 1.

Acknowledgements:- I thank many people at the Applied Physics and Theoretical Physics Groups at the University of Cologne for useful discussions, especially P. Fazekas, E. Holland-Moritz, E. Müller-Hartmann, G. Neumann, R. Pott, B. Roden, J. Röhler and B. Wittershagen, who also calculated the curves of $F_s(\nu,T)$ in Fig. 9. S. Wood and A. Schneider are thanked for their speedy and efficient technical assistance. This work was supported by the Deutsche Forschungsgemeinschaft through SFB 125.

REFERENCES

1. D. Wohlleben, lecture notes at ASI "Physics and Chemistry of Electrons and Ions in Condensed Matter" Cambridge, England, Sept. 5-17 (1983).
2. A. Iandelli and A. Palenzona, "Handbook on the Physics and Chemistry of Rare Earths", K.A. Gschneider and L.R. Eyring, eds. North Holland (1979), Vol. 2, p.1.
3. D. Wohlleben, in "Valence Fluctuations in Solids", L.M. Falicov, W. Hanke & M.B. Maple, eds. North Holland (1981), p.1.
4. E.E. Vainsthein, S.M. Blokhin and Yu.B. Paderno, Sov. Phys. Sol. St. 6, 231 (1965).
5. K.R. Bauchspiess, W. Boksch, E. Holland-Moritz, H. Launois, R. Pott and D. Wohlleben, Ref. 3, p.417.
6. H. Launois, M. Rawiso, E. Holland-Moritz, R. Pott and D. Wohlleben, Phys. Rev. Lett. 44, 1271 (1980).
7. L.C. Gupta, E.V. Sampathkumaran, R. Vijayaraghavan, M.S. Hedge and C.N.R. Rao, J. Phys. C: Sol. St. Phys. 13, L455 (1980).
8. K.R. Bauchspiess, Diplomarbeit, Universität zu Köln (1983).
9. G. Neumann, R. Pott, J. Röhler, W. Schlabitz, D. Wohlleben and H. Zahel, in "Valence Instabilities", P. Wachter and H. Boppart, eds., North Holland (1982), p.87.
10. G.K. Wertheim, W. Eib, E. Kaldis and M. Campagna, Phys. Rev. B22, 6240 (1980).

11. A Berrada, J.P. Mercurio, J. Etourneau and P. Hagenmuller, Mat. Res. Bull. 11, 947 (1976).
12. J.P. Mercurio, J. Etourneau, R. Maslain, P. Hagenmuller and J.B. Goodenough, J. of Sol. St. Chem. 9, 37 (1974).
13. M. Ishii, M. Aono, S. Kawai and S. Muranaka, Solid State Commun. 20, 437 (1976).
14. J. Friedel, Adv. in Phys. 3, 446 (1954).
15. T. Penney, B. Barbara, R.L. Melcher, T.S. Plaskett, H.E. King Jr. and T.J. La Placa, Ref. 3, p. 341.
16. A. Jayaraman, Ref. 2, p. 707.
17. H.E. King, Jr., S. La Placa, T. Penney and Z. Fisk, Ref. 3, p. 333.
18. H. Zahel, Diplomarbeit, Universität zu Köln (1983).
19. R. Pott, R. Schefzyk and D. Wohlleben, Z. Phys. B44, 17 (1981).
20. S. Horn, E. Holland-Moritz, M. Lowenhaupt, F. Steglich, H. Scheuer, A. Benoit and J. Flouquet, Phys. Rev. B23, 3171 (1981).
21. E. Holland-Moritz, D. Wohlleben and M. Loewenhaupt, Phys. Rev. B25, 7482 (1982).
22. L.L. Hirst, Phys. Kond, Mat. 11, 255 (1970).
23. W.C.M. Mattens, Phd. Thesis, University of Amsterdam (1980).
24. P. Wachter, Ref. 9, p. 145.
25. γCe: G. Busch, H.J. Güntherodt, H.U. Künzi and L. Schlapbach, Phys. Letters 31A, 191 (1970); $CeIn_3$: W.H. Dijkman, F.R. de Boer, P.F. de Chatel and J. Aarts, J. of Mag. Mag. Mat. 15-18, 970 (1980); $CePt_2$: Ref. 31; $CeSi_2$ and $CeSn_3$: W.H. Kijkman, PhD. Thesis, University of Amsterdam (1982).
26. P.W. Anderson and S.T. Chui, Phys. Rev. B9, 3229 (1974).
27. E. Umlauf and E. Hess, Physica 108B, 1347 (1981).
28. E. Umlauf and E. Hess, Solid State Commun. 44, 311 (1982).
29. R. Pott, PhD. Thesis, University of Cologne (1982).
30. R. Pott, R. Schefzyk, W. Boksch and D. Wohlleben, Ref. 3, p. 337.
31. P. Weidner, B. Wittershagen, B. Roden and D. Wohlleben, Solid State Commun. 48, 915 (1983).
32. J. Röhler, D. Wohlleben, J.P. Kappler and G. Krill, to be published.
33. D. Wohlleben, E. Cattaneo, D. Müller, S. Hussain and W. Schlabitz, to be published.

NEUTRON INELASTIC SCATTERING AS A PROBE OF MOMENTS IN METALLIC SYSTEMS

T.M. Holden

Atomic Energy of Canada Limited
Chalk River Nuclear Laboratories
Chalk River, Ontario, KOJ 1J0, Canada

INTRODUCTION

Neutron inelastic scattering is one of the most powerful methods for investigating the excitations in solids because the kinetic energy of a thermal neutron falls in the same range as typical excitations in solids. In a scattering event a large relative change usually occurs in the kinetic energy and this is easily measured precisely. The range of frequencies that may be probed is impressively wide with modern methods, ranging from 10^7 to 10^{14} Hz, and wave vector transfers in the range 10^{-3} to 10^2 Å^{-1} are currently accessible.

The correlations measured in neutron scattering at low wave vectors and frequencies in the ordered phase of conventional magnets are periodic; they are known as spin waves. These can be interpreted in terms of the exchange coupling between stable magnetic moments. If the local magnetic moment is unstable the scattering is in the form of a continuum centered on zero frequency whose width depends on the lifetime of the moment. Lifetimes that fall in the range 10^{-10} - 10^{-13} s. match the capabilities of the neutron probe very well.

In the cerium and uranium systems that are the subject of this paper, the magnitude of the lattice parameter is an important variable. As the lattice parameter decreases the net magnetic moment is suppressed and the elementary excitations acquire a lifetime.

The paper is arranged as follows. In the first section the neutron scattering cross sections are given and certain sum rules are introduced. The second section deals briefly with the scattering from conventional metallic magnets to establish the canonical behaviour. The third section deals with the behaviour of Ce dissolved dilutely in La_xTh_{1-x}. This is an interesting system since the magnetism can be fine-tuned by controlling the La concentration. The final section deals with the inelastic scattering in uranium rock-salt structure compounds. In the compounds with the smallest lattice parameter no collective excitations of a spin-wave nature are observed even though the compounds order magnetically. Only a broad continuum of inelastic scattering is observed. A phenomenological explanation of this continuum is discussed.

NEUTRON SCATTERING CROSS SECTIONS

The magnetic neutron scattering cross section from N atoms per unit angular frequency interval and per unit solid angle, is, following the method of Marshall and Lovesey[1],

$$\frac{d^2\sigma}{d\Omega d\omega} = \frac{(\gamma r_0)^2}{4} \frac{k_1}{k_0} S(\vec{Q},\omega) , \qquad (1)$$

where $\vec{k}_0$ is the incident neutron wave vector and $\vec{k}_1$ is the scattered neutron wave vector, r_0 is the classical electron radius and γ is the magnetic moment of the neutron in nuclear magnetons. For the purposes of calculation $(\gamma r_0)^2 = 73$ mb sr^{-1} where $1 mb = 10^{-24} cm^2$. The wave vector transfer, $\vec{Q}$, and energy transfer, $\hbar\omega$, are defined by

$$\vec{Q} = \vec{k}_0 - \vec{k}_1 \text{ and } \hbar\omega = E_0 - E_1$$

where $E_0(E_1)$ is the incident (scattered) neutron kinetic energy. The scattering law $S(\vec{Q},\omega)$ contains the physics of the scattering system and is given by

$$S(\vec{Q},\omega) = g_J^2 |f(Q)|^2 \sum_{\alpha,\beta} (\delta_{\alpha\beta} - \hat{Q}_\alpha \hat{Q}_\beta) S^{\alpha\beta}(\vec{Q},\omega). \qquad (2)$$

The tensor character follows from the dipole-dipole interaction between the neutron nuclear dipole moment and the magnetic dipole moments of the scattering system. Cartesian coordinates with respect to a convenient set of axes are specified by α and β, and for the systems described here, only diagonal terms occur. The spatial distribution of unpaired spin gives rise to the form factor $f(\vec{Q})$ and g_J is the Landé factor for the ions. The function $S^{\alpha\beta}(\vec{Q},\omega)$ is the spectrum of fluctuations in the magnet, the time and space Fourier transform of the spin correlation function $\langle J^\alpha(i,t) J^\beta(j,0)\rangle$. This is the likelihood of finding a spin deviation in the α-component on site i at a time t after a spin deviation in the β-component was impressed on site j. We have

$$S^{\alpha\beta}(\vec{Q},\omega) = \frac{1}{2\pi} \sum_{i,j} \exp[i\vec{Q}\cdot(\vec{r}_i-\vec{r}_j)] \int_{-\infty}^{+\infty} dt\, e^{i\omega t} \langle J^{\alpha}(i,t) J^{\beta}(j,0)\rangle . \qquad (3)$$

If we introduce the spatial Fourier transform of the total angular momentum

$$J^{\alpha}(\vec{Q},t) = \sum_i \exp(i\vec{Q}\cdot\vec{r}_i)\, J^{\alpha}(i,t), \qquad (4)$$

we can write

$$S^{\alpha\beta}(\vec{Q},\omega) = \frac{1}{2\pi} \int_{-\infty}^{+\infty} dt\, e^{i\omega t} \langle J^{\alpha}(\vec{Q},t)\, J^{\beta}(-\vec{Q},0)\rangle . \qquad (5)$$

The integral over all frequencies at constant $\vec{Q}$ is

$$\int_{-\infty}^{+\infty} S^{\alpha\beta}(\vec{Q},\omega)d\omega = \langle J^{\alpha}(Q,0)\, J^{\beta}(-Q,0)\rangle , \qquad (6)$$

and the sum of this function with $\alpha = \beta$ over all wave vectors, $\vec{q} = \vec{Q}-\vec{\tau}$, in a Brillouin zone is simply related to the square of the α-component of the local moment by

$$g_J^2 \sum_{\vec{q}} \int_{-\infty}^{+\infty} S^{\alpha\alpha}(\vec{q},\omega)d\omega = N \langle (M^{\alpha}(0))^2\rangle . \qquad (7)$$

Thus by integrating the magnetic scattering we can find the average square of the local moment. In localized systems this is trivially $g_J^2 J(J+1)$, but in mixed valent systems the moment is essentially unknown. A practical method of calculating the spin correlation function makes use of the retarded Green function

$$G^{\alpha\beta}(\vec{Q},t) = -i\,\theta(t) \langle [J^{\alpha}(\vec{Q},t), J^{\beta}(-\vec{Q},0)]\rangle . \qquad (8)$$

The spin response function, $G^{\alpha\beta}(\vec{Q},\omega)$, is the time Fourier transform of the retarded Green function and is related to $S^{\alpha\beta}(\vec{Q},\omega)$ by the fluctuation-dissipation relation,

$$S^{\alpha\beta}(\vec{Q},\omega) = \frac{-1}{\pi} \frac{1}{(1-e^{-\beta\omega})} \mathrm{Im}\, G^{\alpha\beta}(\vec{Q},\omega) . \qquad (9)$$

The static susceptibility is the limit as $\vec{Q}$ and ω tend to zero of the susceptibility

$$\chi^{zz}(\vec{Q},\omega) = - g_J^2 \mu_B^2 G^{zz}(\vec{Q},\omega) \; . \qquad (10)$$

Practical aspects of neutron scattering have been described in many places, for example by Dolling[2]. There are two ways to obtain the inelastic scattering cross section depending on whether time-of-flight or single-crystal diffraction is used to measure the neutron wave vector and energy. The methods are often complementary since a triple-axis spectrometer is better able to investigate specific regions of wave vector-frequency space, for example close to reciprocal lattice points. Time-of-flight is well adapted, by virtue of large counter-banks, to measure distributions throughout wave vector-frequency space.

CONVENTIONAL METALLIC MAGNETS

Nickel is the archetype of transition metal ferromagnets. Measurements of Im $G(\vec{Q},\omega)$, see for example Mook and Tocchetti[3], and calculations[4] of Im $G(\vec{Q},\omega)$ have been carried out throughout the last two decades. Another example of a typical ferromagnet is manganese silicide studied by Ishikawa et al.[5]. Two points stand out clearly. The first is that the excitations at low temperatures are well-defined collective excitations in both wave vector and frequency. At higher frequencies these spin waves merge with a continuum of single spin flip scattering. The second point is that the continuum extends in frequency up to several times kT_C. This latter range reflects the fact that the magnetic electrons are distributed in energy bands and these are several eV wide in 3d metals.

The scattering from well-localized 4f systems such as rare earth metals and compounds, like Pr and TbSb,[6] is relatively easily understood. The two primary components of the Hamiltonian are the crystal field and the interatomic exchange. The spin response in the random phase approximation (RPA) takes the schematic form[7]

$$G(\vec{Q},\omega) = \frac{G^0(\omega)}{1 - \mathcal{J}(\vec{Q})\, G^0(\omega)} \qquad (11)$$

where $G^0(\omega)$ is the single-ion response in the presence of exchange and crystal fields and $\mathcal{J}(\vec{Q})$ is the Fourier transform of the interatomic exchange. Sharp excitations are observed whose frequencies and intensities can usually be accounted for within the RPA. To introduce lifetime effects, it is necessary to go beyond the RPA. Becker et al.[8] have derived expressions for the intrinsic halfwidths in frequency arising from the coupling between f-states and conduction electrons.

CERIUM IN LANTHANUM-THORIUM ALLOYS

If the Ce ion is in a 3+ valency state, with a single 4f electron localized well below the Fermi level, the crystal field splits the J = 5/2 free ion level into a Γ_7 doublet and a Γ_8 quartet. The dipole matrix element $\langle\Gamma_8|J^z|\Gamma_7\rangle$ is large and the splitting, Δ, between Γ_7 and Γ_8 should easily be observable. If the transition is broadened the Ce ion is not in a long-lived or stable $4f^1$ state.

The static susceptibility of Ce dissolved in La_xTh_{1-x} alloys has been previously studied[9]. The incremental susceptibility per Ce atom in La-rich alloys resembles a Curie-Weiss law suggesting Ce local moments and low spin fluctuation temperatures as gauged by the negative intercepts on the temperature axis. The susceptibility of Ce in Th is much weaker and is almost temperature independent. The inference is that Ce in Th is weakly or non-magnetic.

The local susceptibility of Ce implanted in La and Th has recently been measured by the perturbed angular distribution of γ-rays[10]. The local susceptibility of Ce in La again follows a Curie-Weiss law but the susceptibility of Ce in Th resembles that of an ion whose moment arises from thermal occupation of an excited state. The susceptibility is comparable to that for Ce^{3+} at 800 K but is nearly temperature independent at low temperatures with a weak maximum around 300 K.

Riegel et al.[10] proposed a simple scheme to account for the susceptibility with an f^0 state lying 12 THz below the excited f^1 state. From the damping of the perturbed angular distribution a spin-fluctuation frequency ν_{sf} of 5 THz was deduced which corresponds to a spin fluctuation temperature of 250 K.

Several neutron inelastic scattering experiments on fcc Ce alloys have been performed. For the alloy $Ce_{0.76}Th_{0.24}$, in which a valence transition occurs near 150 K, Shapiro et al.[11] found a broad distribution of scattering centered on zero frequency and extending beyond 17 THz with no evidence for crystal field peaks. The half-width parameter of this Lorentzian distribution was about 5 THz in agreement with the perturbed angular correlation result suggesting that the Ce spin-fluctuation time behaves like a single-ion property. In a further series[12] of experiments in which La was partly substituted for Th, in addition to the peak centered on zero frequency, a peak with an intrinsic width was observed near 3.4 THz which was interpreted as the Γ_7–Γ_8 crystal field transition. In the Chalk River experiments[13] with a low concentration of Ce ions in the ternary system $Ce_{0.1}(La_xTh_{1-x})_{0.9}$ a broadened crystal field excitation was found at the La rich end, and the lifetime and frequency of this transition increased as Th was added.

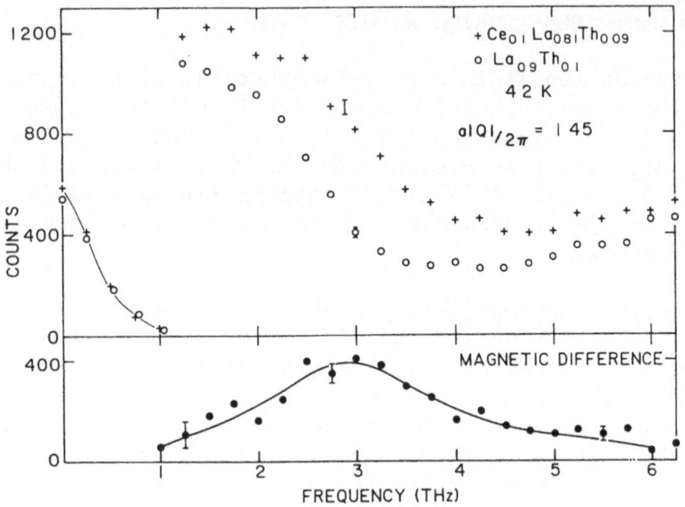

Fig. 1. Inelastic neutron scattering from $Ce_{0.1}(La_{0.9}Th_{0.1})_{0.9}$ and $La_{0.9}Th_{0.1}$ alloys at 4.2 K and the Ce difference scattering.

The results of the measurements are shown in Fig. 1 for x=0.9. for the Ce alloy and a corresponding blank. The difference scattering shows a wave vector dependence which is consistent with the f^1 "magnetic" form factor. Its shape is independent of wave vector in the range $a|Q|/2\pi = 0.8$ to 1.5 confirming the single ion character. The scattering is centered on 2.9 ± 0.2 THz with an intrinisc half-width $\Gamma = 0.8 \pm 0.4$ THz. The magnetic difference counts at 4.2 K for x = 0.0, 0.6 and 0.9 are shown in Fig. 2. There is no scattering for Ce in Th. The result could be consistent with the model of Riegel et al[10] since even with an intrinsic fluctuation temperature of 250 K, the effective population of the f^1 state is very low and this diminishes the scattering by a factor of 20 compared with an f^1 ground state.

The increase in splitting as x changes from 0.9 to 0.6 has a component which is consistent with a crystal field model, since the change in lattice parameter would lead to an increase in Δ of 10%. The increase in Γ to 1.6 ± 0.2 THz keeps step with the apparent Curie-Weiss temperature of this alloy (65 K, or 1.35 THz) suggesting that both have a common origin in the fluctuation time of the magnetic moment.

Since a crystal field transition is observed, the Ce ions are likely to be in a state close to $4f^1$. The presence of intrinsic widths, however, points to some degree of hybridization with the conduction electrons. The effect of decreasing the lattice parameter, through adding Th, is to increase the intrinsic widths and

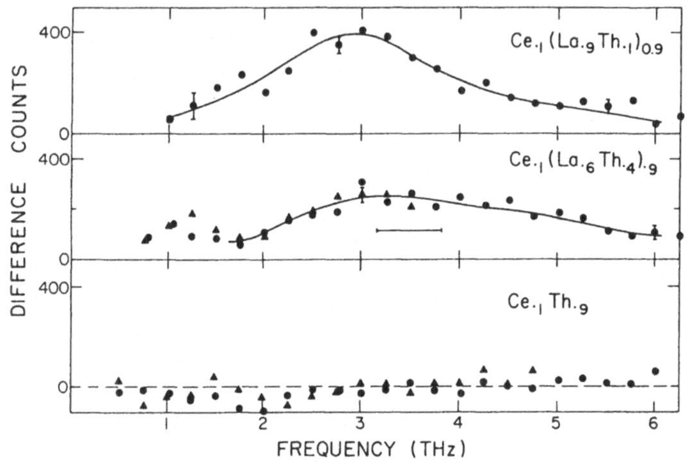

Fig. 2. Cerium difference scattering for La concentrations x=0.9, 0.6, 0.0 at 4.2 K. Circles and triangles correspond to different wave vectors.

suppress the magnetism. On the non-magnetic side of a mixed-valent transition, however produced, the inelastic scattering is highly suppressed.

URANIUM ROCK-SALT STRUCTURE COMPOUNDS

From elementary valence considerations the uranium ions in pnictide compounds should lose one 6d and two 7s electrons to fill the three holes in the pnictogen p-shell and be left in a $5f^3$ state. Likewise uranium ions in chalcogenide compounds should be in a $5f^3 6d$ configuration. The 5f-electrons are in energy bands which cross the Fermi energy[14] and are overlapped by a broader 6d band. Any mixed valent tendencies of uranium would come from two magnetic configurations of the ion unlike the magnetic-non magnetic configurations of Ce. The uranium pnictides order antiferromagnetically[15] but the magnetic moments and the Néel temperatures are small when the lattice parameter is small; the magnetic ground state varies systematically from a 3 $\vec{k}$ type-I structure with spins pointing in [111] directions for USb, through a 2 $\vec{k}$ type-I structure with spins in [110] directions for UAs, to a type-I structure with spins in [001] directions for UN. The chalcogenides are feromagnetic,[15] with [111] easy directions: the magnetic moment decreases with decreasing lattice parameter but T_C increases.

For the compounds with the largest lattice parameters, USb[16] and UTe[17], there exist well-defined branches of spin-waves with a gap in frequency at the magnetic zone-centre. Such an anisotropy gap is not surprising for a 5f electron system and may be a remnant of a crystal field effect. The result for UTe is shown in Fig. 3. The initial wave-vector dependence is quadratic. To date resolution-limited spin waves have not been observed at frequencies above the optic phonon branch. In USb, the collective response is superposed on a wide scattering distribution centred on zero frequency. However, neither of the pnictides or chalcogenides with the smallest lattice parameters exhibit sharp spin-waves even at q = 0, in spite of the magnetic ordering. Thus, UN[18] and US[19] are the first examples of ordered antiferromagnets and ferromagnets which do not possess sharp collective excitations. The magnetic scattering in US has been put on an absolute scale by comparison with the phonon intensity and the response function in units of μ_B^2 THz^{-1} is shown in Fig. 4. The distribution is very broad extending beyond 16 THz with a weak peak around 10 THz at the zone centre. One suspects that at the zone boundary the scattering extends well above the maximum frequency measured. The result is reminiscent of the single particle scattering in nickel, extending as it does to several times kT_C. One way of reconciling the scattering in US with the sharp spin wave in UTe is to suppose[20] that the weak peak at 10 THz represents a crystal field transition but that on-site exchange coupling with the 6d electrons is responsible for the marked

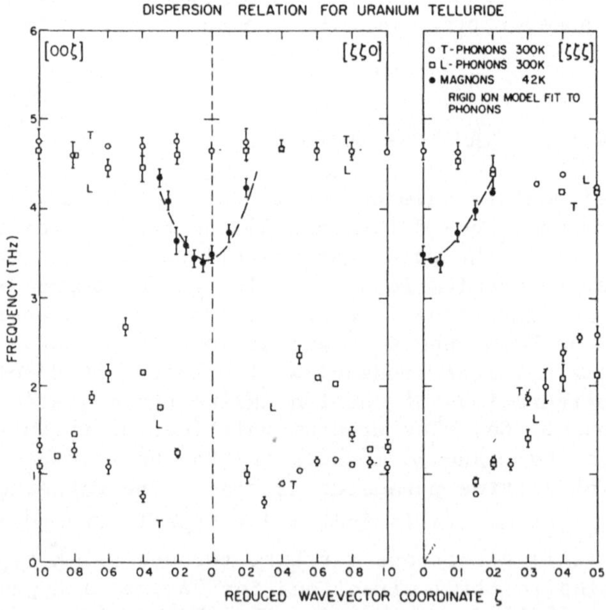

Fig. 3. Dispersion relations for phonons in UTe and for magnons where these are well defined.

damping. If we take a one-pole approximation for the 5f single-ion response,

$$g_f^{-+}(\omega) = \alpha^2/(\omega - \Delta) \tag{12}$$

and a broad d-response

$$g_d^{-+}(\omega) = 1/(i\omega - \gamma) \tag{13}$$

we can obtain the f response by the equation of motion method:

$$G_{ff}^{-+} = g_f^{-+}/\ \{1 - g_f^{-+}(J_{ff} + J_{fd}^2\ g_d^{-+})\}. \tag{14}$$

We have introduced an independent interion coupling J_{ff} as an phenomenological interaction, although it may also be related to the f-d coupling. A broad range of behaviour follows on choosing appropriate magnitudes for J_{fd}, Δ and γ. For example, for $\Delta = 10$, $\gamma = 5$ THz and $J_{fd} = 0$ we recover the sharp excitation near $\omega = \Delta$. On increasing αJ_{fd} to 6 THz, we can spread the excitation out over the full range of observed frequencies. If we

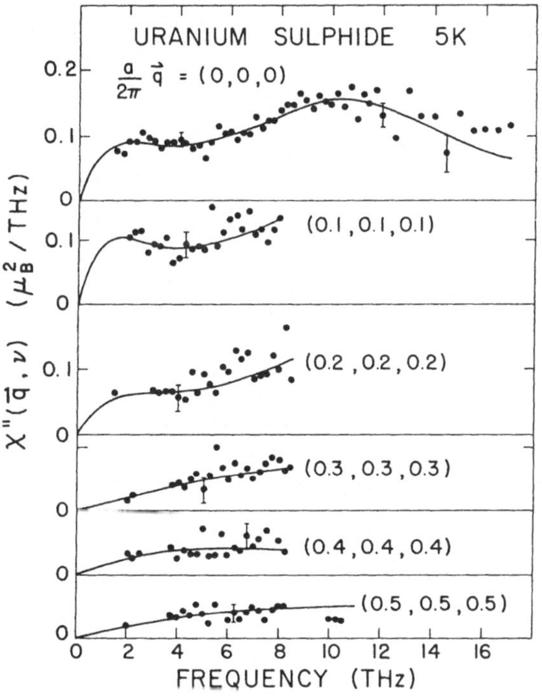

Fig. 4. Dynamic response of US at 4.2 K in the ordered ferromagnetic phase.[19] Lines are a two-Lorentzian fit to the data.

increase γ greatly, the d-response is too weak in the neighbourhood of the f-response and we recover the f-response localized in frequency. The point is that we need γ to be about 5-10 THz, not of order 1 eV, over the whole range of wave vectors studied. Hence we surmise that the damping must come from electron states in a narrow energy region perhaps where hybridization occurs.

When the inelastic scattering at 4.2 K in US[19] is integrated over frequency and wave vector, using Lorentzians to extrapolate to 100 THz, the result is 1.1 μ_B^2, to which must be added the square of the ordered moment, to give a total squared local moment $\langle M^2\rangle$ = 4.2 μ_B^2. This is far short of the squared local moment ≈ 13 μ_B^2 expected for an f^2 or f^3 configuration and suggests either strong depression of the moment or else a strong response at frequencies above the range of the extrapolation.

The paramagnetic scattering in US[19] has also been carefully measured at 300 K, well above the ordering temperature of 180 K. The susceptibility at $\vec{q} = 0$ integrated over frequency $\chi(\vec{q} = 0)$ agreed with the measured static susceptibility to within 10%. This gives a good estimate of the overall accuracy of the measurements and of the reliability of the frequency integration. It was found that $\chi(\vec{q})$ was described empirically by

$$\chi(\vec{q}) = 1/3\ \mu_{eff}^2 / \left\{T - \theta\left|\frac{J(\vec{q})}{J(0)}\right|\right\} \tag{15}$$

where μ_{eff} and θ are the paramagnetic moment and Curie-Weiss temperature. The wave-vector dependence of the exchange is given by

$$\left|\frac{J(\vec{q})}{J(0)}\right| = \frac{1}{Z} \sum_{\vec{\delta}} e^{i\vec{Q}\cdot\vec{\delta}} , \tag{16}$$

where $\vec{\delta}$ runs over the Z nearest uranium neighbours. The wave-vector dependent susceptibility is not therefore so unusual since it is the same as that for a localized system; what is unusual is the wide distribution in frequency. The dynamical susceptibility for wave vectors in the [111] direction was described by

$$\mathrm{Im}\ \chi(\vec{q},\omega) = \chi(\vec{q}) \frac{\omega\Gamma}{\omega^2 + \Gamma^2} , \tag{17}$$

where Γ varies from about 2 THz at the zone centre to 9 THz at the zone boundary. The integrated scattering at 300 K gives the square of the total moment $\langle M^2\rangle \approx 5 \pm 2\ \mu_B^2$ which is again much less than the full localized moment for an f^2 or f^3 configuration.

For uranium compounds we can summarize the systematic effects succinctly by supposing that the mixing between f-states and the conduction electrons increases as the lattice parameter decreases. This has the effect of suppressing the moment below the localized limit. The mixing affects the fluctuation spectrum by introducing strong damping effects which appear to be strongest for those compounds with the maximum moment suppression.

ACKNOWLEDGEMENTS

It is a pleasure to acknowledge the contributions of my colleagues W.J.L. Buyers, G. Dolling, J.A. Jackman, P. Martel and E.C. Svensson to the experiments at Chalk River. The help and enthusiasm of G.H. Lander, P. de V. DuPlessis, M. Loewenhaupt, and A.F. Murray during their visits to Chalk River is also acknowledged.

REFERENCES

1. W. Marshall and S.W. Lovesey "Theory of Thermal Neutron Scattering", Clarendon, Oxford (1971), p. 176.
2. G. Dolling, in "Lattice Dynamics and Intermolecular Forces", Academic Press, New York (1975).
3. H.A. Mook and D. Tocchetti, Phys. Rev. Lett. 43: 2029 (1979).
4. J.F. Cooke and H.L. Davis, in "Magnetism and Magnetic Materials - 1972", C.D. Graham, Jr., and J.J. Rhyne, eds., AIP, Conference Proceedings No. 10, American Institute of Physics, New York, (1972) p. 1218.
5. Y. Ishikawa, G. Shirane and J.A. Tarvin, Phys. Rev. B16: 4956 (1977).
6. J.G. Houmann, B.D. Rainford, J. Jensen and A.R. Mackintosh, Phys. Rev. B20: 1105 (1979); T.M. Holden, E.C. Svensson, W.J.L. Buyers and O. Vogt, Phys. Rev. B10: 3864 (1974).
7. W.J.L. Buyers, T.M. Holden and A. Perreault, Phys. Rev. B11, 266 (1975).
8. K.W. Becker, P. Fulde and J. Keller, Z. Physik B28: 9 (1977).
9. J.G. Huber, J. Brooks, D. Wohlleben and M.B. Maple in "Magnetism and Magnetic Materials - 1975", C.D. Graham, Jr., G.H. Lander and J.J. Rhyne, eds., AIP Conference Proceedings No. 24, American Inst. of Physics, New York (1975) p.475.
10. D. Riegel, H.J. Barth, M. Luszik-Bhadra and G. Netz in "Valence Instabilities", P. Wachter and H. Boppart, eds., North-Holland, Amsterdam (1982) p. 497.
11. S.M. Shapiro, J.D. Axe, R.J. Birgeneau, J.M. Lawrence and R.D. Parks, Phys. Rev. B16: 2225 (1977).
12. S.M. Shapiro, J. Appl. Phys. 52: 2129 (1981).
13. T.M. Holden, W.J.L. Buyers, P. Martel, M.B. Maple and M. Tovar in "Valence Instabilities", P. Wachter and H. Boppart. eds., North-Holland, Amsterdam (1982) p. 325.

14. M.S.S. Brooks and D. Glötzel, Physica 102B: 51 (1980).
15. D.J. Lam and A.T. Aldred in "The Actinides: Electronic Structure and Related Properties. Vol. 1", A.J. Freeman and J.B. Darby, Jr., eds., Academic, New York (1974) p. 109.
16. G.H. Lander and W.G. Stirling, Phys. Rev. B21: 436 (1980).
17. W.J.L. Buyers, T.M. Holden, J.A. Jackman, A.F. Murray P. de V. DuPlessis, J. Magn. and Magn. Mater. 31: 229 (1983).
18. T.M. Holden, W.J.L. Buyers, E.C. Svensson, J.A. Jackman, A.F. Murray, O. Vogt and P. de V. DuPlessis, J. Appl. Phys. 53: 1967 (1982); T.M. Holden, W.J.L. Buyers, E.C. Svensson and G.H. Lander, Phys. Rev. B (1984) to be published (June).
19 J.A. Jackman, P. de V. DuPlessis, A.F. Murray, T.M. Holden and W.J.L. Buyers, to be published.
20. W.J.L. Buyers and T.M. Holden in "Handbook of the Physics and Chemistry of the Actinides", G.H. Lander and A.J. Freeman, eds., North-Holland, Amsterdam, to be published.

ORBITAL EFFECTS IN ACTINIDE SYSTEMS

G. H. Lander

Argonne National Laboratory
Argonne, Illinois 60439

ABSTRACT

Actinide magnetism presents a number of important challenges; in particular, the proximity of 5*f* band to the Fermi energy gives rise to strong interaction with both *d* and *s* like conduction electrons, and the extended nature of the 5*f* electrons means that they can interact with electron orbitals from neighboring atoms. Theory has recently addressed these problems. Often neglected, however, is the overwhelming evidence for large orbital contributions to the magnetic properties of actinides. Some experimental evidence for these effects are presented briefly in this paper. They point, clearly incorrectly, to a very localized picture for the 5*f* electrons. This dichotomy only enhances the nature of the challenge.

I. Introduction

Elsewhere in this volume Holden and Loewenhaupt have discussed properties of actinide magnetism in terms of the magnetic response function $\chi''(Q,\omega)$ as measured by inelastic neutron scattering. They have shown that the situation is complex and that in only one metallic uranium system (UPd_3) studied so far can a simple localized picture of the 5*f* electrons be correct.[1] In this short paper I shall present a few examples of large orbital effects in actinide magnetism. These all point to localized magnetism despite the contradiction of the preceeding remarks.

II. What is a localized moment?

We know that for uranium and beyond the 5*f* electron states are populated. To have some understanding of a localized *f* electron we can turn to lanthanide (4*f*) series. As one progresses across the series one follows Hund's rule and Russell-Saunders coupling such that

$$\vec{S} = \Sigma_i \vec{s}_i , \quad \vec{L} = \Sigma_i \vec{\ell} \quad \text{and} \quad \vec{J} = \vec{L} + \vec{S},$$

where $\vec{s}_i$ and $\vec{\ell}_i$ are the spin and orbital momenta of each 5*f* electron, and the total $\vec{J}$ component defines the magnetic moment, through $\vec{\mu} = g\vec{J}$, where g is the Lande factor. All this is quite conventional and in every textbook on magnetism. To answer the question posed in this section, localized magnetism of *f* electrons, really requires a well-defined spin and orbital component. (Except for a half-filled shell when L = 0). The orbital component implies long-time correlations of the *f* electrons around a particular nucleus.

There are other consequences of localized systems. The orbital degeneracy can be lifted by an electrostatic crystal-field potential. Strong spin-orbit interactions mean that $\vec{L}$ and $\vec{S}$ are frequently not good quantum numbers and we have also to take account of higher L and S states. This has to be treated in a so-called intermediate coupling.[2]

We can see from this what really distinguishes localized from itinerant (or even intermediate valence) systems is the presence of a large orbital moment. Americium systems present a good example. None so far have been found that are magnetic. This is ascribed to there being a J = 0 ground state composed of $\vec{L} = +3$, $\vec{S} = -3$ contributions. Band structures calculations of Ni, for example, do project out an orbital moment but it is small.

III. Magnetic anisotropy and magnetoelastic effects

One consequence of a large orbital moment is magnetic anisotropy and magnetoelastic effects. The actinide ferromagnets have some of the highest anisotropies known. Figure 1 shows the magnetization of UTe along the three principal axes.[3] Note that the easy axis is <111> and that the magnetization in the other directions is exactly the projection of the moment along the nearest <111> in this direction. The anisotropy in UTe, and more recently in Pu ferromagnets PuSb and PuAs, is enormous, much bigger than anything observed in the 3*d* or even the 4*f* series.

Magnetoelastic effects are basically coupling between the lattice and orbital wavefunctions. For example, such effects in an itinerant system such as chromium are very small, but in a system with a large orbital moment, e.g., Er or Dy, they are appreciable. Table I shows some of the lattice distortions of

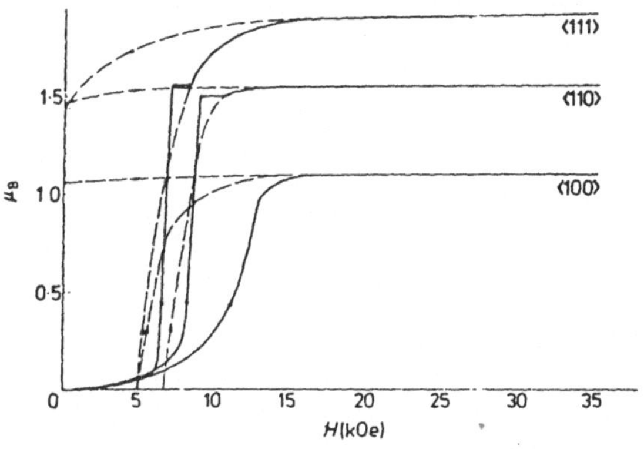

Fig. 1 Magnetization (reduced to μ_B) against applied field for three principal directions in UTe at 1.5 K (Ref. 3).

actinide ferromagnets.[4] These distortions are very large. They show a strong coupling to the lattice, presumably through the orbital moment, and are consistent with the anisotropy illustrated in Fig. 1.

Table I

Lattice distortions in the ordered state of actinide ferromagnetics with the NaCℓ crystal structure; a_o is the lattice parameter, T_c is the ordering temperature, at which the distortion, either rhombohedral (R) or tetragonal (T), begins.[4] In the rhombohedral phase "c" is defined as a unit length along the <111> direction, "a" in a direction perpendicular to this.

Compound	a_o (Å) at 300K	T_c (k)	Distortion	$10^4 \frac{c-a}{a}$
US	5.49	178	R	+ 105
USe	5.75	160	R	+ 81
UTe	6.16	104	R	+ 67
NpC	5.00	220	R	+ 23
NpN	4.90	87	R	- 52
PuP	5.66	126	T	- 31

IV. Neutron elastic scattering-magnetic form factors

The magnetic moment of the neutron intereacts with both the spin and orbital moment of the unpaired electrons around the nucleus. Since these electrons have a spatial distribution (1 - 2Å), comparable to the neutron wavelength then we will see a decrease in the magnetic scattering amplitude for increasing momentum transfer Q. This implies a form factor which is related to the total magnetization density through the Fourier transform

$$f(\vec{Q}) = \int M(\vec{r}) e^{i\vec{Q}\cdot\vec{r}} d\vec{r} \quad .$$

Scattering from orbital moments is quite easily distinguished from spin-only scattering. We show, for example, in Fig. 2 the form factor of the induced magnetic moment in α-U at low temperature[5]. We know that the 5*f* wavefunctions of α-U are very broad and strongly hybridized so that we should not expect an orbital moment, and none is apparent. For such a well-localized system as UO_2 the reverse, of course, is true.[6,7]

Form factors are therefore ideal at showing whether or not an orbital moment is present. A particularly interesting form factor is that of Sm^{3+} or Pu^{3+} in which the $^6H_{5/2}$ configuration is the ground state. Since both L(=5) and S(=5/2) are large and opposed to each other, the ground state J value (=5/2) is small. Furthermore, because the spin distribution is more extended in real space than the orbital contribution, regions of negative (with respect to $\vec{J}$) magnetization density exist around the nucleus. The resulting form factor gives a peak at $Q \neq 0$, and this has been clearly seen in Sm compounds.[8] In Pu systems the situation is complicated by the need to take into account the large spin-orbit coupling, but the two form factors measured so far are essentially flat for small Q values.[9,10] This is a clear signature of an orbital contribution.

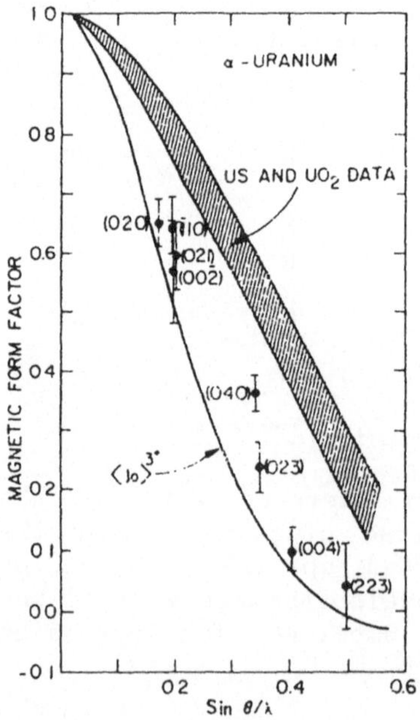

Fig. 2 Experimental magnetic form factor for α-uranium metal compared to $\langle j_o \rangle$ and combined results on uranium in compounds - hatched region (Ref. 5)

Another important aspect of form factors is that they can be thought of as measuring the time-averaged localized moment. The modifier "time-averaged" is important. Nickel is not usually thought of as a localized system; however, the form factor agrees well with atomic calculations.[11] The reason is that the hopping time of the itinerant d electrons is much less than their resident time at a particular site, so the "time-average" misses the hopping aspect. On the other hand, if electrons have wavefunctions that are spread out in real space, then the Fourier transform of this will be a sharp peak in Q space. Since the most accurate form factors are measured at Bragg peaks (with polarized neutrons utilizing the interference between magnetic and nuclear scattering), a diffuse electron distribution is often missed. Its presence is then inferred from the difference between the μ_{local} extrapolated from the neutron work at finite Q and μ_{total} measured by bulk susceptibility and by definition at Q = 0. The difference is then called the "conduction electron polarization", μ_{cep}. This quantity has been the subject of much interest in iron, nickel, gadolinium,[12] and in lanthanide compounds, and also exists in actinide systems.[3] Unfortunately, neutrons have difficulty in measuring the small Q (< 1 $Å^{-1}$) region because in normal systems (with moderate lattice parameters) there are no Bragg peaks. Other methods of measuring f(Q) through the spin waves, the magnetovibrational scattering , or the diffuse scattering in the paramagnetic phase are difficult and subject to rather large uncertainties.

In a few systems "extra", i.e. non-localized, scattering can be seen. Good examples are the work on $CeSn_3$[13] and $CePd_3$,[14] where an additional d contribution can be seen. Note should also be made of the fact that the d polarization is parallel to $\vec{J}$, whereas in all other light lanthanide or actinide systems it is antiparallel. In the actinides an interesting magnetization density was obtained in UGe_3, in which the hybridization results in a polarized spin density observable near the Ge site.[15]

V. Hyperfine field-magnetic moment correlations

We now turn to quite a different aspect. One of the terms that may be derived from Mossbauer spectroscopy is the hyperfine field - or strength of the magnetic interaction, H_{hf}. We show in Fig. 3 the relationship between hyperfine fields in Np systems and the ordered moments derived by neutron scattering.[16] The relationship in the Np systems is as linear as in the Yb salts. Note some of the small values of the ordered moment found in the Np systems, less than 1 μ_B. Moreover, because of the very large values of H_{hf} the orbital component must contribute a major component. In contrast the hyperfine field in ordered iron is 340 kOe. The fact that H_{hf} comes so close to the free-ion Np^{3+} value suggests an essentially atomic character with this configuration. The proportionality between H_{hf} and $\langle r^{-3} \rangle$, where this

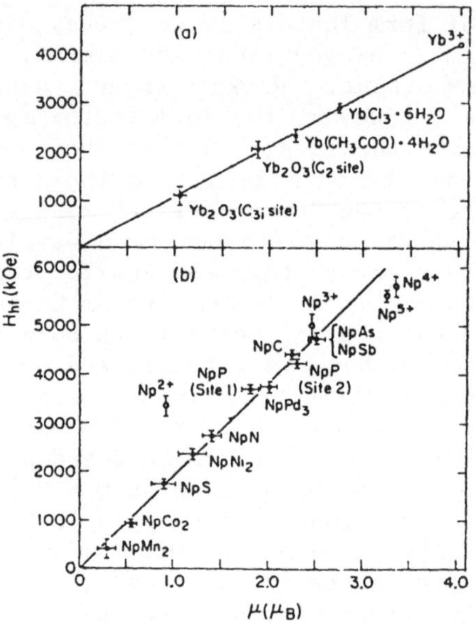

Fig. 3 Comparison between magnetic hyperfine field and electronic magnetic moment for a number of (a) Yb salts and (b) Np intermetallics. Experimental points are shown by solid circles and calculated free-ion values by open circles.

is the average value of $1/r^3$ for the open-shell electrons, remain roughly constant and only $\langle J_z \rangle$, the average value of the angular momentum, is altered in the various materials. Thus, even for systems with small moments a proportional orbital must exist.

VI. Conclusions

I have tried in this short paper to draw attention to the importance of orbital magnetism in actinide systems. The present NATO School has focussed attention on intermediate valence, unstable moments, and actinide systems. These classes have many things in common. It is interesting, for example, that recent neutron experiments on intermediate valence systems such as $CeSn_3$ and YbCuAℓ have claimed to find evidence for crystal-field transitions. The latter imply an orbital degeneracy lifted by the electrostatic field of the surrounding ions. We do not easily find crystal-field transitions in U systems, but much evidence for strong orbital contributions exists.

None of the properties discussed in this paper for actinide systems are by themselves too surprising. Similar magnetization, magnetostrictive, and form factor properties exist in the heavy rare-earths, which also have large orbital moments. What is surprising is that these localized properties are observed in systems with unusual photoemission (see the paper by Y. Baer in this volume) and neutron inelastic scattering (Holden and Loewenhaupt in this volume) measurements. In the neutron inelastic scattering, for example, it has proved very difficult to observe propagating spin-wave modes, and yet all the properties discussed in the present article would point to their existence, as indeed is the case in normal rare-earth materials. It is this conjunction of properties that is the most interesting.

The challenge to theory is then to include both strong spin-orbit coupling and hybridization in a realistic way. Recent work by Brooks and Kelly[17] on the form factors of the actinide nitrides is a step in this direction, as are the most recent calculations of Edwards in this volume, both of which start with a band description and proceed to determine the extent of f-d hybridization.

Acknowledgements

I am grateful to Mike Brooks, Barry Cooper, and Art Freeman for interesting discussions on this topic over a period of some years.

References

1. W. J. L. Buyers, A. F. Murray, T. M. Holden, E. C. Svensson, P. de V. DuPlessis, G. H. Lander, and O. Vogt, Physica 102B 291 (1980).
2. B. G. Wybourne, "Spectroscopic Properties of Rare Earths", Interscience Publishers, New York, 1965.
3. G. Busch, O. Vogt, A. Delapalme, and G. H. Lander, J. Phys. C 12 1391 (1979).
4. H. W. Knott, G. H. Lander, M. H. Mueller, and O. Vogt, Phys. Rev. B 21 4159 (1980).
5. R. Maglic, G. H. Lander, M. H. Mueller, and R. Kleb, Phys. Rev. B 17 308 (1978).
6. A. J. Freeman, J. P. Desclaux, G. H. Lander, and J. Faber, Phys. Rev. B 13 1168 (1976).
7. G. H. Lander, J. Faber, A. J. Freeman, and J. P. Desclaux, Phys. Rev. B 13 1177 (1976).
8. J. X. Boucherle, D. Givord, and J. Schweizer, J. de Physique 43 C7-199 (1982).
9. G. H. Lander and D. J. Lam, Phys. Rev. B 14 4064 (1976).
10. G. H. Lander, A. T. Aldred, B. D. Dunlap, and G. K. Shenoy, Physica 86-88B 152 (1977).
11. H. A. Mook, Phys. Rev. 148 495 (1966).
12. R. M. Moon, W. C. Koehler, J. W. Cable, and H. R. Child, Phys. Rev. B 5 997 (1972).
13. C. Stassis, C.-K. Loong, B. N. Harmon, and S. H. Liu, J. Appl. Phys. 50 7567 (1979).
14. C. Stassis, C.-K. Loong, J. Zarestky, O. D. McMasters, R. M. Moon, and J. R. Thompson, J. Appl. Phys. 53 7890 (1982).
15. G. H. Lander, J. F. Reddy, A. Delapalme, and P. J. Brown, Phys. Rev. Letters 44, 603 (1980).
16. B. D. Dunlap and G. H. Lander, Phys. Rev. Letters 33 1046 (1974).
17. M. S. S. Brooks and P. J. Kelly, to be published (1983).

MAGNETIC RESPONSE OF UNSTABLE MOMENTS MEASURED BY NEUTRON SCATTERING

M. Loewenhaupt

Institut für Festkörperforschung
KFA-Jülich, Postfach 1913
D-5170 Jülich, W. Germany

INTRODUCTION

The instability of magnetic moments in metallic systems is due to the strong coupling (or even hybridization) between the local (3d, 4f or 5f) electrons and the conduction electrons. The magnetic response and its temperature dependence as measured by inelastic magnetic neutron scattering will be used as a guide to classify the different behavior of stable, Kondo, mixed valent or spin fluctuation systems. Ideally, a stable moment would be a local moment (built up by intra-atomic correlations of a partly filled 3d, 4f or 5f shell) without any interactions with its surroundings. Its response would be a δ-function at zero frequency (ω =o) and possible sharp excitations at finite frequencies (e.g. transitions to crystal field levels in a 4f system). A more realistic definition, however, would be a local moment with some weak interactions to its surroundings. The sharp structure of the magnetic response should be conserved; the meaning of "sharp", however, depends now on the type of experiment and the corresponding frequency or energy resolution. In neutron scattering the accessible energy transfer goes from μeV through the typical meV range up to the eV range available only at spallation sources, with resolutions of 1% to 10% of the incident neutron energy. In addition to the energy spectrum neutron scattering also gives information about the spatial extent and spatial correlations of the magnetic moments through its Q dependence (Q=momentum transfer). For a detailed description of the magnetic scattering cross-sections see the article by T.M. Holden in this book.

Sum rules can be used to extract

(i) the local susceptibility $\chi'(Q, \omega=o, T)$ which can be compared for $Q \to 0$ with the static or bulk susceptibility $\chi(T)$ and

(ii) the average of the square of the magnetic moment which can be compared with the effective paramagnetic moment as measured by the Curie constant or with theoretical expectations, e.g. for the $4f^1$ configuration of Ce^{+3} ions.

Kinematical restrictions of the scattering process and the rapid decrease of the magnetic form factor with Q (see the article by G.H. Lander in this book) confine the observation of magnetic scattering to a certain area in the Q-ω- diagram. An example is shown in Fig.1 for a rather high incident neutron energy of 790 meV (lower incident energies enlarge the inaccessible area even further). Applying the sum rules (i) and (ii) requires integration from $-\infty$ to $+\infty$ of ω. This may yield ambigous results if the right extrapolation into the inaccessible regions of the Q- ω-plane is not known.

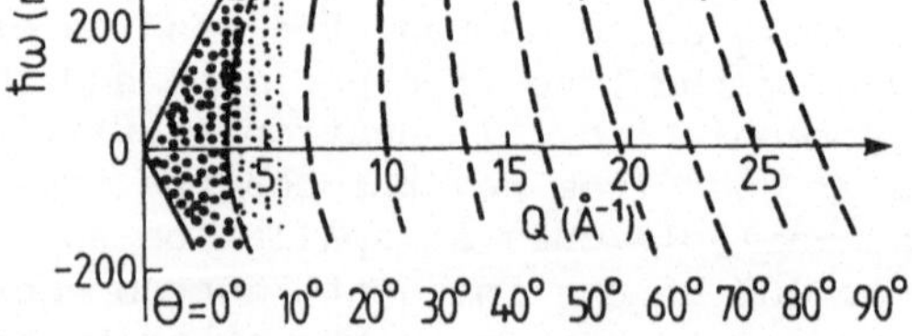

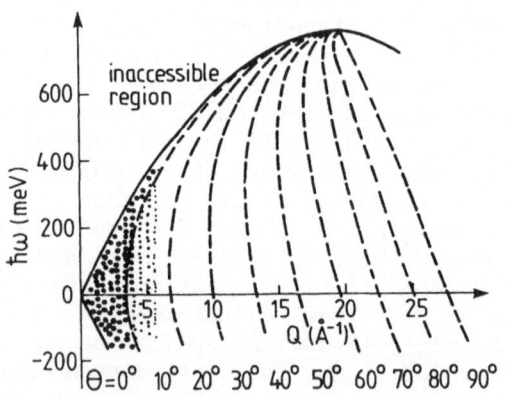

Fig.1 Kinematical conditions for inelastic neutron scattering with an incident energy of E =790 meV. Magnetic scattering is confined to dotted area because of the rapid decrease of the formfactor with Q (magnetic intensity $\sim f^2(Q)$).

The following article is meant to give a brief review of experiments performed by the author and others during the last years. Detailed descriptions are published elsewhere[1,2,3,4,5]. Also included are results from recent experiments with high incident neutron energies performed at the IPNS, Argonne, in collaboration with J.M. Carpenter and C.K. Loong.

MAGNETIC RESPONSE OF RELAXING MOMENTS

If spin-spin interactions are negligible (dilute limit, systems with no or very low magnetic ordering temperatures), it turns out that the dynamical behavior of the investigated systems can be described by a simple relaxation process. The scattering law (in zero external field and for unpolarized neutrons) can be written as

$$S(Q,\omega)=\frac{1}{1-e^{-\omega/T}}\cdot\chi'(Q,\omega=0,T)\cdot\omega\cdot\frac{\Gamma/2}{\omega^2+(\Gamma/2)^2} \qquad (1)$$

where sum rule (i) has already been applied to replace the imaginary part χ'' by the real part χ' and a spectral function (Kramers-Kronig relation). The line width is a direct measure of the relaxation rate or of the inverse of the lifetime of the moment originating from the coupling to the conduction electrons. For stable moments the line width goes to zero if $T\rightarrow 0$ because the Korringa law gives $\Gamma/2=\alpha T$ with $\alpha\sim 10^{-1}$ for 3d impurities and $\alpha\sim 10^{-3}$ for 4f impurities in a metallic environment. If there are quasielastic and inelastic transitions the line width of the central (quasielastic) line always tends to zero if $T\rightarrow 0$ while the line width of the inelastic lines may be finite for $T=0$ (BFK theory[6]).

This behavior changes drastically if we look at systems which show the Kondo effect . There is a cross-over from the high temperature region ("stable-moment-like") to the low temperature region ("non-magnetic)". The quasielastic line width stays finite for $T\rightarrow 0$! Its value is a measure of the Kondo temperature T_K. The cross-over takes place where $\Gamma/2\simeq k_BT$. In the high temperature region we have $\Gamma/2\ll k_BT$, in the low temperature region $\Gamma/2\gg k_BT$ the corresponding spectra are shown schematically in Figs. 2a and 2b, respectively.

The underlying scattering law, however, is the same in both regions and of the relaxational type as discussed above. The change from "magnetic to non-magnetic" behavior of a Kondo system thus depends on the energy or frequency window that we use to look at the system. A measurement on an energy scale that corresponds to k_BT "sees" the full moment only at high temperatures and a reduced or no moment at low temperatures. The main part of the magnetic response at low temperatures is lying at high energies and can only be reached by measurements with $\omega\gg\Gamma/2$.

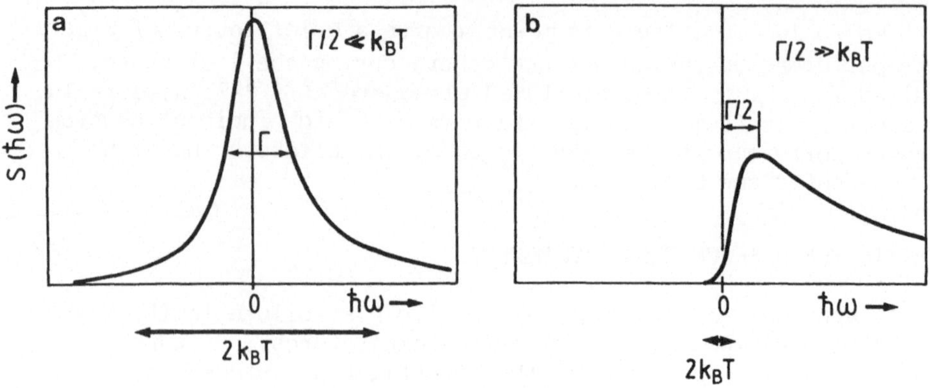

Fig.2 Scattering spectrum for a relaxing moment
a) in the classical high temperature limit; the slightly asymmetric Lorentzian is due to the Bose factor. (left)
b) in the quantum low temperature limit; for T=0 the Bose factor is zero at the neutron energy gain side ($\omega < 0$) and equal to one at the energy loss side yielding there a curve of the form ω *Lorentzian. (right)

Similarly a Van Vleck system with splitting Δ shows two regions $\Delta \ll T$ and $\Delta \gg T$. The magnetic response at T=0 would only be a sharp excitation at Δ for the Van Vleck system in contrast to the broad response of the Kondo system, starting with a linear term in ω at $\omega = 0$, a broad maximum at $\omega \simeq \Gamma/2 \simeq k_B T_K$ and a slow decrease at higher frequencies. At very high frequencies there must be a cut-off, however, otherwise sum rule (ii) would give a diverent moment. A typical example for the temperature dependence of the line width of the quasielastic line is shown in Fig.3 for a series of Ce alloys where increasing lattice pressure yields an increase of the Kondo temperature and hence of the residual quasielastic line width.

Similar curves are also obtained for the "Kondo compounds" $CeAl_2$[2], CeB_6[3], $CeCu_2Si_2$[4] and $CeAl_3$[7]. All spectra exhibit crystal field transitions with a doublet as ground state. $CeAl_2$ and CeB_6 show magnetic ordering at low temperatures, CeCu Si_2 becomes superconducting, and $CeAl_3$ does not order.

TmSe, a mixed valent compound which orders magnetically, has a relatively sharp excitation spectrum at low temperatures[1,8] with an inelastic line at 10 meV which may originate from a crystal field transition (see the article of P. Schlottmann in this book). With increasing temperature central and inelastic lines broaden considerably and merge around T ~ 100 K into one broad quasielastic line.

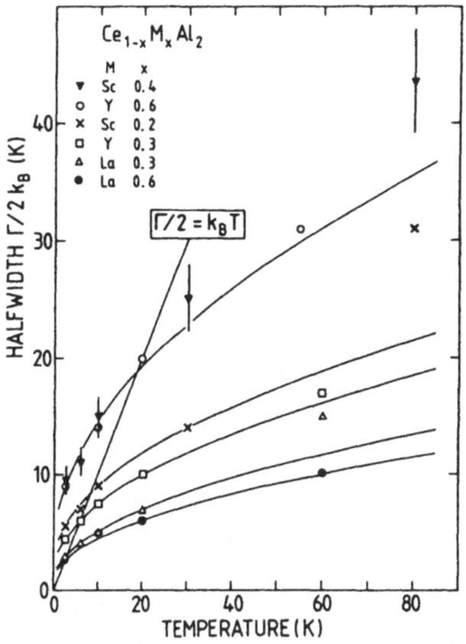

Fig.3 Temperature dependence of the quasielastic line width for the Kondo series $Ce_{1-x} M_x Al_2$ with M=La, Y or Sc. Increasing lattice pressure (going from La to Y to Sc) yields increasing residual line widths or Kondo temperatures.

Between 100 K and 300 K the line width ($\Gamma/2 \sim 7$ meV) stays practically constant. The latter behavior is also observed for mixed valent Ce and Yb compounds[5] . Fig. 4 shows the temperature dependence of the line width for $CePd_3$, $CeSn_3$ and $CeBe_{13}$.

In the observed temperature region the spectra consist of just one broad quasielastic line ($\Gamma/2$~20-25 meV), interpreted as a fast quantum mechanical relaxation process. Recent experiments on $CeSn_3$ show an additional inelastic contibution (at 40 meV) at low temperatures[9]. This reduces the quasielastic line width to a residual width of only 11 meV, a factor of two smaller than the high temperature value. This raises the question of crystal field transitions in mixed valent compounds which seem to be observable when $\Gamma/2<\Delta$ (as also seen in $YbCu_2Si_2$[5]). More experiments, however, are necessary to reveal the details of the low temperature magnetic response of the mixed valent Ce compounds.

At the end of this article the magnetic reponse of two intermetallic U compounds (UAl_2[10], USn_3[11]) , which do not order magnetically, will be discussed briefly. The energy spectra again can be fitted by a simple relaxational model which seems to be valid in the whole experimentally accessible region up to 200 meV (UAl_2 , using high incident neutron energies).

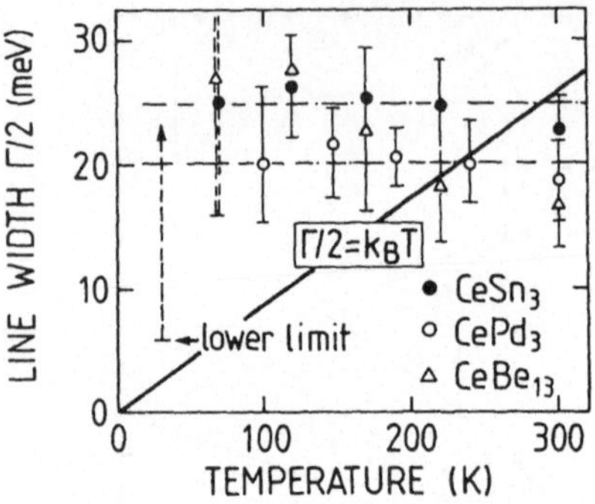

Fig. 4. Temperature dependence of the quasielastic line width single line fit) of some mixed valent Ce compounds. Experimental conditions restrict the determination of the line width to $T \geq 100$ K.

This supports the idea that the cut-off energy, as discussed earlier, is of the order of eV (band width of the conduction electrons as used in Ref. 5 to deduce the 4f occupation in Ce mixed valent compounds. The residual line widths of UAl_2 and USn_3 are 25 meV and 4 meV, respectively, in strikingly good agreement with the Curie-Weiss temperatures (-310K and -58K) as measured by the high temperature susceptibility.

In conclusion, we find that the instability of a magnetic moment shows up as a finite residual line width of the quasielastic part of the magnetic response function. The inelastic structure seems to be conserved if $\Gamma/2 \ll \Delta$ or wiped out if $\Gamma/2 > \Delta$. Kondo systems with $T_K \simeq$ 5K to 10K show a rather strong temperature dependence of the quasielastic line width; in mixed valent Ce systems with $T_{SF} \simeq$ 100K to 300K the temperature dependence is considerably weaker.

REFERENCES

1. M. Loewenhaupt and E. Holland-Moritz, J. Appl. Phys. 50:7456 (1979) and J. Magn. Magn. Mater. 14:227 (1979).

2. S. Horn, F. Steglich, M. Loewenhaupt and E. Holland-Moritz, Physica 107B:103 (1981) and M. Loewenhaupt, S. Horn and F. Steglich, Solid State Commun. 39:295 (1981).

3. S. Horn, F. Steglich, M. Loewenhaupt, H. Scheuer, W. Felsch and K. Winzer, Z. Physik B 42:125 (1981) and M. Loewenhaupt and J. M. Carpenter, Bulletin of Amer. Phys. Soc.28:286 (1983).

4. S. Horn, E. Holland-Moritz, M. Loewenhaupt, F. Steglich, H.Scheuer, A. Benoit and J. Flouquet, Phys. Rev. B 23:3171 (1981).

5. E. Holland-Moritz, D. Wohlleben and M. Loewenhaupt, Phys. Rev. B 25 :7482 (1982).

6. K. W. Becker, P. Fulde and J. Keller, Z. Physik B 28:9 (1977) and article by P. Fulde in this book.

7. A. P. Murani, K. Knorr, K. H. J. Buschow, A. Benoit and J. Flouquet, Solid State Commun. 36:523 (1980).

8. E. Holland-Moritz and M. Prager, J. Magn. Magn. Mater. 31-34: 395 (1983).
9. A. P. Murani, preprint (1983)

10. M. Loewenhaupt, S. Horn, F. Steglich, E. Holland-Moritz and G. H. Lander, J. de Physique 40:C4-142 (1979) and M. Loewenhaupt and C. K. Loong, Bulletin Amer. Phys. Soc. 28:341 (1983).

11. Preliminary results from a combined experiment on a single crystal and a polycrystal performed on triple-axis (Chalk River) and time-of-flight (IPNS) spectrometers by T. M. Holden, W. J. L. Buyers, C. K. Loong and M. Loewenhaupt.

INHOMOGENEOUS MIXED VALENCY: PARAMAGNETIC DIMERS

C.M. Hurd

National Research Council of Canada

Ottawa, Ontario, KIA OR9, Canada.

INTRODUCTION

We shall review the evidence for what is apparently a new magnetic species: ordered magnetic dimers intrinsic to a crystalline lattice. The dimers have been detected in a class of inhomogeneous, mixed-valence compounds known as the Magnéli titanium oxides.

INHOMOGENEOUS MIXED VALENCY

'Mixed valency' means that a given element is present in more than one level of oxidation. It is a widespread phenomenon that can involve most of the d- or f-block elements[1] and so frequently concerns magnetism. The level of oxidation of an ion is deduced from the measured bond lengths. Consequently, the question of experimental time scales arises. Consider, for example, the coexistence of Fe^{2+} and Fe^{3+} ions in an oxide. Oscillation between these states occurs by electron transfer typically more than 10^8 times /sec. If the 'observation time' of the crystallographic probe is much less than 10^{-8} sec we would distinguish Fe^{2+} and Fe^{3+} ions in the compound. This is the so-called 'inhomogeneous mixed-valence' case. But if the observation time is much longer than 10^{-8} sec we would then see a smeared result and would assign the same nonintegral valence to all the cations. This is the 'homogeneous mixed-valence' case. It is typical of compounds containing rare earth elements, such as Ce. However, the inhomogeneous case is our present interest.

These are families of inhomogeneous mixed-valence oxides with composition M_nO_{2n-1}, where M is typically Ti or V, and $4 \leq n \leq 9$. We shall concentrate on the Ti-based series. This series, which is essentially one of oxygen-deficient rutile (TiO_2), has homologous structures consisting of parallel chains of edge-sharing TiO_6 octahedra arranged as in rutile. The main difference from rutile is that the chains are interrupted every n octahedra along the chain direction by a shear plane lying in what would be the rutile (121) plane.[2,3]

The compounds are generally semiconductors, and they have attacted attention because — for some at least — one or more transitions occur below room temperature.[4] These transitions are commonly ascribed to 'charge ordering', as is the canonical Verwey transition[5] in magnetite (Fe_3O_4), but it is important to emphasize that presently there is no unequivocal explanation of their microscopic origins.

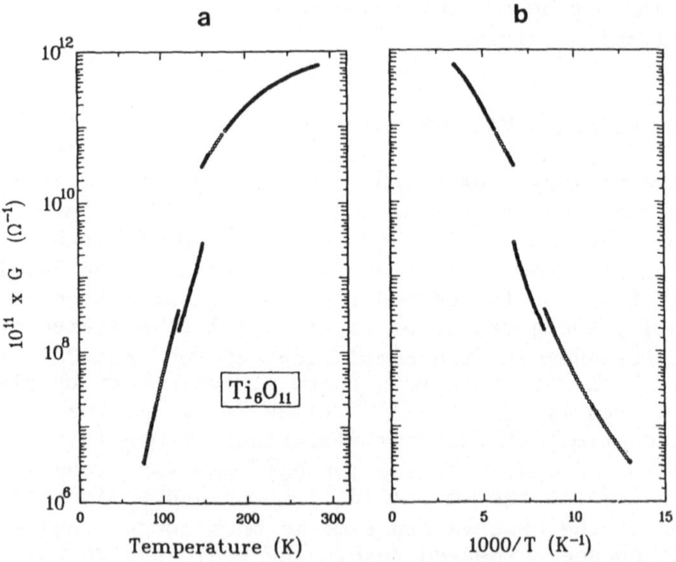

Fig. 1. Showing the temperature dependence of the electrical conductance of Ti_6O_{11}. (From Ref. 4.)

CHARGE-ORDERING TRANSITIONS

Consider Ti_6O_{11}, since it has been most thoroughly studied. It was originally thought to show one phase transition below room temperature[6], but recently two have been identified at about 119K and 147K from electrical[4], magnetic , X-ray[3] and EPR[8] studies. Figure 1 shows the electrical conductance (G) versus temperature[4]. The data are plotted in two ways for comparison: conventionally as in the Arrhenius plot of logG vs T^{-1} and alternatively as logG vs T. The linearity of the latter below the transition of 119K may be interpreted in terms of electron tunneling between $Ti^{3+} \longrightarrow Ti^{4+}$ cations with the Bethe overlap of these wave functions being temperature dependent due to the ions' thermal oscillations.[4,9]

Two different superstructures have been identified[3], one in the range 147K - 119K and the other below 119K. The Ti-O bond lengths change on passing though either transition, and by relating these lengths to effective ionic charges it has been shown that ordering of the valence charge occurs in each of the superstructures below 147K. Above 147K there is no valence ordering on the crystallographic time scale.

Essentially three types of cation are found in the ordered phases:

(1) Ti^{4+} ions,

(2) $Ti^{3.5+}$ paired with $Ti^{3.5+}$ (i.e. Ti_2^{7+}),

(3) Ti^{3+} paired with Ti^{3+} (i.e. Ti_2^{6+}).

(The original papers[7,8] give tabulated values of the numbers of ions and dimers in each phase).

EPR IN Ti_6O_{11}

The first measurements of EPR in single crystals of Ti_6O_{11} were made recently.[8] A single, intense EPR line was obtained below 147K; none was detectable above about 150K. From the spectra, it is clear that the number of spins decreased considerably on cooling through 119K. The g-tensor was determined to be orthorhombic in both phases with the minimum principal value g(min) aligned along the rutile c-axis.

Previous analysis[10] of EPR in polycrystalline Ti_6O_{11} attributed the signal to an unpaired electron located on an oxygen vacancy, but it was impossible to determine the principle directions of the g^2 tensor to confirm this assignment. The results from the single crystal samples, however, suggest strongly that the unpaired electron is associated with more than one Ti ion.

Given this feature, we turn to the crystallographic data. The symmetry of the g-tensors allows us to dispense[8] with the $Ti^{3+} - Ti^{3+}$ pairs [(3) above] as possible paramagnetic centres. Hence the EPR spectrum most probably arises from the Ti_2^{7+} dimers [(2) above], which lie parallel to the rutile c-axis in accord with the symmetry of the EPR site. The spin count shows that the sites exibit an appropriate concentration change at 119K, in accord with the magnetic[7] and crystallographic studies[3]. According to the X-ray results, the Ti_2^{7+} dimers are located at the corners of the rutile unit cell, as shown in Fig. 2, consistent with the EPR requirement that g(min) be parallel to the rutile c-axis. Note that the Ti_2^{7+} dimers occupy the rutile-like regions between the shear planes. In the lower temperature phase there is one Ti_2^{7+} dimer per four rutile chains; in the intermediate temperature range there are four such dimers per six chains.

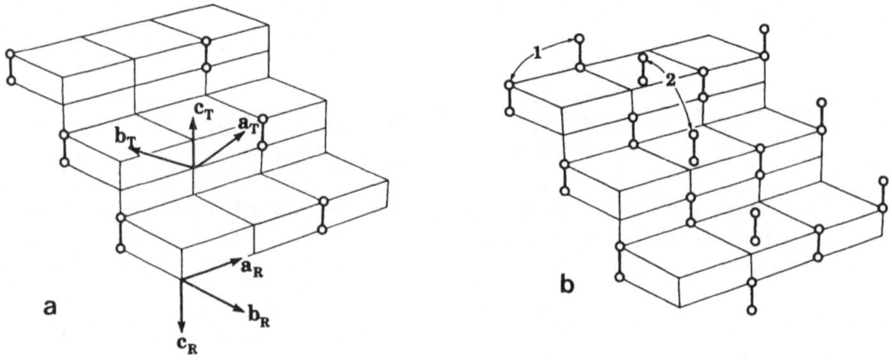

Fig. 2. Showing the location of Ti_2^{7+} dimers in Ti_6O_{11}. (From Ref. 8.) (a) 90K, (b) 130K. R relates to the pseudo-rule axes and T to the triclinic set.

From the structure shown in Fig. 2, it can be deduced that the number of Ti_2^{7+} dimers will increase by a factor 8/3 on warming through the transition at 119K. Good agreement with this expectation is obtained from a quantitative analysis of the EPR results.[8]

Two different Ti_2^{7+} dimers exist in the structure (refer to Fig. 2): (1) at the corners of the rutile unit cell, and (2) at its body centre. It has not been possible, however, to distinguish experimentally between these types. They are possibly indistinguishable on the EPR time scale.

EPR IN OTHER MAGNELI Ti OXIDES

More recently, a powerful EPR spectrum has been observed[11] between 90K and 150K in single crystals of Ti_nO_{2n-1} for $5 < n < 8$. Again, this spectrum is apparently due to the same dimer Ti_2^{7+}, which also occurs intrinsically in these compounds. This conclusion derives from the similarity of the principal g-factors, and from the fact that in each compound the g(min) lies along the rutile c-axis.

In Ti_4O_7 and Ti_9O_{17}, however, the EPR signal is very much weaker and implies that a paramagnetic defect is responsible. This is tentatively identified[11] as Ti^{3+} in Ti_9O_7, but a different (and unresolved) defect is responsible for EPR in Ti_4O_7.

BIPOLARONS AND PARAMAGNETIC DIMERS

Earlier structural studies [12,13] of Ti_4O_7 were interpreted in terms of bonded Ti^{3+} - Ti^{3+} pairs in the low temperature phases. These pairs (called bipolarons) were believed to be 'frozen' in the low temperature phase (below about 130K) but to be active, mobile electron pairs in the intermediate phase until they break up at about 150K. The idea of such 'bipolaron conductors' was made attractive because of the analogy between the bipolaron and the Cooper pair[14], and it has been applied to several systems particularly by the Grenoble group.[15,16] Is there a connection between this bipolaron picture and that of paramagnetic dimers?

It seems there is not. Conceptually, the bipolaron is associated with a mobile, bonded pair where, by electron transfer and lattice relaxation, the entity appears to move through the solid. The dimer, on the other hand, is a fixed, intrinsic part of the structure located on ordered sites.

The future of the bipolaron hypothesis is in any case uncertain because recent work[17] on the superstructure of Ti_4O_7 at 140K, to which mobile polarons were previously ascribed, shows that the structure is radically different from that previously reported. These new results are incompatible with the hypothesis of independently mobile bipolarons.

CONCLUSIONS

There is much indirect evidence to support the view that paramagnetism below about 147K in the series Ti_nO_{2n-1} for $5 < n < 8$ comes from Ti_2^{7+} oxygen-bridged dimers present as intrinsic components that are ordered in the crystal structure. Such a magnetic complex, which is an intrinsic part of the structure rather than a randomly-placed defect, is apparently a new magnetic species.

Acknowledgements. I am very grateful to J.R. Morton, A.D. Inglis, Y. LePage and S.P McAlister for helpful advice and discussion of their unpublished results.

REFERENCES

1. P. Day, Sci. Prog. Oxf. 68, 83 (1982).
2. L.A. Bursill and B.G. Hyde, Prog. Solid State Chem. 7, 177 (1972).
3. Y. LePage and P. Strobel, J. Solid State Chem. 43, 314 (1982); 44, 273 (1982); 47,6 (1983).
4. A.D. Inglis, Y. LePage, P. Strobel and C.M. Hurd, J. Phys. C. Solid State Phys. 16, 371 (1983), and references cited there.
5. N.F. Mott, Philos. Mag. 42, 327 (1980).
6. See, for example, R.F. Bartholomew and D.R. Frankel, Phys. Rev. 187, 828 (1969).
7. S.P. McAlister & A.D. Inglis, S.S. Commun. 47, 931 (1983).
8. S.A. Fairhurst, A.D. Inglis, Y. LePage, J.R. Morton and K.F. Preston, Chem. Phys. Lett. 95, 444 (1983).
9. W.R. McKinnon and C.M. Hurd, J. Phys. Chem. 87, 1283 (1983).
10. J.F. Houlihan, W.J. Danley and L.N. Mulay, J. Solid State Chem. 12, 265 (1975).
11. S.A. Fairhurst, A.D. Inglis, Y. LePage, J.R. Morton and K.F. Preston, J. Mag. Res. (in press).
12. J.L. Hodeau and M. Marezio, J. Solid State Chem. 23, 253 (1979); 29, 47 (1979).
13. M. Marezio, D.B. McWhan P.D. Dernier and J.P. Remeika, J. Solid State Chem. 6, 213 (1973).
14. B.K. Chakraverty, J. Physique 42, 1351 (1981); Nature 287,393 (1980).
15. B.K. Chakraverty, Phil. Mag. B42, 473 (1980).
16. B.K. Chakraverty and C. Schlenker, J. Physique C4, Suppl. 10, 37, C4-353 (1976).
17. Y. LePage and M. Marezio, J. Solid State Chem. (in press).

MOMENT FORMATION IN $TmSe_{1-x}Te_x$: FROM IV SEMICONDUCTORS TO IV METALS

Heinz Boppart and Peter Wachter

Laboratorium für Festkörperphysik, ETH Zürich

8093 Zürich, Switzerland

INTRODUCTION

The TmSe-TmTe system is unique among the rare earth monochalcogenides as it shows a compositionally induced semiconductor to metal transition (SMT). TmTe on the one end of the system is an ionic semiconductor with an energy gap of about 0.3 eV and the Tm ions are in a divalent state. TmSe on the other end is an intermediate valent metal with a Tm valence of about 2.8. For $x > 0.4$ the compounds are semiconducting and for $x < 0.2$ metallic. Both phases are separated by a miscibility gap in the composition range $0.2 < x < 0.4$ [1].

To start from semiconducting TmTe one observes a reduction of the energy gap with increasing incorporation of Se. This gap is determined by the separation in energy of the $4f^{13}$ and $4f^{12}$ 5d configurations. The substitution of Te by Se reduces the energy gap from 300 meV in TmTe to about 40 meV in $TmSe_{0.60}Te_{0.40}$.

For the semiconducting compositions ($x > 0.4$) a SMT can also be induced by external pressure. This SMT is driven by an increase of the ligand field splitting of the 5d conduction band as the anion-cation distance decreases. Consequently, the $4f^{13}$-$4f^{12}5d$ energy gap is reduced and finally the 4f electrons flow into the conduction band.

The reduction of the energy gap either by chemical composition or by external pressure leads to an increasing f-d mixing long before the energy gap is reduced to zero and the transition to the metallic state takes place. This mixing or hybridization in the semiconductors leads to the same effects as known from intermediate valent metals such as "gold" SmS, TmSe or SmB_6. The ionic radii of the rare earth ions are intermediate between the divalent and the trivalent radii, the compressibilities are very large, negative elastic constants c_{12} are observed and a phonon mode typical for a breathing deformation occurs[2,3,4].

EXPERIMENTAL

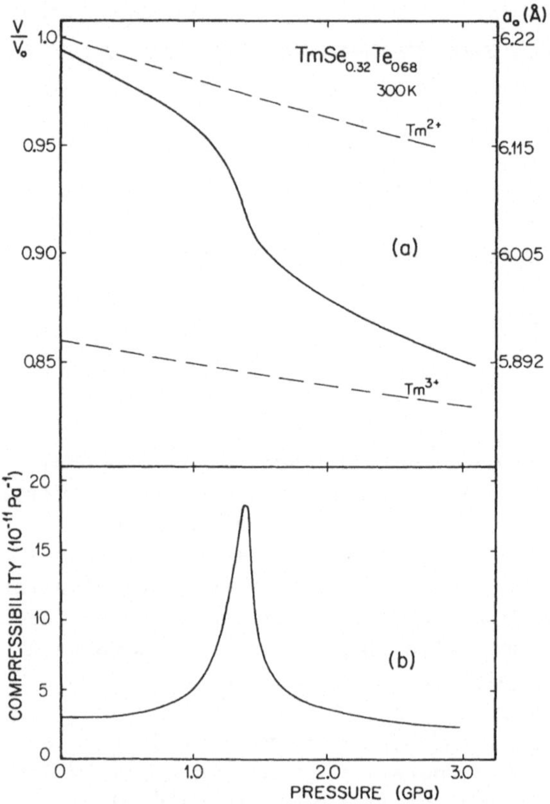

Fig.1: Pressure dependence of a) the specific volume and b) the compressibility of $TmSe_{0.32}Te_{0.68}$.

In Fig. 1a we show for example the volume change with pressure of a sample of $TmSe_{0.32}Te_{0.68}$ as measured with a strain gauge technique [2]. Also shown are theoretical curves indicating the expected change of the volume when the valency remains integer di- or trivalent. Already at zero pressure the volume according to the lattice constant is less than the one for a divalent compound and the compressibility, shown in Fig. 2b is nearly as large as for intermediate valent TmSe. At 1.4 GPa the volume curve is practically at a valence of 2.5 and the compressibility has a maximum.

The electrical resistivity versus pressure curve shows an exponential decrease of the resistivity with pressure, indicating a linear closing of the energy gap with pressure. When the gap is closed the resistivity becomes practically pressure independent, the SMT is achieved, which for this sample is observed at 1.4 GPa. Thus the maximum of the compressibility and the inflexion point of the volume change agree with the SM transition pressure and they are observed at a valence of about 2.5. This is a general phenomenon and it holds for all investigated semiconducting compositions. It is then important to realize that while still in the semiconducting range the volume or the lattice constant has decreased appreciably compared to an integer valent state and the compressibility has reached very large values.

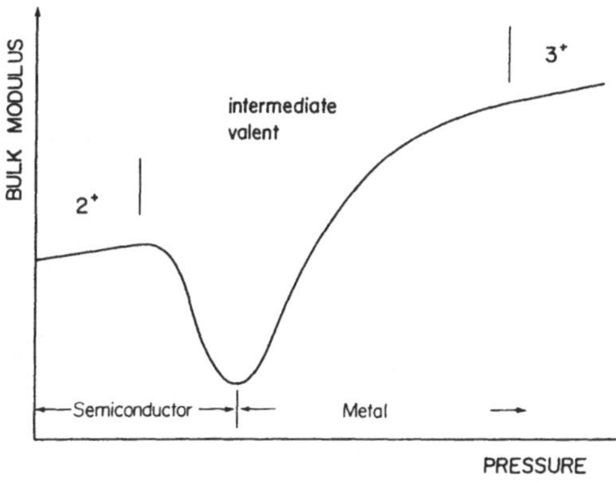

Fig.2: Behavior of the bulk modulus in a SMT.

In Fig. 2 we give an overview of the behavior of the bulk modulus, which is the inverse of the compressibility. The bulk modulus exhibits a strong softening at the SMT and for a critical composition near $x = 0.5$ the bulk modulus is even zero. The figure indicates that, as judged from the volume, lattice constant, compressibility or bulk modulus, due to the strong proximity of f- and d states and their mutual mixing something like an intermediate valent semiconductor may exist.

It then comes as no surprise that other properties which are connected with the lattice also show anomalies for these hybridized semiconductors. Ultrasonic measurements in various crystallographic directions permit the derivation of the elastic constants c_{11}, c_{12} and c_{44}. For $TmSe_{0.32}Te_{0.68}$ these measurements have been performed even under pressure up to the SM transition at 1.4 GPa[3]. The interesting feature in this connection is that c_{12} is positive at ambient pressure, but becomes negative at about 0.5 GPa long before the SMT transition, i.e. in the semiconducting range. An explanation for a negative c_{12} can only be given through a breathing deformability of the rare earth ion due to a valence instability [4]. c_{12} has a minimum at the SM transition at 1.4 GPa.

The breathing deformability mentioned above leads also to a renormalization of the phonon dispersion inasmuch as the LO phonons soften in the [111] direction at the L point and the LA phonons show a flattening near the middle of the zone also in the [111] direction. The consequence is a new peak in the weighted phonon density of states at about 60 cm^{-1} in the TA branch. This phenomenon has been observed in intermediate valent TmSe and is now found also

Table I.

Semiconducting $TmSe_{1-x}Te_x$ $(x>0.40)$

x	0.40	0.55	0.68	1.0
lattice constant (Å)	6.02	6.14	6.20	6.36
energy gap (meV)	40	120	180	300
p_{eff} (μ_B)	4.56	4.51	4.61	4.73
Θ (K)	-2	-2	-2	-1.5
T_N (K)			0.20	0.235

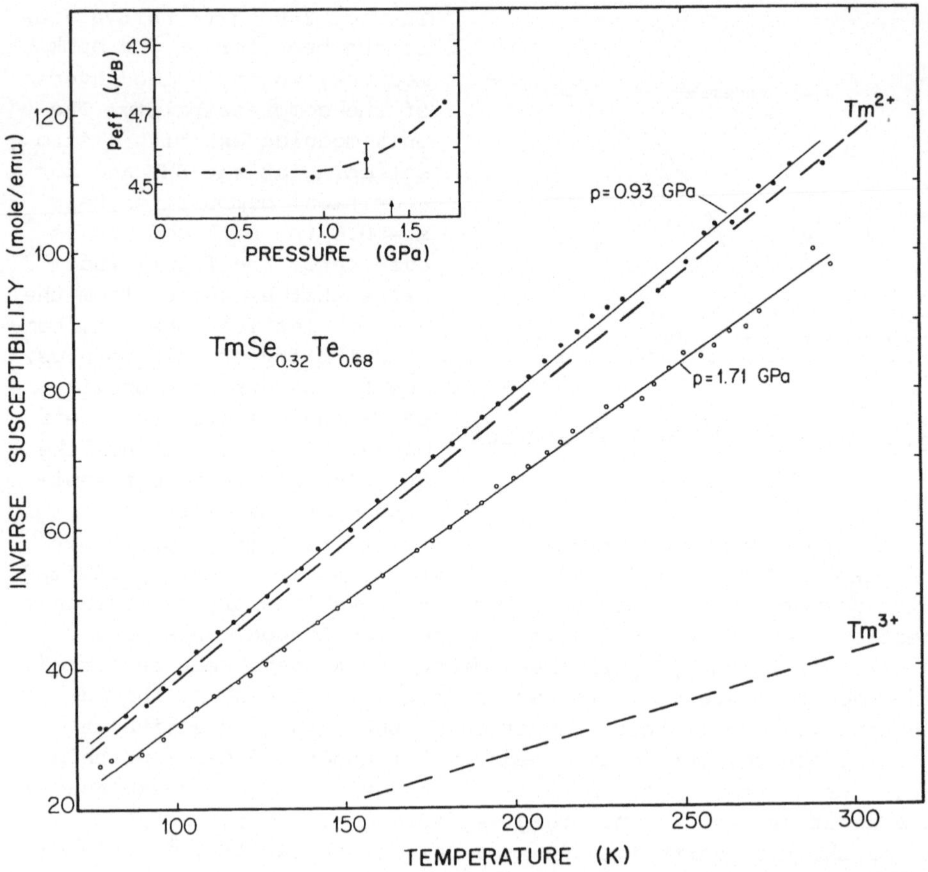

Fig.3: Inverse magnetic susceptibility as a function of temperature for 0.93 and 1.71 GPa. in the insert p_{eff} is shown as a function of pressure. The arrow at 1.4 GPa indicates the SMT.

for the semiconducting compounds with strong f-d mixing, i.e. for $0.4<x<0.68$, but not for TmTe, where the f-d separation in energy of about 300 meV is too large[5]. We thus realize that every physical property connected with the lattice of the strongly mixed semiconductors is highly anomalous and points to a strongly hybridized state.

We then have measured the magnetic properties of these semiconductors. The data obtained at ambient pressure and at temperatures up to 500 K are collected in Table I. From the effective moments (p_{eff}) the divalent character of Tm in $TmSe_{1-x}Te_x$ has to be inferred. These values are very close to the free ion value of 4.54 μ_B for Tm^{2+} except the one of TmTe. Attributing the deviation from pure divalency

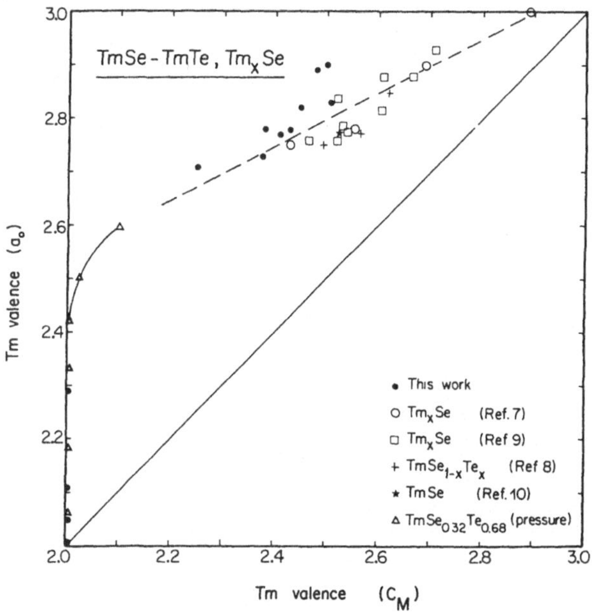

Fig.4: Tm valencies determined by the lattice - or the molar Curie constant.

in TmTe to a fraction of trivalent Tm we compute for the lattice constant of 6.36 Å a Tm^{3+} contamination of 6%, leading to a theoretical value of $p_{eff} = 4.72\ \mu_B$, which is in excellent agreement with the experimental value of $4.73\ \mu_B$.

The divalency of the semiconducting $TmSe_{1-x}Te_x$ compounds at ambient pressure comes as a surprise since we have shown above that all lattice connected properties show strong f-d mixing effects. Also as $TmSe_{0.60}Te_{0.40}$ has at normal pressure a formal valence of 2.35, deviations of p_{eff} from divalency should be easily detectable.

To increase the f-d mixing we then have performed susceptibility measurements under pressure. In Fig. 3 the inverse susceptibility as function of temperature is shown for the example of $TmSe_{0.32}Te_{0.68}$ for two representative pressures, 0.93 GPa and 1.71 GPa. The dashed lines represent the Curie behavior for the free ion values 4.54 and 7.56 μ_B for Tm^{2+} and Tm^{3+}, respectively. We recall that the transition pressure for the SMT for this compound is 1.4 GPa. We see at once that the points for 0.93 GPa agree well with the free ion line for Tm^{2+}, whereas the points for 1.71 GPa clearly lie between the Tm^{2+} and Tm^{3+} line. In the insert of Fig. 3 p_{eff} as calculated from the slope of the corresponding Curie-Weiss lines is shown for five different pressures. From this it is obvious that the Tm valence remains divalent up to the transition pressure of 1.4 GPa. Qualitatively the same behavior was observed in a similar investigation on TmTe under pressure [6]. We conclude from our measurements on different compositions of semiconducting $TmSe_{1-x}Te_x$ that it is a general behavior that p_{eff} stays divalent up to the SM transition pressure and becomes intermediate valent only at pressures in the metallic range.

We then realize a clear discrepancy in determining the valence from lattice properties (for simplicity we use the lattice constant) or from the molar Curie constant C_M. This incompatibility does not only exist for strongly f-d mixed semiconductors, but also for the metallic compositions Tm_xSe. In fact there is a systematic difference between both methods.

In Fig. 4 the valency as determined by the lattice constant on the one hand and by the molar Curie constants on the other hand is shown for different compounds from the series $TmSe_{1-x}Te_x$ and Tm_xSe, including pressure variations. The full line would correspond to valence values being the same for both methods. The full circles in Fig. 4 represent the valence values of Tm in the $TmSe_{1-x}Te_x$ system. The values indicated by triangles represent the results of the pressure variation in $TmSe_{0.32}Te_{0.68}$. All the other points result from investigations by various authors [7,8,9,10] and, in good agreement with each other lie near the broken line, which is thought to be a guide line for the eye. From our experiments we have clear evidence that both valence determinations cannot agree.

DISCUSSION

The most striking evidence taken from Fig. 4 is the fact that up to a valence of about 2.5, as judged by the lattice properties, the effective magnetic moment remains divalent and only in the metallic state, which is achieved for valencies above 2.5, the effective moment is intermediate. In other words, as long as the $4f^{13}$ electron is still localized we measure divalency with the magnetic moment. However, the lattice properties suggest that the $4f^{13}$ character is no longer retained and at least part of the charge is outside the original 4f shell, in fact even outside the 5s-and 5p shells, thus reducing the electron screening of the core charge on the 5s-and 5p shells, which in turn results in a reduced diameter of the ion. We must conclude that with strong f-d mixing the $4f^{13}$ charge distribution gets renormalized in such a way as to reduce the charge inside the 5s-5p shell and to increase the charge outside the 5s-5p shell. The quantum numbers L,S and J, however, must remain the same until the 4f electron is really delocalized (SM transition), from when on in the high temperature limit, which is probed by a p_{eff} measurement, with thermally driven valence fluctuations and for T larger than the mixing energy, one observes a mixture between a $4f^{13}$ and a $4f^{12}$ configuration and thus a molar Curie constant intermediate between the Tm^{2+} and the Tm^{3+} values.

Up to now we have compared only measurements which transmit to the 4f system either practically no excitation energy or one definitely less than the hybridization energy. It will be very interesting to extend our measurements to high energy experiments measuring only excited states such as UPS, XPS or L edge absorption experiments.

At the moment it is an open but interesting question whether such a high energy experiment will yield divalency or intermediate valence for the strongly f-d mixed semiconductors, for which lattice properties indicate intermediate valence, but p_{eff} indicates divalency. These measurements are under way. In any case it has become evident that the lattice properties are much more sensitive to the f-d mixing effects than the magnetic moment.

ACKNOWLEDGEMENT

The authors are very grateful to Dr.E. Kaldis for the growth and chemical characterization of the single crystals and to K. Mattenberger and S. Kiener for magnetic measurements.

REFERENCES

1. E. Kaldis, B. Fritzler, H. Spychiger and E. Jilek, in: Valence Instabilities, P. Wachter and H. Boppart eds., North-Holland, Amsterdam, (1982) p.131.
2. H. Boppart, P. Wachter, B. Batlogg and R.G. Maines, Solid State Commun. 38, 75 (1981).
3. H. Boppart, W. Rehwald, E. Kaldis and P. Wachter, in Ref. 1, p.81.
4. H. Bilz, G. Güntherodt, W. Kleppmann and W. Kress, Phys. Rev. Letters 43, 1998 (1979).
5. I. Mörke and P. Wachter, Solid State Communications 48, 441 (1983).
6. D. Wohlleben, J.G. Huber and M.B. Maple, AIP Conf. Proc. 5, 1478 (1972).
7. B. Batlogg, H.R. Ott, E. Kaldis, W. Thöni and P. Wachter, Phys. Rev. B 19, 247 (1979).
8. U. Köbler, K. Fischer, K. Bickmann and H. Lustfeld, J. Magn. Material 24, 34 (1981).
9. F. Holtzberg, T. Penney and R. Tournier, J. de Physique 40-C5, 314 (1979).
10. P. Wachter, in Ref. 1, p.145.

OPTICAL PROPERTIES OF CERIUM AND URANIUM COMPOUNDS

J. Schoenes

Laboratorium für Festkörperphysik, ETH Zürich
CH 8093 Zürich, Switzerland

ABSTRACT

New or very recent results of the optical properties of CeS, CeN, US, UAs and $U_xY_{1-x}Sb$ are presented. A correlation is established between the oscillator strength of f→d and d→f transitions and the delocalization of the f states. For CeN good agreement is obtained with a band structure calculation. In the pseudobinary system $U_xY_{1-x}Sb$ a valence transition near x = 0.15 is found.

INTRODUCTION

The intensive study over the last decade of f-state materials with unstable valence has emphasized a great conflict between many spectroscopic results and available theories to interpret these experiments.[1] Today's band structure calculations describe the ground state; however, spectroscopies measure by definition excited states. The knowledge of excited states is of undisputed importance for the understanding of the electronic structure and the problem can not be solved by discarding spectroscopies. Instead spectroscopic methods should be well selected for the problem in question and the results have to be critically analyzed.

In this paper we present optical data for a few cerium and uranium compounds with the intention of showing some of the typical features which point either to integral valent, intermediate valent or band-like f electrons. One of the advantages of optical spectroscopy compared, for example, with some photoelectron spectroscopy techniques is related to the fact that the input energy in optics is generally close to the binding energy. Thus optical spectroscopy may in many respects be classified with the so-called "soft" spectroscopies.

CeS and US

For the sake of brevity we will focus on the optical study of electronic interband transitions. Typical excitation energies are 0.1 to 10 eV, and the most general method for the study of the optical properties are reflectivity measurements.[2] Fig. 1 displays

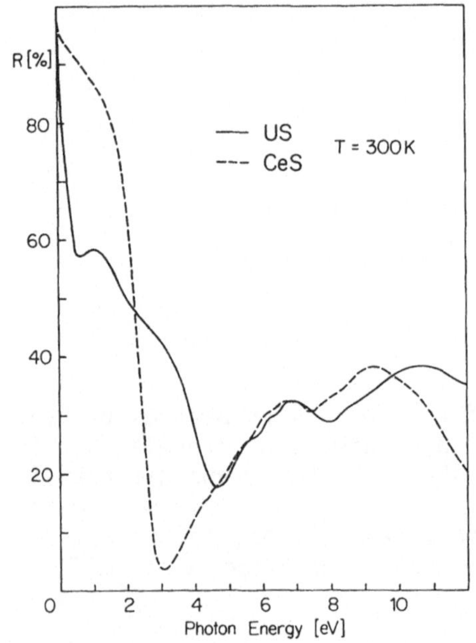

Fig. 1. Optical reflectivity of CeS and US single crystals.

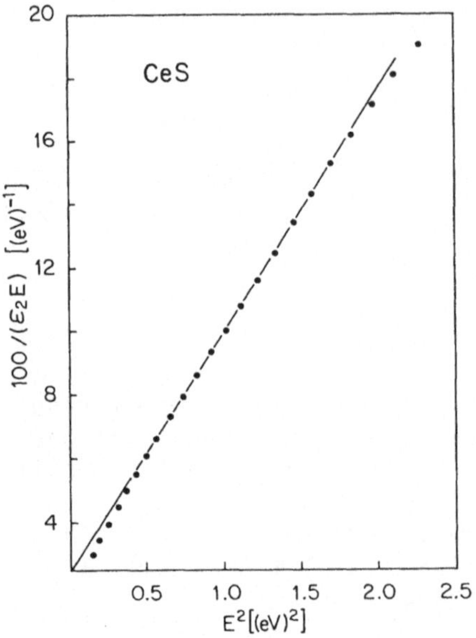

Fig. 2. Drude fit of the optical conductivity of CeS at low energies.

the near-normal-incidence reflectivity of CeS[3] and US as obtained on single crystals cleaved in high or ultrahigh vacuum.[4] The strong increase of the reflectivity at low energies indicates a metallic conductivity for both compounds. Yet, the energy dependence of the reflectivity from $\hbar\omega_p \approx 0$ up to the so-called plasma minimum located at 3 eV in CeS and at 4.6 eV in US follows a free electron behavior only in CeS and not in US. Fig. 2 shows for CeS a Drude fit of the absorptive part of the dielectric function. The fitting parameters are an energy independent damping $\hbar\gamma$ = 0.56 eV and the unscreened plasma energy $\hbar\omega_p = (4\pi Ne^2/m^*)^{1/2}$ = 4.83 eV. From the latter value one can either derive the carrier concentration N or the optical effective mass m*. Assuming 1 conduction electron per formula unit, as one expects for trivalent cerium, we compute m* = 1.25m in good agreement with m* = 1.3m derived previously for LaS and GdS.[5] In this context we note that CeS and LaS show very similar reflectivity spectra, except for the absolute values at high energies which were too low in LaS because of reflectivity losses caused by elastic light scattering on polished samples.

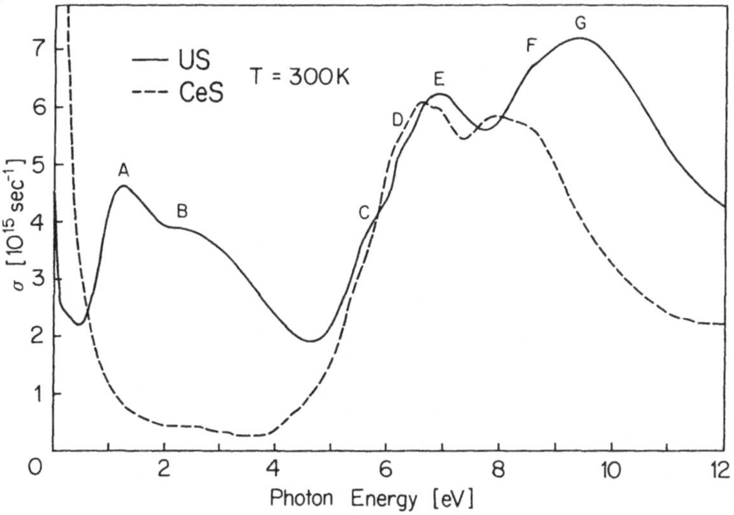

Fig. 3. Optical conductivity of CeS and US.

From the reflectivity spectra the optical conductivity $\sigma_1 = \varepsilon_2 \cdot \omega/(4\pi)$ has been derived via a Kramers-Kronig transformation (Fig.3). The CeS spectrum is quite simple and easily interpreted. The dominant structures occur between 5 and 10 eV and they contribute nearly six electrons per formula unit to

$$n_{eff} = m/(2\pi e^2 N_M) \int_{\hbar\omega=5}^{\hbar\omega=12} \omega\varepsilon_2(\omega)d\omega .$$

where N_M is the number of molecules per cm^3 in the material. Six electrons are expected from sum rule considerations for p(S)→5d, 6s(Ce) transitions and we assign the two peaks near 6.5 and 8 eV to p-valence to d-conduction band transitions, the splitting being presumably an indication of the crystal field splitting of the d band. Between the above mentioned free electron absorption below 2 eV and the p→d transitions we find very weak structures near 2.5 and 4.5 eV which are indicative of transitions originating from localized f states.

The results of this short discussion may be summarized in an energy level scheme as shown in Fig. 4. Because spectroscopies measure energy differences, transitions involving localized and

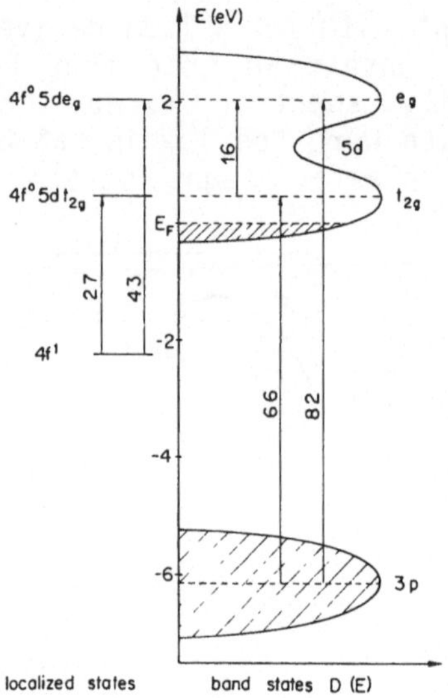

Fig. 4.
Energy level scheme for CeS.

correlated states, for which the energy is occupation dependent are shown separated from band to band transitions. To give a rough idea of the relative energies, we have placed the final state energies of the $4f^1 \rightarrow 4f^0 5d^1 (t_{2g}, e_g)$ transitions at the energies of the maxima of the densities of states of the $5dt_{2g}$ and $5de_g$ subbands. With the p and d bandwidths estimated from the width of the p→d transitions and assuming a filling of the conduction d band of 1/10 of its total width, one derives an apparent binding energy of the f state of 2.1 eV which is in reasonable agreement with an apparent binding energy of 2.4 eV found by photoemission techniques.[6,7] However, it should be emphasized again that these estimates neglect possible relaxation and correlation effects. As long as these effects are not quantitatively understood, we feel it is more important to note the extreme weakness of the f→d transition in CeS and the very small perturbation of the free electron conductivity by f-d interactions. Both observations point to a localized 4f state which is well separated from the conduction electrons at E_F.

The situation is very different in US. While above 5eV the p→d transitions give similar structures, except for a larger crystal field splitting of the 6d states in US than the 5d states in CeS, below 5eV interband and intraband transitions show marked differences. In US we observe two strong interband transitions denoted by A and B in Fig. 3 and only weak intraband transitions. A fit of the low energy part of the spectrum with Lorentzians gives the values $\hbar\omega_A$ = 1.15eV, $\hbar\gamma_A$ = 1.1eV, f_A = 0.85; $\hbar\omega_B$ = 2.75eV, $\hbar\gamma_B$ = 2.8 eV and f_B = 1.56.[8] From these oscillator strengths and the fact that peak B is strongly depressed in the uranium monopnictides, where formally no d electrons exist, one assigns peak A and B to f→d and d→f transitions, respectively.[23] If we compare the oscillator strengths f_A for the f→d transition in US with that in CeS, we find an increase of f_A by one order of magnitude. The square of the f-d dipole transition strength $R^2_{fd} = |\int R_f(r)R_d(r)r^3dr|^2$ computed with atomic wave functions[9] is less than a factor two larger in US than in CeS and the f occupation may at maximum be three times larger in US than in CeS. One concludes then that the f delocalization goes along with a substantial increase of the oscillator strength and, in fact, a computation of the f-d radial overlap integral with LMTO wave functions[10] gives for US a nearly three times larger value than the calculation using atomic wave functions. The situation appears to be even more dramatic for d→f transitions. To the author's knowledge this transition has not been observed optically in any rare earth compound with localized f states. On the contrary in US it gives rise to the large peak B in Fig. 3. Going from US to UTe the oscillator strength f_B decreases from 1.56 to 0.7 to 0.3,[11] indicating the increasing localization of the final state of this transition with increasing lattice constant, without reaching a degree of localization comparable to that in CeS for which no d→f transition is discernable.

Besides the different oscillator strengths for f→d and d→f transitions in localized (CeS) and itinerant (US) f systems, we also observe in the latter material a strong f-d interaction in the free electron behavior. Thus the "free electron" contributions to the optical conductivity are much weaker in US than in CeS and the damping is larger. This latter fact manifests itself in the shallower plasma minimum of US compared to the deep minimum in CeS (Fig. 1). A fit of optical and magnetooptical spectra gives for the free electron damping in US at room temperature $\hbar\gamma = 3.5$ eV,[8] which is six times the value derived in CeS.

CeN, UAs and USb

CeN belongs to the classical intermediate valent compounds.[1] Its optical reflectivity has been studied previously on polished samples and the intraband part of the optical conductivity has been analyzed in detail to derive the number of d conduction electrons n_d, which was then correlated to the valence of cerium assuming $n_d + n_f = 1$.[12] For the interband part of the spectra only a tentative interpretation was given which appears to be very unlikely. We present new reflectivity measurements in the fundamental absorption region for CeN single crystals cleaved under ultrahigh vacuum conditions and measured in situ. Fig. 5 displays these new data together with the reflectivity spectra of UAs and USb.[4] As expected, the reflectivity values above ≈ 5 eV are substantially higher for cleaved than for polished CeN samples. Yet we should add that the CeN samples did not cleave as well as UAs and USb and the true reflectivity above 5 eV of CeN may still be higher than shown in Fig. 5. Disregarding this difference in the

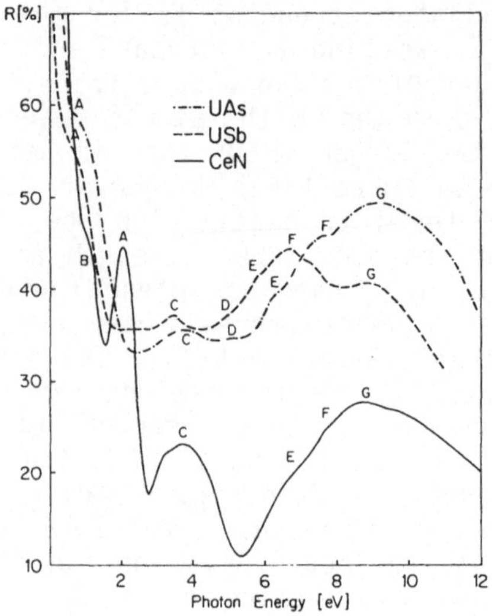

Fig. 5. Optical reflectivity of CeN, UAs and USb single crystals.

magnitude of the p→d transitions, one recognizes large similarities between the cerium and the uranium pnictides. In particular: i) nearly a one-to-one correspondence of interband transitions as visualized by the letters A to G, ii) no well defined plasma minimum, iii) small intraband contributions.

To discuss CeN in more detail. Fig. 6 displays the optical conductivity obtained from a Kramers-Kronig transformation of the reflectivity. The empirical assignments of the various structures rely mainly on the similarities with the uranium monopnictides and the observed changes of the reflectivity spectrum between 0.5 and 4.5 eV on exposure for a few minutes of the samples to normal atmosphere. In the latter case, the structure B disappears, the peak C disappears (or shifts to above 4.5eV) and peak A increases somewhat in intensity as well as in energy (0.1-0.2eV). Peak A is assigned to an $f^1 \to d$ transition, peak B to a $d \to f^1$ transition and peak C to a (pd)→f transition. Obviously, by the short oxygen exposure Ce becomes integer trivalent, so the f→d transition increases in intensity and in energy, while transitions into f states just above E_F are not possible any more due to the absence of empty $4f^1$ states.

The above assignments are also corroborated by a comparison with band structure calculations. Fig. 7 displays the band dispersion as obtained in a self-consistent APW calculation by Pickett and Klein.[13] From this calculation strong f→d transitions are expected to occur between the bands having the symmetries Γ'_2 and Γ'_{25} at the zone center and extending along the ΓX direction. The

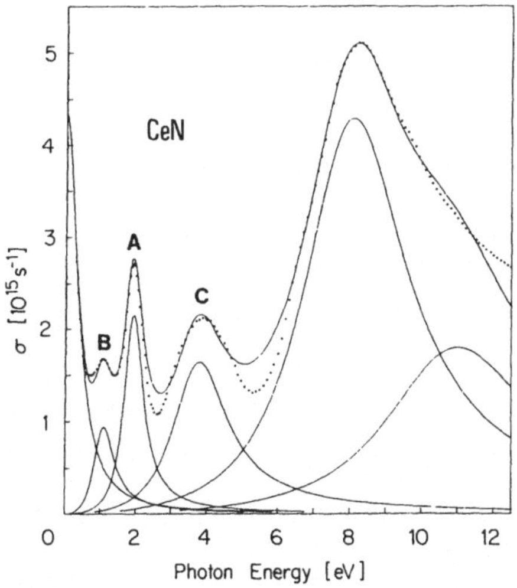

Fig. 6. Optical conductivity of CeN. Experimental data:•••••. The six lower full lines are single Lorentzians. The upper full line, which fits the experimental results, is the sum of these six Lorentzians.

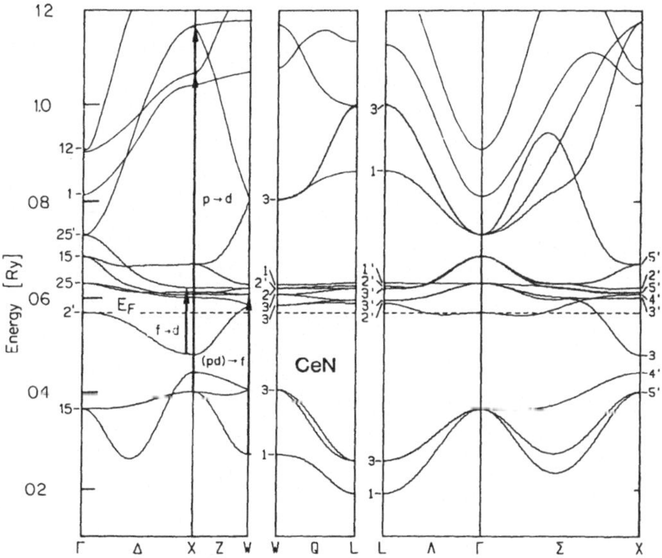

Fig. 7. APW energy bands of CeN (Pickett and Klein[13]).

energy separation derived from Fig. 7 is ≈ 2eV in perfect agreement with the experiment. The d→f transitions can not be deduced directly from the band dispersion. However, the partial density of states[13] shows strong d character just below E_F and a peak in the density of empty f states 0.8 eV above E_F. Transitions from the bonding (pd) states into empty f states are expected between 3.5 and 4.5 eV and the major p→d transitions are predicted to occur between 7.5 and 10.7 eV. Thus, an astonishingly good agreement between the one-particle calculation and the optical results exists. This might appear very surprising for a 4f material. However, because the 4f state in CeN is occupied by less than one electron and it is pinned at E_F, the one-particle description provides satisfying results.

Fig. 6 shows also a fit of the experimental conductivity spectrum with six Lorentzians. Usually such fits are not unambiguous in solids and therefore not very meaningful. Yet, the CeN spectrum in Fig. 6 displays four well separated and pronounced maxima, allowing a definite decomposition of the spectrum. The first oscillator with resonance frequency zero represents the free electrons, the oscillators 2,3 and 4 correspond to the interband transitions B,A and C respectively, and the oscillators 5 and 6 approximate the large and broad p→d transition peak extending from 6 to at least 12 eV. The fit parameters are listed in table 1. Comparing the oscillator strength of the f→d transition in CeN with that in CeS we note an increase of the oscillator strength in CeN by a factor of 3.5 despite a smaller f occupation. Similarly to the case of US, one has to consider overlap integrals as computed with wave functions from band structure calculations in order to account for the strong f→d transition in CeN.

Table 1. Energy, damping and oscillator strength of the six Lorentzians used to fit the optical conductivity of CeN. The errors are less than half the last digit quoted.

i	$\hbar\omega_i$ (eV)	$\hbar\gamma_i$ (eV)	f_i
1	0	0.4	0.33
2	1.1	0.7	0.125
3	1.93	0.66	0.27
4	3.8	1.9	0.59
5	8.1	3.5	2.85
6	11.0	5.0	1.7

With the decomposition of the optical conductivity into Lorentz oscillators one can try to estimate the valence of Ce in CeN from the intensity ratio of the f→d to the d→f transition. Because intercation optical transitions into an f^2 state are very unlikely, we use a localized model for the f state with an f^1 final state for the d→f transition and set $n_{fe} = 1 - n_{fo}$, but $n_{de} = 10 - n_{do}$ (e=empty, o=occupied). Assuming also $n_{fo} + n_{do} = 1$ and scaling the oscillator strengths with the product of occupied initial and empty final states for the two considered transitions, one obtains an f-occupation of 0.52±0.005. The corresponding valence of 3.48 for Ce compares favorably with values of 3.46 and 3.4 derived from the lattice constant[14] and a Drude fit to the intraband part of the optical conductivity[12], respectively.

In the uranium monopnictides the oscillator strength of the f→d transition (peak A) is similar to that in the uranium monochalcogenides, pointing therefore to a substantial delocalization of the f electrons. This is further corroborated by the absence of any fine structure in either the optical or the magneto-optical spectra which could be related to final state splittings typically observed in localized f systems such as EuS[15], TmSe[16] and UO_2[17]. The main difference between CeN and the uranium monopnictides is the absence of peak B in the optical spectra of the uranium monopnictides, manifesting the small d occupation in these latter materials.

$U_xY_{1-x}Sb$

USb and YSb form pseudobinary solid solutions with very interesting magnetic properties[18]. For x > 0.45 (U,Y)Sb orders antiferromagnetically. Between an uranium concentration of 45% and ≈20% the compounds show ferromagnetic order and below ≈20% no magnetic order occurs. Magnetization and elastic neutron scattering measurements also showed a strong decrease of the magnetic moment per uranium atom for x<0.5.[18] Fig. 8 displays optical conductivity spectra for $U_xY_{1-x}Sb$ single crystals with x= 0.08, 0.15, 0.3, 0.7 and 1.[19] Above 4eV the spectra are governed by p→d transitions for which we note interesting changes along the series reflecting the different crystal field splittings and band widths of the 4d and 6d derived conduction bands. Our main interest, however, focusses on the low energy part of the spectra in which f→d transitions occur. Going from x = 1 to x = 0.15 we observe a continuous shift of peak A to lower energies, leading for x = 0.15 to the disappearance of this peak. Because in USb we have identified peak A as an f→d transition with nearly three f electrons in the initial state, we conclude that on dilution of USb with YSb the $f^3 \to f^2 d$ transition energy vanishes and one f electron is promoted into the conduction d-band. The remaining f^2 state gives rise to $f^2 \to d$ transitions which we assign to the shoulder occurring near

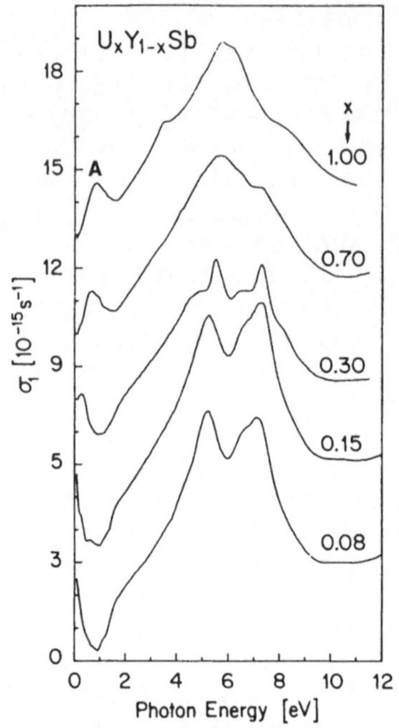

Fig. 8. Optical conductivity of $U_xY_{1-x}Sb$ single crystals for five different values of x.

1.6 eV in the uranium-poor solid solutions. Thus the optical data indicate a valence transition from nearly trivalent to tetravalent uranium between x = 0.3 and x = 0.15. The occurrence of a valence transition is corroborated by the concentration dependence of the lattice constant and the magnetic moment per formula unit.[19] Both quantities show an abrupt change of their slope near x = 0.15 and x = 0.18, respectively. From the magnetic data one derives a magnetic moment per uranium of approximately $0.5\mu_B$ for $x < 0.18$ and of $3.3\mu_B$ for $x > 0.18$. The latter value is close to the free ion value of the $f^3(^4I_{9/2})$ Russell-Saunders ground state, but the former value is too small to be explained within an f^3 configuration if one considers reasonable crystal field parameters. Instead, a value of $0.5\mu_B/U$ can be easily obtained if one assumes an f^2 state which is split by the octahedral crystal field in a nonmagnetic Γ_1 groundstate and a first excited state Γ_4 at a suitable energy.[20]

As a driving mechanism for the valence transition we propose the anisotropic p-f mixing which lowers the f-state energy and increases the valence band energy.[21] From a comparison of optical data for U_3P_4 and Th_3P_4 the p-f mixing has been derived to be 0.85 eV,[22] which is, within the accuracy of the measurements, of the necessary size to shift peak A centered between 0.85 and

0.9 eV in USb to zero energy when the p-f mixing has dropped to a negligible value in the very yttrium rich sample.

CONCLUSIONS

By comparison of the optical spectra for a few selected cerium and uranium compounds the delocalization of f electrons has been correlated to a substantial increase of the oscillator strength of the f→d and d→f transitions. For the first time an estimate of the valence has been performed using the relative intensities of f→d and d→f transitions in CeN. In the pseudobinary system $U_xY_{1-x}Sb$ optical data are presented giving evidence for the formation of a magnetic ground state for $x \geqslant 0.15$.

ACKNOWLEDGEMENTS

The author is indepted to P. Wachter for his support and for valuable discussions. The various crystals used in this work were grown by F. Hulliger, E. Kaldis and O. Vogt to whom the author expresses his gratitude. He is also grateful to his former or present collaborators, B. Frick, M. Küng, J. Neuenschwander and W.Reim for contributions to this investigation.

REFERENCES

1. See for example: P. Wachter and H. Boppart, eds. "Valence Instabilities", North-Holland, Amsterdam (1982).
2. J. Schoenes, Optical and Magneto-optical Properties, in: "Handbook of the Physics and the Chemistry of the Actinides", A.J. Freeman, G. Lander and C. Keller, eds., North-Holland, Amsterdam (1984).
3. M. Küng, J. Schoenes and F. Hulliger, Helv. Phys. Acta 53:578 (1980) and unpublished results.
4. J. Schoenes, Physics Report 66:187 (1980).
5. W. Beckenbaugh, J. Evers, G. Güntherodt, E. Kaldis and P. Wachter, J. Phys. Chem. Solids 36:239 (1975).
6. W. Gudat, M. Campagna, R. Rosei, J.H. Weaver, W. Eberhardt, F. Hulliger and E. Kaldis, J. Appl. Phys. 52:2123 (1981).
7. M. Croft, J. Weaver, A. Franciosi, D. Peterman and A. Jayaraman, in: "Valence Fluctuations in Solids", L.M. Falicov, W. Hanke, M.B. Maple, eds., North-Holland (1981) p. 401.
8. W. Reim, J. Schoenes and O. Vogt, Solid State Commun. 47:597 (1983).
9. F. Herman and S. Skillman, "Atomic Structure Calculation", Prentice Hall, New-York (1963).
10. M.S.S. Brooks, private communication.
11. W. Reim, J. Schoenes and O. Vogt, to be published.
12. A. Schlegel, E. Kaldis, P. Wachter and Ch. Zürcher, Phys. Letters 66A:125 (1978).

13. W.E. Pickett and B.M. Klein, J. Less-Com. Met. 93/94:.... (1983) and priv. commun.
14. Y. Baer and Ch. Zürcher, Phys. Rev. Letters 39:956 (1977).
15. J. Schoenes, Z. Physik B20: 345 (1975).
16. O.E. Hüsser, Diplomarbeit ETH Zürich (1983), unpublished.
17. W. Reim and J. Schoenes, Solid State Commun. 39:1101 (1981).
18. J. Rossat-Mignod, P. Burlet, S. Quézel, O. Vogt and H. Bartholin, in "Crystalline Electric Field Effects in f-Electron Magnetism", Guertin, Suski and Zolnierek, eds. Plenum Publ. Corp, New-York (1982) p. 501.
19. B. Frick, J. Schoenes, F. Hulliger and O. Vogt, Extended Abstract "13èmes Journées des Actinides", Elat, Israel (1983) p. B3, and to be published.
20. B.R. Cooper, O. Vogt and R. Siemann, Physica 102B:41 (1980).
21. K. Takegahara, H. Takahashi, A. Yanase and T. Kasuya, J.Phys.C 14:737 (1981).
22. J. Schoenes, M. Küng, R. Hauert and Z. Henkie, Solid State Commun. 47:23 (1983).
23. J. Schoenes, O. Vogt and J. Keller, Solid State Commun. 32:873 (1979).

REMARKS ON EFFECTS OF LARGE ORBITAL DEGENERACY AND ON TRANSPORT PROPERTIES OF MIXED VALENT SYSTEMS

T.V. Ramakrishnan

Department of Physics
Indian Institute of Science
Bangalore 560012, India*
and
Institute for Theoretical Physics
University of California
Santa Barbara, California 93106

I. INTRODUCTION

I discuss here the many ways in which large orbital degeneracy, N (of one of the two relevant ionic configurations) plays a role in the mixed valence and Kondo problems. Some unusual transport properties of mixed valent systems are then described, namely the large electrical resistivity of mixed valent metals, their anomalous temperature dependent Hall coefficient, and the giant negative magnetoresistance of SmS and TmSe. Explanations are proposed for these effects.

II. LARGE ORBITAL DEGENERACY (MIXED VALENCE REGIME)

We consider first the old question of the ground state of a single mixed valent impurity,[1] emphasizing the effect of large N.[2] The suggestion that large N is an essential simplifying feature of the mixed valence problem was made by Anderson[3] and a detailed theory with (1/N) as expansion parameter was worked out in Refs. 4 and 2. Recently, Zhang and Lee[5] have analyzed extensively the convergence of the (1/N) expansion. Coleman[6] has proposed a new Boson Fermion Hamiltonian specially suitable for perturbation theory.

The two ionic configurations, namely f^{n-1} with energy ε_0 and f^n with energy ε_f with respect to the Fermi level admix by hybridization with conduction electrons. The matrix element for this is

*Permanent address.

$(V_{km}/\sqrt{N})$ where k describes the conduction electron state and m is one of the N orbital states f^n. The basic energy scale in the problem is the hybridization energy or width $\Delta = N^{-1} \sum_m \sum_k |V_{km}|^2 \delta(\tilde{\varepsilon}_k)$. The matrix element is normalized by a factor $1/\sqrt{N}$ so that if a perturbation expansion in powers of V exists, the successive terms will be of higher order in (1/N). Now consider the special case $\varepsilon_0 = \tilde{\varepsilon}_f = 0$, so that the two configurations have the same energy. The state f^{n-1} (f^n) is lowered in energy by admixture with f^n (f^{n-1}) and a conduction hole (electron). For a band symmetric about the Fermi level, the intermediate states in the two cases have exactly the same energies; the only difference is that intermediate states f^n are N times as many as those with f^{n-1}, so that the nondegenerate singlet is stabilized N-fold in comparison to the multiplet. It is easy to see that to lowest order in admixture, the renormalized energies E_0 and $\tilde{E}_f$ are

$$E_0 = -\Delta \ln(D/\Delta); \quad \tilde{E}_f = -(\Delta/N) \ln(D/\Delta) \tag{1}$$

where 2D is the bandwidth and a flat density of states has been assumed. Since the renormalized singlet has the configuration f^n admixed in it, the ground state is mixed valent; the valence n_v (zero for pure f^{n-1} and unity for pure f^n) is

$$n_v = \Delta/\{\Delta + |E_0 - \tilde{E}_f|\} = (1 + \ln(D/\Delta))^{-1} \tag{2}$$

the last equality in Eq. (2) being true only when $\varepsilon_0 = \tilde{\varepsilon}_f = 0$. Thus the valence can be strongly nonintegral and much less than the value (N/N+1) (which is the limit for zero hybridization and iso-energetic configurations).

The main qualitative consequence of large N is the strong biasing of one set of states with respect to others. We have seen above that this greatly stabilizes the nondegenerate singlet. Some further consequences of large N are now discussed below.

(i) Increased stabilization energy for mixed valent singlet. The microscopic hybridization matrix element is $\langle m|H|k\rangle = V_{km}/\sqrt{N}$, so that the Δ used above is N times the conventional virtual level width value $\Gamma_m = \sum_k |\langle m|H|k\rangle|^2 \delta(\tilde{\varepsilon}_k)$. The size of the latter is not well known, but estimates for Ce are in the range .03 to .05eV. The singlet stabilization energy is N ln(D/.03N) times this value (for $\varepsilon_0 = \tilde{\varepsilon}_f = 0$), i.e. $\simeq 15 \times .03$eV for D = 2eV and $\Gamma = 0.03$eV. We thus see that large stabilization energies are possible with relatively small virtual level widths.

(ii) Perturbation theory. Very crudely, the expansion parameter in the theory is $(\Delta/N) \times$ (energy denominator). The latter is typically ($E_0 - \tilde{E}_f$) which has a value $\sim \Delta$ when $\tilde{\varepsilon}_f - \varepsilon_0 = 0$ and decreases smoothly as $\tilde{\varepsilon}_f = \varepsilon_0$ falls below the Fermi level. Thus, when the two configurations have nearly the same bare energy, i.e. in the mixed valence regime, the expansion parameter is $(\Delta/N)(1/\Delta) \sim (1/N)$. This is confirmed by calculating higher order terms in Brillouin Wigner perturbation theory.[2,4,5] For the case $N = 6$ and $D = 33\Delta$ the next order correction to the ground state energy is 10% for $(\varepsilon_0 - \tilde{\varepsilon}_f) = -3\Delta$ (errors for susceptibility are much worse).[5]

At this energy the impurity is nearly high valent, $n_v \simeq 0.8$. Thus one has a convergent perturbation theory for much of the mixed valent regime, covering $0.85 \gtrsim n_v > 0$.

(iii) Expansion of the mixed valence regime. The f level can be as much as $\Delta \ln(D/\Delta)$ below the Fermi energy, with the valence still nonintegral. Thus there is an N-fold expansion of the probability of mixed valence, because there are N channels for valence mixing. If a first order transition can occur (as in Ce) due to an appropriate balance between elastic energy cost and the hybridization or mixed valence gain, the level can lie well below the Fermi energy in the integral valent (γ - Ce like) phase and be unstable towards the formation of a mixed valent (α - Ce like) phase. This further expands the mixed valent regime. A simplified theory along these lines is described in Ref. 4; a detailed phenomenological theory is given by Wohlleben.[7] This situation is to be contrasted for example with early theories of the γ-α transition in which $\tilde{\varepsilon}_f \simeq -.05$eV in the γ phase (the 'experimental' values are in the range 1 to 2eV).

(iv) Separation of single-site and intersite energy scales. It can be shown[4] that for the mixed valent lattice, the intersite energy is (1/N) times the one-site singlet stabilization energy which is of order Δ. Thus even in concentrated systems a dense impurity model is accurate for energetics. For quantities like susceptibility it is accurate provided thermal excitations [at $T \gtrsim (\Delta/N)$] or alloying disorder break up lattice coherence. At lower temperatures, in pure systems, the system is a homogeneous Fermi liquid with a small characteristic energy scale[3] of order (Δ/N). This could be the reason why in many mixed valent systems the susceptibility has a plateau and then rises at lower temperatures.

(v) Local Fermi liquid model. The ground state of a single mixed valent impurity is nonmagnetic, with a finite $T = 0$ susceptibility χ (with a positive T^2 curvature) and a specific heat c_v linear in temperature. The Wilson ratio $W = (\pi^2/3)(T/c_v)(3\chi/\mu^2_{eff}) = \{1 + \Delta/N|E_0 - \tilde{E}_f|\}$ i.e. it has the noninteracting local Fermi gas value 1 with small corrections of order (1/N).[2,5] Thus the local Fermi liquid has weak self-interaction, and the ground state is essentially a local potential, scattering conduction electrons. An attractive form is a resonant f level,[2,8] a model developed by Newns and Hewson[9] on the basis of results for the N orbital Kondo impurity problem, and used successfully to describe susceptibility of intermediate valence systems. Since many-body wave function renormalization effects are small in the mixed valence regime, the resonant level model can be used literally for the impurity and for the lattice; in the Kondo regime these renormalization effects are large, and affect different quantities(e.g. χ and density of states) very differently so that care is needed in using such a model.

(vi) Green's functions. These can be calculated with systematically improvable accuracy as a power series in (1/N), as shown for single particle spectral density by Gunnarsson and Schönhammer[10] and by Coleman[6] using a new Boson Fermion Hamiltonian. Kuramoto,[11] as well as Schlottmann[11] present results for response functions in these proceedings.

One of the important consequences of the large N is that simple, reliable calculations can be done for physical properties of mixed valent systems. Several examples have already been mentioned above. The magnetic susceptibility $\chi(T)$ is discussed in Ref. 2, where the dominant energetic constraints on valence are included in a simple way.

III. LARGE ORBITAL DEGENERACY (KONDO REGIME)

In the Kondo regime, i.e. for $(\tilde{\varepsilon}_f - \varepsilon_0)$ negative and much larger in magnitude than Δ, the Kondo temperature T_K or singlet energy increases greatly for large N, i.e. $T_K = D_{eff} \exp(-1/NJ\rho)$ where $J\rho = (\Gamma/|E_0 - \tilde{E}_f|) = (\Delta/N|E_0 - \tilde{E}_f|)$. The Kondo resonance is relatively broad. The RKKY intersite coupling is reduced by a factor of N.[12] The Wilson ratio $W = (N/(N-1))$ so that for large N, the Fermi liquid corrections are small.

Though the single Kondo impurity problem has been solved exactly and numerically using RG methods, a perturbative (1/N) expansion is of great interest for several reasons, and is the center of much current activity.[13,6,14] First, many systems, e.g., those based on Yb and Ce, are near the Kondo-mixed valent crossover, perhaps to keep the valence integral, reducing the elastic energy cost, but at the same time maximizing singlet stabilization energies. A theory which can simply and accurately describe both regimes is needed. Secondly, while most theories focus on and solve the unique infrared problem, many quantities such as valence, ground state energy, and charge susceptibility are not entirely infrared determined, and a direct "brute force" theory is of interest. Such a theory would also be useful for correlation functions and for the Kondo lattice.

The first question of interest is the expansion parameter,

which is naively $(\Delta/N) \times$ (energy denominator). The latter is of order $(1/T_K)$, so that one may be expanding in powers of $(1/N)(\Delta/T_K)$ which is not necessarily small. However, because of wave function renormalization, only a fraction (T_K/Δ) of states have energy denominator $(1/T_K)$, so that the expansion parameter could be $(1/N)$. Only an explicit higher order calculation of physical quantities such as χ and c_v in powers of $(1/N)$ can decide this question. In addition to Gaussian fluctuations,[13] quartic fluctuations, which are formally $O(1/N)$ and have these dangerous energy denominators, have to be considered.[14]

Another problem concerns the $N = \infty$ limit. In this limit, there is a phase transition for a single impurity at a nonzero temperature, so that the crossover from high temperature to the nonmagnetic state at low temperature is not smooth. Fluctuation effects, of higher order in $(1/N)$, are necessary for a smooth crossover. In a poor man's scaling picture, higher order effects, retarded coupling etc. become important when $\tilde{J}_{eff} \sim N$ so that assuming $N = \infty$ in the beginning leads to a spurious fixed point. Is a systematic, perturbative, large but finite N, crossover theory possible? Quite likely yes, but we don't have it yet.

IV. TRANSPORT ANOMALIES

(i) Resistivity. The temperature dependent resistivity of many mixed valent metals and alloys is very high; $\Delta\rho(T) = \rho(T) - \rho(0)$ can be 200 $\mu\Omega$cm or so,[15] a factor of ten or more higher than normal electron phonon resistivities. Now the size of the latter is limited since $\rho_{e-ph} \lesssim (\hbar\pi^2/e^2k_F)(\langle x^2\rangle/a^2)$ where the first factor is the Mott maximum metallic resistivity, ρ_{Mott}, and the second is the dimensionless mean squared ion displacement causing scattering. From say Lindemann's law, $\langle x^2\rangle/a^2 \lesssim .01$ and $\rho_{Mott} \sim 10^3$ $\mu\Omega$ cm so that $\rho^{max}_{e-ph} \sim 10$ $\mu\Omega$ cm. Correspondingly, the inverse lifetime is $(h/\tau_{e-ph}) \simeq \epsilon^0_F(\langle x^2\rangle/a^2) \simeq \hbar\omega_D$. The observed resistivities are thus far too high, and need explanation. Other unusual features are the following[15]: (a) saturation behavior at high resistivity, (b) extreme sensitivity to disorder at low resistivity and low temperature, (c) large departure from Matthiessen's rule, (d) T^2 dependence at low temperature, and (e) upturn at very low temperature in the highly resistive systems. Though $CePd_3$ is the best studied example, many Ce and Yb containing systems, as well as Np and nearby actinides show similar behavior.

We now consider a resonant level lattice model for transport properties.[8] The f channel phase shift, η_m, at each site is tied to the valence; crudely $(\eta_m/\pi) \simeq (n_v/N)$. Thus the scattering phase shift fluctuates with the local valence which can vary due to thermal or disorder induced strains. Connecting the valence change with volume change for the γ-α transition in Ce, for example, one finds $\delta\eta_m \simeq 2.5\, e_{\alpha\alpha}$ where $e_{\alpha\alpha}$ is the local bulk strain. Thus sizable local phase shift fluctuations are possible. It does not take much of these to randomly dephase scattering centers from each other since the bandwidth or intersite coherence energy is small, of order (Δ/N). Further, scattering from one site to another is off-diagonal in channel index and there are N channels at each site. The off-diagonal matrix elements are randomized by longitudinal as well as transverse strain fluctuations, and the disorder effect is multiplied N-fold.

The resistivity saturates when each site scatters randomly due to thermal or disorder caused strains. Their effect is thus highly nonlinear. In this limit, the resistivity depends only on valence, and has the value $(\pi^2\hbar/e^2k_F)(\pi^2n_v^2/N)$. This works out to about 200 μΩ cm for $CePd_3$, close to the T = 0 saturation limit. The low temperature T^2 behavior could be due to electron-phonon scattering because the normal $(v_S/v_F) \simeq (\theta_D/T_F^*)$ phase space suppression may be ineffective in these narrow band systems with $T_F^* \sim (\Delta/N)$.

(ii) Hall Anomaly. In many metallic glasses, in liquid Ce, and in $CePd_3$, the Hall effect has the wrong sign (i.e. it is positive).[16] In $CePd_3$, the Hall constant R_H is also very large ($\sim$ 10-20 times larger than in YPd_3) and strongly temperature dependent, becoming negative at low temperatures. In Yb compounds, the sign is not anomalous, but R_H depends on temperature.

A plausible explanation for these effects is spin orbit coupling, well known as giving rise to anomalous Hall effect. In most simple metals and alloys, this is a few percent correction but in the above systems, it can be very large. Crudely, the scattering rate $W_{kk'}$ for an electron from a state $\vec{k}$ to a state $\vec{k}'$, to the lowest order in spin-orbit coupling, λ, is $W_{kk'} \sim i\,\lambda\,\vec{k} \times \vec{k}' \cdot \vec{S}$. If there is a magnetic field B in the z direction, $\vec{S} = \chi\, B\, \hat{e}_z$. A net drift in the y direction due to an electric field means that $\delta n_{k'} \sim k'_y E_y$, so that the spin-orbit term leads to an additional Hall current $\langle\vec{k}\rangle$ in the x direction. One has, approximately,

$$(\delta R/R_0) \simeq (\tilde{\lambda})(\rho/\rho_{sat})(\chi/\chi_{Pauli}) \,. \qquad (3)$$

In rare earth mixed valent systems, spin orbit coupling is large, so that $\tilde{\lambda} \simeq (\lambda N/\Delta)$ is of order unity. Indeed, locally, scattering takes place in channels with definite m_J, i.e. one is in the strong spin-orbit coupling limit. The resistivity at high temperatures is close to the saturation limit, and the susceptibility can be much larger than the Pauli value, so that all factors help to make a large anomaly possible. The sign of λ in Yb is opposite that in Ce; this is the basic reason why Ce has positive R_H.

(iii) Giant negative magnetoresistance. We briefly mention now the small gap (semiconducting) mixed valent systems. One striking feature of TmSe and low pressure, semiconducting SmS ($p < 5$ k bar) is the gigantic negative magnetoresistance (Ref. 17). For example, in SmS, a field of 100 kG reduces ρ by more than a factor of ten, and the resistivity is metallic for $H \gtrsim 250$ kG. The authors explain this as being due to extrinsic effects, i.e. magnetic Sm^{3+} impurities in the gap. However, unrealistic parameter values are needed, e.g. a g factor of ~ 13. A simpler explanation is that the effect is intrinsic, namely that the magnetic field reduces the excitation gap and finally changes the ground state to a metallic one. If the gap E_g in semiconducting SmS is connected with the $(E_0 - \tilde{E}_f)$ difference discussed in Sec. II, it will decrease with field, i.e. $E_g(H) \sim E_g^0(1 - \frac{1}{2}\chi H^2)$ so that the resistivity $\rho(T) \sim \rho_0 \exp(E_g/T)$ has a Gaussian field dependence, close to what is observed. One could make a more serous theory if the ground state of SmS, SmB_6 etc. were better understood; as it is, the nature of mixed valence and small gap in SmS (6 k bar $< p <$ 20 k bar), its metallization above 20 k bar and its small magnetoresistance in the mixed valent regime, are also not understood.

These three example illustrate how the unusual transport properties of mixed valent systems raise questions regarding their nature.

ACKNOWLEDGMENTS

I would like to thank P.W. Anderson, P. Coleman, and H.R. Krishnamurthy for discussions. This material is based upon research supported in part by the National Science Foundation under Grant No. PHY77-27084, supplemented by funds from the National Aeronautics and Space Administration.

REFERENCES

1. C.M. Varma and Y. Yafet, Phys. Rev. B 13:2950 (1976); H.R. Krishnamurthy, J.W. Wilkins, and K.G. Wilson, Phys. Rev. B 21:1044 (1980); F.D.M. Haldane, Phys. Rev. Lett. 40:416 (1978); A. Bringer and H. Lustfeld, Z. Phys. B 22:213 (1977).
2. T.V. Ramakrishnan and K. Sur, Phys. Rev. B 26:1798 (1982).
3. P.W. Anderson, in "Valence Fluctuations in Solids," L.M. Falicov, ed., North Holland, Amsterdam (1981), p. 451.
4. T.V. Ramakrishnan, ibid., p. 13.
5. F.C. Zhang and T.K. Lee, Phys. Rev. B 28:33 (1983).
6. P. Coleman, to be published and these proceedings.
7. D. Wholleben, these proceedings.
8. T.V. Ramakrishnan, in "Valence Instabilities," P. Wachter and H. Boppart, eds., North Holland, Amsterdam (1982), p. 351.
9. D.M. Newns and A.C. Hewson, J. Phys. F 10:2429 (1980).
10. O. Gunnarsson, these proceedings.
11. Y. Kuramoto, ibid., and P. Schlottmann, ibid.

12. P. Coleman, to be published; N. Read, D.M. Newns, and S. Doniach, to be published.
13. N. Read and D.M. Newns, to be published.
14. H.R. Krishnamurthy and T.V. Ramakrishnan, to be published.
15. P. Scoboria, J.E. Crew, and T. Mihalisin, J. Appl. Phys. 50:1895 (1979); H. Schneider and D. Wohlleben, Z. Phys. B 44:193 (1981).
16. E. Cattaneo, U. Häfner, and D. Wohlleben, in Ref. 8, p. 451.
17. P. Haen and F. Lapierre, Ref. 3, p. 313 (TmSe); M. Konezykowski et al., Ref. 8, p. 447 (SmS).

LOCAL FERMI LIQUID THEORY OF INTERMEDIATE VALENCE SYSTEMS AND ITS ABINITIO DERIVATION

D.M. Newns* and N. Read

Institut Laue-Langevin
156X, 38042 Grenoble Cedex, France

and

A.C. Hewson

Department of Mathematics
Imperial College
London SW 7 2BZ, U.K.

I. INTRODUCTION

Intermediate Valence (IV) materials are rare earth (RE) materials in which electrons are *partially* free to hop in and out of the f shell, in contrast to their core-like behaviour in normal RE compounds. They include many normal metals (YbCuAl, $CeSn_3$) and a few normal insulators (SmB_6, gap 3 mev[1]), and we shall discuss only such nonmagnetic compounds here.

Some of the questions (all but the last for metallic IV's) we want to answer are :

- Why in a plot of $T = 0$ susceptibility χ_0 vs. linear coefficient of specific heat $\gamma = \partial C_V/\partial T|_{T=0}$ the points separate onto different straight lines for different RE ions (Fig. 1a), independent of structure or composition[2,3].
- χ_0 and γ go up to huge values for some IV's - why so large?
- Why χ_0 versus mean valence gives a very strong systematic dependence (Fig. 1b of Ref. 4, Ref. 5) despite variation in structure and composition.

*Permanent address: Dept. of Mathematics, Imperial College, London.

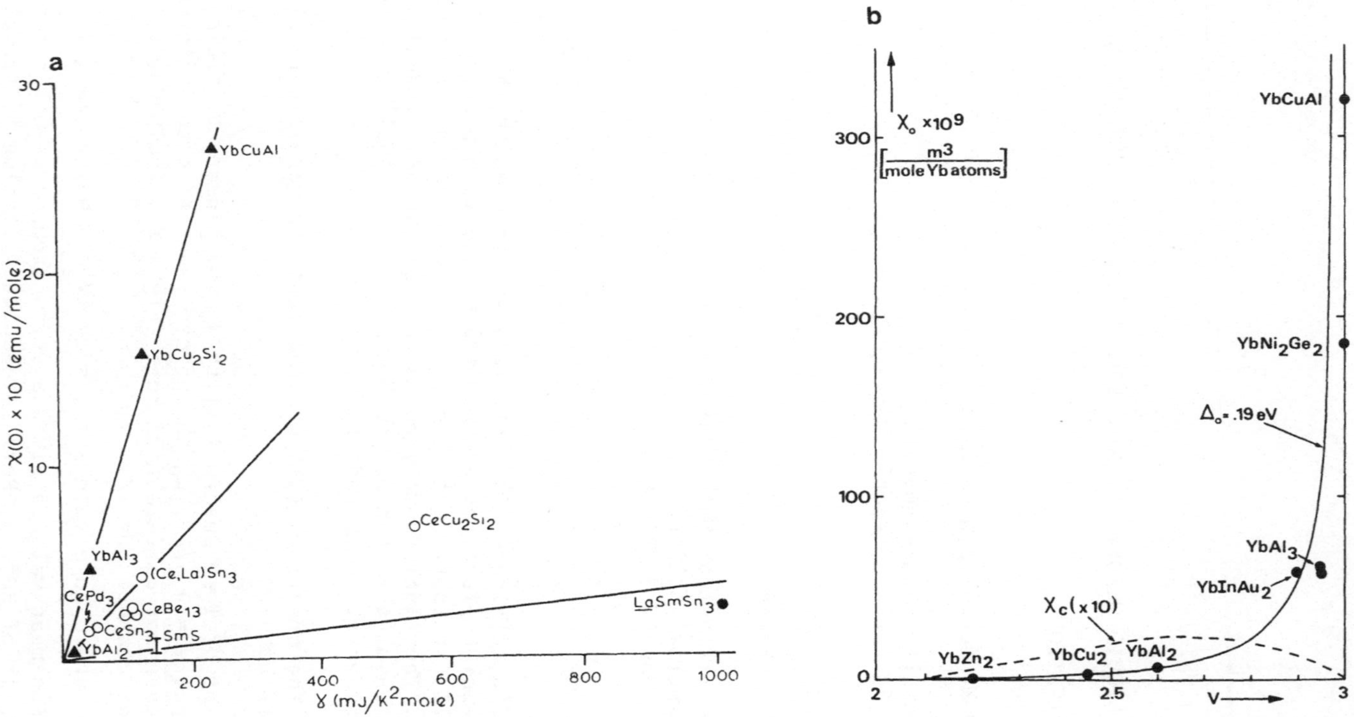

Fig. 1. (a) Plot of χ_o vs. γ for IV compounds. Straight lines are Eq. (33). (b) T = 0 magnetic susceptibility χ_o of Yb compounds plotted vs. valence after Ref. 4. Assuming $\Delta° = 0.19$ eV, full curve is χ_o (Eq. (35)), broken curve $\mu^2_{eff}\ \chi_c$ (Eq. (36)), increased by factor 10.

- Why $\chi(T)$ curves almost always have a maximum, i.e. at low T $\chi(T) = \chi_0 (1+\kappa T^2)$, $\kappa > 0$. Similarly the magnetization $M(H) = \chi_0 H (1+\mu H^2)$, and electronic specific heat $C_V(T) = \gamma T (1+B_e T^2)$ with B_e, $\mu > 0$, at least for YbCuAl[6].
- What is the weight of the "quasiparticle" peak in the spectral density near the Fermi level, as seen in Ultra-Violet Photoemission (UPS) spectroscopy?
- How can we arrive at an explanation of transport properties compatible with a resistivity $\rho(T) \alpha T^n$, $n \sim 2$, at low temperatures[7], and a thermopower that is positive for Ce, but negative for Yb - based IV's[8]?
- What is the origin of the gap in SmB_6, and near-gap in SmS?

II. THE MODEL-1 IMPURITY

Table I, which excludes Tm, shows the allowed valencies for the 4 main IV RE ions. For each, one valence is nonmagnetic, e.g. f° for Ce. For each, one valence is magnetic with total angular momentum j, e.g. f' has $j = 5/2$ for Ce. The Table also shows the degeneracy of the magnetic configuration,

$$N = 2j + 1, \tag{1}$$

which plays a big role in IV theory.

Table I

	Ce^{4+}	Ce^{3+}	Sm^{2+}	Sm^{3+}	Eu^{3+}	Eu^{2+}	Yb^{2+}	Yb^{3+}
j =	0	5/2	0	5/2	0	7/2	0	7/2
	e	6	h	6	e	8	h	8

Now we shall introduce, and stick to for a while, a model of one impurity at the origin in an electron gas. Hopping from the f-shell is only strong into a subspace of the electron gas states $|k\rangle$ given by $|\ell = 3mk\rangle$, i.e. states of angular momentum $\ell = 3$, magnetic quantum number m, and radical momentum k. Including spin, these states may be coupled to form states $|jmk\rangle$, where $j = \ell + 1/2 = 7/2$ or $j = \ell - 1/2 = 5/2$, and $m = -j, -j + 1, \ldots j$.

When the valence state of the ion changes, say from $|jm\rangle$ to $|0\rangle$ a particle is emitted into the electron gas which by conservation of angular momentum is in any available state $|jmk\rangle$. The nature of the particle, electron or hole, is listed in Table I. Let the matrix element for the process be V (its k-dependence is hereby ignored).

We write the Hamiltonian following Coleman[9]

$$H = \sum_{km} \varepsilon_k C^+_{km} C_{km} + E_o \sum_m f^+_m f_m + V \sum_{km} (C^+_{km} f_m b^+ + f^+_m C_{km} b) \quad (2)$$

Here C_{km} and f_m are Fermion operators for the states $|jmk\rangle$ of energy eigenvalue ε_k and for an f-orbital of magnetic quantum number m and energy eigenvalue E_o respectively. The sums over m run over the N values (1). The first term in H describes the energy of the electron gas, and the second that of the ion. The third describes the coupling between them.

So far we have not said anything about the mysterious boson creation and destruction operators b^+,b. They are a brilliant contrivance introduced by Coleman[9] to prevent the system going into forbidden multiple occupation states, such as f^2, f^3 for Ce. Suppose the system were in a state $|1mo\rangle$ with 1 f electron and no boson. Action of H on this state can generate a state $|o1\rangle$ with no f electron and one boson, but no higher f-occupations because the boson has no negative energy states.

Thus (2) will work provided

$$Q = n_f + b^+b = 1 \quad (3)$$

where

$$n_f = \sum_m f^+_m f_m \quad . \quad (4)$$

Q commutes with H, but (3) nevertheless provides a constraint on the problem since we always want to work with an open system at constant μ.

We conclude this section by noting two approximations we have made. The first is the neglect of crystal field splitting, not always negligible in IV systems (e.g. $YbCu_2Si_2$[10]). The second is the neglect of higher excited states of the two valencies of the ion. This assumption is good for Yb. But as we go down the series of decreasing excited state energy

Yb > Ce > Sm, Eu

it becomes more debatable. We think it is adequate for most cases at least for systems with large γ, but one should watch for its breakdown.

Finally, we note that (2) has been written for particles and one must make an electron-hole inversion to apply it to the hole cases (Table I).

III. FUNCTIONAL INTEGRALS

Actually, the Hamiltonian (2 + 3) is exactly soluble[11], for linear dispersion $\varepsilon_k \sim k$. The solution is not as far as we know extensible to the lattice, even in one dimension. Rather, we choose to proceed by formulating a mean field solution[12] to (2 + 3), together with a procedure for calculating corrections to it when necessary. The mean field solution is the Local Fermi Liquid theory[2,3]. This approach seems to be readily extensible to the lattice[13].

A key justification for such a mean field approach is the 1/N expansion. That is, corrections to it are believed to come in as powers of 1/N. This has been demonstrated explicitly in the Kondo limit by two of us[14]. Since N is fairly large (6 or 8), this should give adequate results for practical purposes, and can be systematically improved as long as the expansion holds.

We write the partition function as

$$Z = \int_{-\pi/\beta}^{\pi/\beta} \frac{\beta d\lambda}{2\pi} \, \mathrm{Tr} \exp\,[-\,\beta H - i\beta\lambda\,(n_f + b^+b - 1)] \tag{5}$$

We have here dealt with the constraint (3) by introducing an integral over a parameter λ, exactly the prescription used by Read and Newns (RN)[14] to put the constraint $n_f = 1$ into the SU(N) Kondo problem. This integral over λ produces a Kronecker delta inside the trace, so that nonzero contributions come only from states satisfying (3), as required.

We may rewrite (5) as a functional integral

$$Z = \int_{-\pi/\beta}^{\pi/\beta} \frac{\beta d\lambda}{2\pi} \, \mathcal{D}b\mathcal{D}b^+\mathcal{D}f\mathcal{D}f^+\mathcal{D}C\mathcal{D}C^+ \exp\,[-\int_0^\beta d\tau\, L(\tau)] \tag{6}$$

where

$$L = b^+ \frac{d}{d\tau} b + \sum_m f_m^+ \left(\frac{d}{d\tau} + E_o\right) f_m + \sum_{km} C_{km}^+ \left(\frac{d}{d\tau} + \varepsilon_k\right) C_{km}$$

$$+ V \sum_{km} (C_{km}^+ f_m b^+ + f_m^+ C_{km} b) + i\lambda(n_f + b^+b - 1). \tag{7}$$

Here $\mathcal{D}f$ is shorthand for $\pi_m \mathcal{D}f_m$, etc. The f_m's and C_{km}'s are anticommuting Grassman numbers[15].

We now make a gauge transformation, analogous to that in RN. Writing $b(\tau) = r(\tau) \exp(i\theta(\tau))$, whence

$$db\, db^+ = r\, dr\, d\theta\ , \tag{8}$$

we set

$$f'_m(\tau) = f_m(\tau)\, e^{-i\theta(\tau)} \tag{9}$$

$$C'_{km}(\tau) = C_{km}(\tau) \tag{10}$$

and

$$\lambda'(\tau) = \lambda + \dot{\theta}(\tau) \quad , \tag{11}$$

to obtain a new functional integral (dropping normalising factors and the primes)

$$Z = \int \mathcal{D}\lambda\, (\prod_\tau r(\tau))\, \mathcal{D}r \mathcal{D}f \mathcal{D}f^+ \mathcal{D}C \mathcal{D}C^+ \exp\, [-\int_0^\beta L'(\tau)d\tau] \tag{12}$$

where

$$L' = \sum_m f_m^+ (\frac{d}{d\tau} + E_0 + i\lambda)\, f_m + \sum_{km} C_{km}^+ (\frac{d}{d\tau} + \varepsilon_k)\, C_{km}$$

$$+ rV \sum_{km} (C_{km}^+ f_m + f_m^+ C_{km}) + i\lambda\, (r^2 - 1) \quad . \tag{13}$$

Now $\lambda(\tau)$ is τ-dependent and becomes a new field, replacing $\theta(\tau)$ which has disappeared. A term $\int_0^\beta r\,\dot{r} d\tau$ appeared, which integrates to zero because of the periodicity of b on $o < \tau < \beta$.

Eq. (12-13) are exceedingly similar to the formulation by RN of the SU(N) Kondo problem. It may be used to generate a Feynman diagram expansion with a linked cluster theorem. In this respect, Coleman's original approach to inserting the constraint (3), though it may be criticised[16], seems also to lead to a linked cluster theorem.

Comparing with RN, (12-13) has two new features. It has an explicit energy E_0 for the f-level. It also has a new three-boson field vertex, $i\lambda r^2$.

IV. STATIC APPROXIMATION - LOCAL FERMI LIQUID THEORY

Effective Hamiltonian

In the SU(N) Kondo problem, RN found that the static approximation to the FI was valid to leading order in 1/N, giving for example the correct ground state energy, susceptibility and specific heat for this exactly solvable model.

We shall assume that this is true also for (12-13). We thus replace $i\lambda(\tau)$ and $r(\tau)$ by static quantities $\varepsilon_f - E_0$ and r_0. At low temperatures, in which region the gauge (12-13) is in any case most useful, the Boltzmann factor will prevent excursions away from the stationary values. Therefore, at low-T our problem becomes that of determining ε_f and r_o so as to minimize the free energy of the single particle Hamiltonian

$$H_{eff} = \sum_{km} \varepsilon_k C^+_{km} C_{km} + \sum_m \varepsilon_f f^+_m f_m + r_o V \sum_{km} (C^+_{km} f_m + f^+_m C_{km}) + (\varepsilon_f - E_o)(r_o^2 - 1) \qquad (14)$$

The parameter ε_f is seen to become an effective "renormalized" f-level, while $r_o V$ is a "renormalized" hopping matrix element.

To see first what (14) means, let us obtain the f-propagator by means of the Dyson equation (Fig. 2). Since in the following we are using only one-electron theory, we may employ retarded

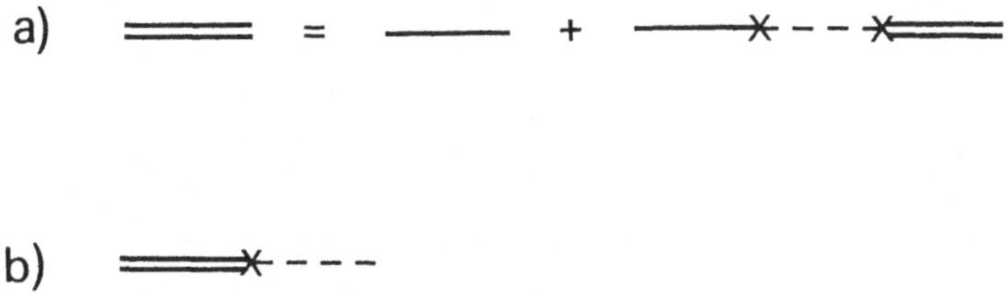

Fig. 2. (a) Dyson equation for G_f. Double line represents complete propagator; single line $(\varepsilon - \varepsilon_f + is)^{-1}$, broken line $\sum_k (\varepsilon - \varepsilon_k + is)^{-1}$; cross, the vertex $r_o V$. (b) Propagator G_{fc}.

progagators as is conventional in this field. Then Fig. 2a gives

$$G_f(\varepsilon) = \frac{1}{\varepsilon - \varepsilon_f + is} + \frac{1}{\varepsilon - \varepsilon_f + is} \sum_k \frac{r_o{}^2 V^2}{\varepsilon - \varepsilon_k + is} G_f(\varepsilon) \qquad (15)$$

Now in working with the non-interacting Anderson model (14), it is customary to put

$$\sum_k \frac{V^2}{\varepsilon - \varepsilon_k + is} = -i\pi V^2 \sum_k \delta(\varepsilon - \varepsilon_k) = -i\pi V^2 \rho \equiv -i\Delta^\circ \qquad (16)$$

The approximation involved is to assume Δ° to be energy-independent and to neglect the real part of (16). We shall also define

$$\Delta = r_o{}^2 \Delta^\circ \qquad (17)$$

Then (15) has the solution

$$G_f(\varepsilon) = (\varepsilon - \varepsilon_f + i\Delta)^{-1} \quad . \qquad (18)$$

We see that there is a scattering resonance on the RE site located at energy ε_f and of width Δ. Δ takes on the meaning of a renormalized resonance width, with r_o^2 the renormalization factor multiplying the bare width $\Delta°$.

Stationary Point

Now to determine ε_f and r_o, we use the Hellman-Feynman theorem to differentiate (14) w.r.t. ε_f and r_o, and put expectation values round the results :

$$\partial F/\partial \varepsilon_f = \langle n_f \rangle + r_o^2 - 1 = 0 \tag{19}$$

or using (17)

$$\Delta/\Delta_o = 1 - \langle n_f \rangle \tag{20}$$

At T = 0 we obtain $\langle n_f \rangle$ from

$$\langle n_f \rangle = -N\pi^{-1} \mathrm{Im} \int_{-D}^{o} G_f(\varepsilon) d\varepsilon \tag{21}$$

where we take μ as energy zero and define a lower conduction band cut-off D. Whence doing (21) using (18), (20) becomes

$$\langle n_f \rangle = \frac{N}{\pi} \tan^{-1} \frac{\Delta}{\varepsilon_f} = 1 - \frac{\Delta}{\Delta°} \quad . \tag{22}$$

The left hand equality is the Friedel Sum Rule of Newns and Hewson[2].

Similarly,

$$\partial F/\partial r_o = 2VN\langle C^+_{km} f_m \rangle + 2r_o(\varepsilon_f - E_o) = 0 \quad . \tag{23}$$

Evaluating $\langle C^+_{km} f_m \rangle$ from the propagator G_{fc} in Fig. 2b we get, using (16) and (18)

$$\langle C^+_{km} f_m \rangle = -\pi^{-1} \mathrm{Im} \int_{-D}^{o} (\varepsilon - \varepsilon_f + i\Delta)^{-1} \; (-i\pi r_o V\rho) \quad . \tag{24}$$

Doing the integral in (24) and inserting into (23)

$$\ln \frac{(\varepsilon_f^2 + \Delta^2)^{1/2}}{D} = -\frac{(\varepsilon_f - E_o)\pi}{N\,\Delta^o} \quad . \tag{25}$$

Eqs. (22) and (25) are the key results.

From (22), at large N $\Delta \simeq \pi N^{-1} \langle n_f \rangle \varepsilon_f$ is small w.r.t. ε_f, enabling (25) to be written approximately

$$\varepsilon_f \simeq E_o - \frac{N\,\Delta^o}{\pi} \ln \frac{\varepsilon_f}{D} \tag{26}$$

an equation which has been obtained by numerous authors from Brillouin-Wigner perturbation theory[17] or otherwise[18].

In the Kondo limit (25) yields the result of RN.

$$(\varepsilon_f^2 + \Delta^2)^{1/2} = T_K = D\exp[\pi E_0/N\Delta^0] \simeq \varepsilon_f \quad . \tag{27}$$

Density of States, Susceptibility, Specific Heat

From (18), the f density of states per channel is a Lorentzian

$$\rho_f(\varepsilon) = \pi^{-1}\,\Delta\,/\,[(\varepsilon - \varepsilon_f)^2 + \Delta^2] \quad . \tag{28}$$

When E_0 lies well above the Fermi level, $\langle n_f \rangle$ is small, and Δ is near Δ^0 from (20). Also from (26) ε_f is near E_0. We have a broad nearly empty resonance just above E_0 of width $\sim \Delta^0$. But when E_0 is large and negative, ε_f lies just above the Fermi level, being given by (27). Δ is small according to (20), since $\langle n_f \rangle$ is near to unity. We have a narrow resonance at ε_f close to the Fermi level, with $\varepsilon_f \simeq N\Delta/\pi$ from (22).

The magnetic susceptibility χ_0 is obtained by adding a Zeeman term $-g\mu_B Hm$ (g = g factor) to ε_f in each channel m, giving a channel magnetization $g^2\mu_B{}^2Hm^2\rho_f(o)$. Summing over channels using the identity

$$\sum_m m^2 = \frac{1}{3} N\, j\, (j + 1) \tag{29}$$

gives the susceptibility

$$\chi_0 = \frac{1}{3}\,\mu_{eff}^2\, N\rho_f(o) \tag{30}$$

where

$$\mu_{eff}^2 = g^2\mu_B{}^2\, j(j + 1) \quad . \tag{31}$$

The linear coefficient of specific heat is by standard Sommerfeld arguments

$$\gamma = \frac{1}{3}\,\pi^2\, k_B^2\, N\, \rho_f\,(o) \quad . \tag{32}$$

Comparison with Experiment

To sustain us, let us try comparing this impurity theory with experiments on real IV materials. First, the χ_0/γ ratio, from (30,32) obeys

$$R \equiv \pi^2\, k_B{}^2\, \chi_0/\mu_{eff}^2\,\gamma = 1 \quad . \tag{33}$$

We plot the straight lines (33) on Fig. (1a) : amazingly, they go through the points rather well. Actually, an exact theorem bounds R by [19]

$$1 < R < N/(N-1) \quad , \tag{34}$$

showing that (33) is in error by O(1/N), which is fine as we think we are doing a 1/N expansion.

Next, rewrite χ_0 in terms of $<n_f>$ to give

$$\chi_0 = \frac{N\mu_{eff}^2}{3} \sin^2 \left(\frac{\pi<n_f>}{N}\right) / (1 - <n_f>) \Delta^o \tag{35}$$

This is plotted on Fig. (1b) for $\Delta^o = 0.19$ eV. The fit is not at all bad, considering the same Δ^o is assumed for all materials. Eq. (35) has also been obtained from perturbation theory [3,18].

We have now partially answered, in terms of the impurity model, the first three questions in § 1. Eqs. (33, 35) answer the first and third. The answer to the second, for Ce and Yb, is the $(1 - <n_f>)$ denominator in (35). For Eu and Sm, it is mostly that the Clebsch-Gordon coefficients defined into V make Δ^o small to begin with.

We have also plotted on Fig. 1b the charge susceptibility

$$\chi_c \equiv - d <n_f>/dE_o \simeq \pi N^{-1} \Delta^{o-1} <n_f>^2 (1-<n_f>) \quad , \tag{36}$$

whose spectacular difference from χ_0 as $<n_f> \to 1$ is notweorthy. To calculate the derivative (36), one must take into account the change in stationary point to order dE_o.

Particle Propagator

Finally, the valence fluctuation f-propagator[9] $<Tf_m(\tau)b^+(\tau)b(o)f_m^+(o)>$ becomes in static approximation $|<b>|^2$ $<Tf_m(\tau)f_m^+(o)>$, or

$$r_o^2/(\varepsilon - \varepsilon_f + i\Delta \text{ sgn } \varepsilon) \quad ,$$

which (apart from being τ-ordered) differs from the quasiparticle propagator (18) by the factor $r_o^2 \cong Z = \Delta/\Delta_o$, which is indeed exactly correct[3,20]. It gives a density of states $Z\rho_f(\varepsilon)$, thus answering the fifth point in § 1.

Thermal and Magnetic Field Effects

In the mixed valent regime, $<n_f>$ varies with T and H. To calculate this, one at least needs to take account the following two effects[2]. Firstly, particle conservation effectively adds to the energy a term $\frac{1}{2} \rho^{-1} (<n_f> - \nu)^2$, which tends strongly to peg the f-occupation at a constant value. Secondly, screening, when the RE-RE separation exceeds twice the screening length, tends completely to counteract this effect by localizing the change in number of conduction electrons, equal to the change in $\langle n_f \rangle$, around the RE sites. We shall here stick to the Kondo limit where this unquantifiable net coupling drops out. One then calculates, taking into account the shift of the stationary point with H or T, the quantities defined in § 1 as collected in Table II.

Table II

	Large-N	N=8	Exact N=8	YbCuAl
κ	$\pi^2 S_3/6S_1$	3.97	∿5.4	>0
B_e	$9\pi^2 S_3/10S_1$	21.4	∿35	18
μ	$S_3 S_1^{-1}\,[n_4/3n_2 - n_2/2]$	1.11	1.9	1.9

The notations $n_r = \Sigma\, m^r/N$, $S_r = \sin(r\pi/N)$ are used and T_K is taken as unity. Our results are compared to what is known for the exact solution[21] and with data for YbCuAl[6]. Interestingly, *all* coefficients are predicted to be positive for N > 3, in agreement for κ, μ with exact results[21]. Notice that our large-N theory is only in fair agreement with exact N = 8 results for these T^2 and H^2 coefficients. We guess that, as we go away from T = H = 0, larger N is needed for it to be reasonable.

This gives as answer to the fourth question in § 1 : because the number of channels is greater than 3.

V. THE LATTICE

Model and Static Approximation

We consider, for simplicity, a free electron gas conduction band interacting with $\ell = 3$ impurities and ignore spin, so $N = 2\ell + 1 = 7$. Coleman's Hamiltonian becomes[9]

$$H = \sum_{\vec{k}} \varepsilon_{\vec{k}}\, C^+_{\vec{k}} C_{\vec{k}} + E_o \sum_{im} f^+_{im} f_{im} + V \sum_{im} (C^+_{im} f_{im} b^+_i + f^+_{im} C_{im} b_i) \qquad (37)$$

where i = site at $\vec{R}_i$, $-\ell \leqslant m \leqslant \ell$, and

$$C^+_{im} = \sum_{\vec{k}} (4\pi)^{1/2}\, Y^m_\ell (\Omega_{\vec{k}})^* \, e^{-i\vec{k}.\vec{R}_i}\, C^+_{\vec{k}} \quad . \qquad (38)$$

Here, we seek to generalize the work of Read, Newns and Doniach[13] on the Kondo Lattice. We proceed as in (5), introducing a set of λ_i, to (13). In making the static approximation, an additional assumption that lattice symmetry is not broken is introduced resulting in two parameters $i\lambda_i = \varepsilon_f - E_o$, $r_i = r_o$, so that

$$H_{eff} = \sum_{\vec{k}} \varepsilon_{\vec{k}} C^+_{\vec{k}} C_{\vec{k}} + \varepsilon_f \sum_{i,m} f^+_{im} f_{im} + r_o V \sum_{im} (C^+_{im} f_{im} + h.c.)$$
$$+ \sum_i (\varepsilon_f - E_o)(r_o^2 - 1) \quad . \qquad (39)$$

Equation (19) and (23) remain valid at each site i, but the difficulty comes in the diagrams for the propagators $G_{fim,im}$ and $G_{fcim,im}$ needed to get the expectation values, because the unperturbed conduction band propagator is now a matrix $g_{im,jm'}(\varepsilon) = g_{mm'}(\vec{R}_{ij},\varepsilon)$

$$g_{mm'}(\vec{R},\varepsilon) = 4\pi \sum_{\vec{k}} \frac{Y^m_\ell(\Omega_{\vec{k}}) Y^{m'*}_\ell(\Omega_{\vec{k}})}{\varepsilon - \varepsilon_{\vec{k}} + is} e^{i\vec{k}.\vec{R}} \quad . \qquad (40)$$

To solve the Dyson Equation (Fig. 2a) exactly leads us into the whole problem of the renormalized band structure of (39) - which is something one needs eventually to do. But we can make some progress analytically by 1/N methods. We start expanding Fig. 2a as a perturbation series[13] (Fig. 3), giving for $\varepsilon < \mu$

$$\text{Im } G_{fim,im} = \frac{r_o^2 V^2}{(\varepsilon - \varepsilon_f)^2} \{ \text{Im } g_{mm}(\vec{0},\varepsilon)$$
$$+ \text{Im} \sum_{jm'} g_{mm'}(-\vec{R}_{ij},\varepsilon) \frac{r_o^2 V^2}{\varepsilon - \varepsilon_f} g_{m'm}(\vec{R}_{ij},\varepsilon) + \ldots \} \qquad (41)$$

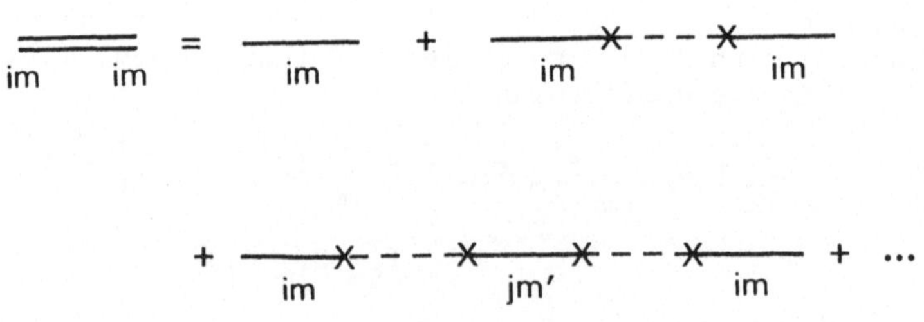

Fig. 3. Perturbation series for propagator $G_{fim,im}$ (double line). Single line, $(\varepsilon - \varepsilon_f + is)^{-1} \delta_{ij} \delta_{mm'}$; broken line, propagator $g_{im,jm'}(\varepsilon)$; cross, the vertex $r_o V$.

If in the second term in (41) we quantize around the direction of the bond $\vec{R}_{ij}$, only m' = m contributes. Now one may make an asymptotic large-ℓ evaluation[13] of the spherical harmonics in (40). For $\vec{R}$ along OZ, and m small (which are the largest terms), (40) becomes in the large-ℓ limit

$$g_{mm}(R,\varepsilon) = \int_0^\infty \frac{\rho(k)k\,dk}{\varepsilon-\varepsilon_k+is} J_0(kR) \quad . \tag{42}$$

In (42) J_0 is the zeroth order Bessel function, and $\rho(k)$ the free electron density of states. The essential point is that (42) is ℓ-independent.

Now the second term in (41) is down on the first by a factor no greater than (for $\varepsilon \leqslant \mu$)

$$\sum_{\vec{R}} \frac{\Delta}{\delta\pi(\varepsilon_f-\mu)} \operatorname{Re} g_{mm}(\varepsilon,\vec{R}) \quad . \tag{43}$$

But, assuming that, as before, $\Delta/\pi(\varepsilon_f-\mu) \sim \langle n_f\rangle/N$, and given that (42) is N-independent, (43) is of O(1/N). The same result may be demonstrated for the next and higher diagrams, so that, to leading order in 1/N, we have only the first, *on site*, term in (41).

The same result may be demonstrated for G_{fc}.

Thus we may, working to leading order in 1/N, *reproduce the results of § 4.* These results are themselves only reproduced to leading order in 1/N, i.e. in (22) $\tan^{-1}(\Delta/\varepsilon_f)$ is replaced by Δ/ε_f, and in (25) the Δ^2 is absent.

Our picture is compatible with unrenormalized band structure calculations, e.g. by Yanase for $CeSn_3$[22], which show that the Fermi level lies below the bottom of the f-band in a region dominated by f-d hybridization, just as would be expected from the single impurity picture

The observation that intersite effects are of O(1/N) is not entirely new. It was first found in the high temperature perturbation theory of Hewson[23], and noted by Ramakrishnan et al[17]. Very recently, it has surfaced again in the work of Coleman[24] and Read et al.[13] on the Kondo Lattice.

Consequences

- As long as the 1/N expansion is good, we get back to leading order in 1/N the thermodynamic properties of the impurity, thus answering finally questions 1-5 of §1.

- Nevertheless there exists a renormalized band structure and the lattice will have zero residual resistance. The transport properties at low-T are calculable from consideration of fluctuations around the stationary point, answering in principle question 6.

  Actually, some Yb materials (YbCuAl, $YbCu_{4.5}$) seem to be in a "dirty" limit where the mean free path of conduction electrons is less than the "impurity radius" v_F/T_K. For these we are tempted to apply impurity transport theory, leading to the exact low-T result for thermopower[3,25]

$$Q = \frac{1}{3} \pi^2 k_B^2 T |e|^{-1} \partial \ell n \rho_f(\mu)/\partial\mu \quad .$$

  This works quite well, and explains the signs of thermopower being positive for Ce and negative for Yb (question 6), since ρ_f' indeed has these signs.

- The 1/N expansion of Fig. 3 breaks down when the Fermi level lies in a gap, which happens for certain fillings. In Fig. 4 we sketch what would happen in SmB_6 within our assumptions. The hybridization gap E_g^h turns out to be $\simeq (\varepsilon_f - \mu)$. The observed, direct gap E_g of 3 meV is less than half E_g^h, making $E_g^h > 6$ meV. In SmS we found,[2] for comparison, $(\varepsilon_f - \mu) = 10$ meV, treating it as a metal.

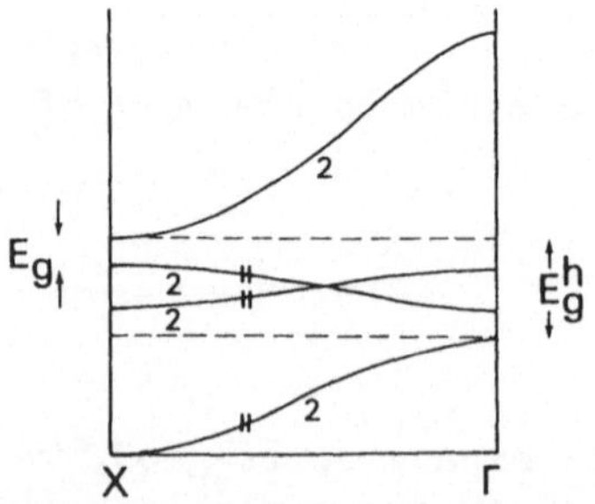

Fig. 4. Schematic band structure of Sm B_6.

At high pressures, the tight-binding band inside the gap (whose width ∿ (lattice constant)$^{-7}$) would tend to widen relative to E_g^h, thus closing the gap. It is suggested that this is what happens in SmB_6 at pressures above ≃ 60 kbar[26].

CONCLUSION

We have set out a mean field theory for the quasiparticle properties of Intermediate Valence systems extending the work of Read and Newns[12], and Read, Newns and Doniach[13] on the Kondo Lattice by means of Coleman's Hamiltonian[9]. The theory leads to a renormalized band structure. For metallic IV materials we nevertheless get back the thermodynamic properties of the local theory which is supported by the data, plus the transport properties of a normal metal. When the renormalized band structure gives a gap, an explanation of insulating materials such as SmB_6 is obtained.

ACKNOWLEDGEMENTS

N. Read has been supported by an SERC studenship.

REFERENCES

1. S. von Molnar, T. Theis, A. Benoit, A. Briggs, J. Flouquet, J. Ravex and Z. Fisk, in "Valence Instabilities," eds. P. Wachter & H. Boppart (North-Holland, Amsterdam 1982).
2. D.M. Newns and A.C. Hewson, J. Phys. F 10, 2429 (1980); in "Valence Fluctuations in Solids", eds. L.M. Falicov, W. Hanke and M.B. Maple (North-Holland, Amsterdam 1981).
3. D.M. Newns, A.C. Hewson, J.W. Rasul and N. Read, J. Appl. Phys. 53, 7877 (1982).
4. J.C.P. Klaase, F.R. de Boer and P.F. De Chatel, Physica 106B, 178 (1981).
5. T. Mihailisin, P. Scoboria and J.A. Ward, in "Valence Fluctuations in Solids", eds. L.M. Falicov, W. Hanke and M.B. Maple (North-Holland, Amsterdam 1981).
6. W.C.M. Mattens, thesis, University of Amsterdam 1980.
7. J.R. Thompson, S.T. Sekula, C.K. Loong and C. Stassis, J. Appl. Phys. 53, 7893 (1982).
8. D. Jaccard and J. Sierro in "Valence Instabilities", eds. P. Wachter and H. Boppart (North-Holland, Amsterdam 1982).
9. See chapter by Piers Coleman.
10. E. Holland-Moritz, D. Wohlleben and M. Loewenhaupt, Phys. Rev. B 25, 7482 (1982).
11. P. Schlottman, Phys. Rev. Lett. 50, 1697 (1983).
12. N. Read and D.M. Newns, J. Phys. C16, L1055 (1983).
13. N. Read, D.M. Newns and S. Doniach, submitted to Phys. Rev. Lett.

14. N. Read and D.M. Newns, J. Phys. C 16, 3273 (1983)
15. P. Ramond, "Field Theory, a Modern Primer", Benjamin/Cummings, Reading Mass. (1981).
16. H. Keiter, Z. Physik 213, 466 (1968).
17. A. Bringer and H. Lustfeld, Z. Physik B28, 213 (1977); T.V. Ramakrishnan and K. Sur, Phys. Rev. B26, 1978 (1982); J.W. Rasul and A.C. Hewson, J. Phys. C16, L933 (1983).
18. O. Gunnarsson and K. Schonhammer, Phys. Rev. B28, 4315 (1983).
19. A. Yoshimori,Progr. Theor. Phys. (Kyoto) 55, 67 (1976).
20. D.C. Langreth, Phys. Rev. 150, 712 (1966).
21. A.C. Hewson, J.W. Rasul and D.M. Newns, Solid State Commun. 47, 59 (1983); V.T. Rajan, Phys. Rev. Lett. 51, 308 (1983).
22. A. Yanase, J. Magnetism and Magnetic Materials 31-34, 453 (1983).
23. A.C. Hewson, J. Phys. C 10, 4973 (1977).
24. P. Coleman, Phys. Rev. B28, 5255 (1983).
25. B. Horiutic and V. Zlatič, preprint.
26. M.B. Maple, private commun.

SELF-CONSISTENT DYNAMICAL THEORY FOR

VALENCE FLUCTUATIONS

Yoshio Kuramoto

Department of Applied Physics
Tohoku University
Sendai 980, Japan

INTRODUCTION

The valence fluctuation phenomena in certain rare-earth systems involve a difficult many-body problem. The source of the difficulty is traced to subtle interaction effects between the localized 4f electrons and itinerant conduction electrons. The interaction often leads to the Kondo effect, but the outcome has several variants depending on each rare-earth system. In this paper we put forward a theory which can take detailed account of realistic 4f-shell structures and is expected to work both for the intermediate-valence (IV) and nearly integral-valence (Kondo) regimes. Since valence fluctuations in dilute rare-earth alloy systems are very similar to those in dense rare-earth compounds,[1] we concentrate here on single-site fluctuation processes and neglect the rare-earth intersite interactions.[2,3] We outline the theoretical formulation[4] and report some numerical results[5] on dynamical properties of Ce systems.

RENORMALIZATION SCHEME

Consider a rare-earth impurity in a metallic matrix whose Hamiltonian is H_c. The rare-earth 4f states have the Hamiltonian H_f if there were no interaction with the environment. We take $H_0 = H_f + H_c$ as the unperturbed Hamiltonian, and assume that H_0 has been diagonalized. The perturbation Hamiltonian H_1 is taken to be the hybridization between 4f and conduction-band states. In dealing with the total system $H = H_0 + H_1$, we respect the fact that there is one and only one rare-earth electronic state at a given instant, and use the time-ordered Goldstone-type perturbation expansion from the atomic limit. The strategy of our formulation is to integrate out first the conduction-band degrees of freedom. Then we are left with renormalized 4f states each of which has a finite lifetime. The dynamics of renormalized 4f states are put into the 4f-resolvent matrix in terms of which 4f-electron density of states and dynamical susceptibilities are derived.

Let us factorize the grand partition function Z of the system as $Z = Z_c Z_f$ where $Z_c = Tr_c \exp(-\beta H_c)$ with the trace over the conduction-electron states. The renormalized 4f part Z_f is given by[4]

$$Z_f = Tr_f \bar{R}(\beta) = \int_C \frac{dz}{2\pi i} e^{-\beta z} Tr_f R(z), \quad (1)$$

where the trace is over the 4f-shell states and the contour C encircles counterclockwise all singularities of the integrand. The propagator $\bar{R}(\beta)$ has in general the argument $\tau > 0$, and reduces to $\exp(-\tau H_f)$ if the hybridization is turned off. The resolvent R(z) is the Laplace transform of $\bar{R}(\tau)$ and has the form $R(z) = [z - H_f - \Sigma(z)]^{-1}$ where the self-energy matrix $\Sigma(z)$ embodies all the effects of H_1.

The hermiticity of the Hamiltonian imposes that (i) R(z) is analytic except on the real axis of z, and that (ii) the spectral intensity matrix $\eta(\varepsilon) = -\mathrm{Im}R(\varepsilon + i\delta)/\pi$ is positive definite with δ being a positive infinitesimal. Furthermore, the boundary condition $\bar{R}(\tau \to 0) = 1$ leads to the sum rule that integration of $\eta(\varepsilon)$ over the whole ε gives the unit matrix. In the simplest self-consistent scheme the self-energy components are related to the resolvent elements by

$$\Sigma_\alpha(z) = \frac{1}{N} \sum_{k\nu\sigma} \sum_\beta | \langle\beta|V|\alpha,k\nu\sigma\rangle |^2 f(\varepsilon_{k\nu}) R_\beta (z + \varepsilon_{k\nu}), \quad (2)$$

$$\Sigma_\beta(z) = \frac{1}{N} \sum_{k\nu\sigma} \sum_\alpha | \langle\beta|V|\alpha,k\nu\sigma\rangle |^2 [1-f(\varepsilon_{k\nu})] R_\alpha (z - \varepsilon_{k\nu}), \quad (3)$$

where N is the number of lattice sites in the metal, α and β denote a zeroth-order $4f^{n-1}$ configuration and a $4f^n$ one, respectively, and $f(\varepsilon_{k\nu}) = [\exp(\beta\varepsilon_{k\nu}) + 1]^{-1}$. We note that Σ and R are diagonal in this basis. The hybridization matrix element $\langle\beta|V|\alpha,k\nu\sigma\rangle$ corresponds to the change $\alpha \to \beta$ with absorption of a ν-th band conduction electron with momentum k, spin σ and energy $\varepsilon_{k\nu}$.

The approximation leading to (2) and (3) is equivalent to summation of all the diagrams for Z_f without crossing of conduction-electron lines, and hence is called non-crossing approximation (NCA).[4] We notice that the NCA can incorporate the realistic 4f-shell structure of each rare-earth with crystalline-electric-field (CEF) effects. In addition, diagrams neglected in the NCA become less important when the 4f-shell degeneracy is large.[6,7] Being a self-consistent approximation, the NCA has further advanges:[4] (i) the resolvents have positive definite spectral intensities which obey the sum rule; (ii) response functions in the NCA are consistent with all available sum rules and conservation laws, so that their static limit agrees with that derived by the thermodynamic derivative of Z_f; (iii) there are no divergence difficulties in (2) and (3) for all temperatures.

RESULTS FOR Ce SYSTEMS

We consider a model Ce system in which the $Ce^{3+}(4f^1)$ ground level with total angular momentum J = 5/2 competes in stability with the $Ce^{4+}(4f^0)$ nondegenerate level. Spherical symmetry around the Ce ion is assumed. Then the hybridization strength, defined by

$$\frac{1}{N}\sum_{k\nu\sigma}|\langle\beta|V|\alpha,k\nu\sigma\rangle|^2\delta(\varepsilon-\varepsilon_{k\nu}) \equiv W_0(\varepsilon), \tag{4}$$

does not depend on β which is reduced to J_z in the case of spherical symmetry. We have solved (2) and (3) numerically[5] using a rectangular model of $W_0(\varepsilon)$: $W_0(\varepsilon) = W_0\theta(D - |\varepsilon|)$ where D corresponds to the half-width of the conduction band. The parameters are taken to be W_0 = 30K and D = 10^4K. We denote by ε_f the energy difference between the $4F^1$ state and the $4f^0$ one.

The 4f-electron density of states $\rho_{4f}(\varepsilon)$ is given in our theory by

$$\rho_{4f}(\varepsilon) = Z_f^{-1}(2J+1)(1+e^{-\beta\varepsilon})\int_{-\infty}^{\infty} d\nu\, e^{-\beta\nu}\,\eta_0(\nu)\eta_J(\varepsilon+\nu), \tag{5}$$

where the spectral intensities η_0 and η_J correspond to $4f^0$ and $4f^1$ configurations, respectively. Figure 1 shows the numerical results. In the IV case $\varepsilon_f = 0$, the peak of $\rho_{4f}(\varepsilon)$ is shifted above the chemical potential owing to the hybridization effect. We note that the peak position is almost independent of temperature while the peak height increases with decreasing temperatures.

The most notable result in the Kondo case $\varepsilon_f = -1200$K is the appearance of a spike at the Fermi level at low temperatures. The presence of the spike has long been suspected on the basis of the Fermi-liquid theory.[8,9] At finite temperatures our theory has derived for the first time the quantitative result for $\rho_{4f}(\varepsilon)$ over the whole range of ε. The origin of the spike in the NCA is that the singlet $4f^0$ acquires sharp spectral intensity around $\varepsilon \sim \varepsilon_f$ as a result of hybridization. We also note that the lower peak in $\rho_{4f}(\varepsilon)$ has a considerably larger width than in the IV case.

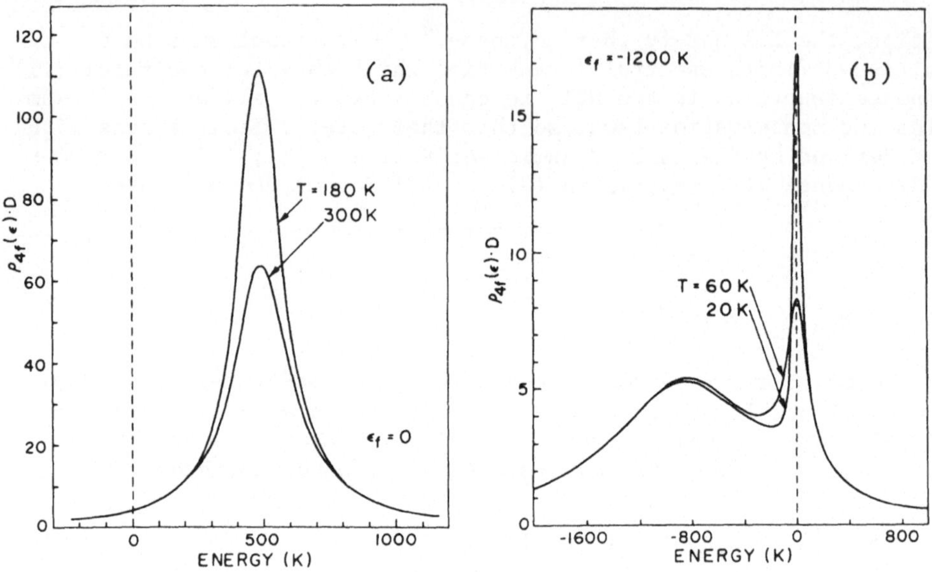

Fig. 1. 4f-electron density of states for (a) the IV case and (b) the Kondo case.

There exists a sum rule

$$\int_{-\infty}^{\infty} \rho_{4f}(\varepsilon)\, d\varepsilon = n_f + (2J+1)(1-n_f), \tag{6}$$

which the NCA correctly obeys.[4] Here n_f is the averaged 4f-electron number. In the IV case, the increase of the peak in $\rho_{4f}(\varepsilon)$ with lowering temperatures should then correlate with decrease of n_f. Numerical calculation of n_f shows[5] that this is indeed the case; $n_f = 0.64$ at $T = 300$K and $n_f = 0.28$ at $T = 40$K. The substantial decrease of n_f is specific to our impurity model since in concentrated rare-earth systems the decrease should be severely restricted by the requirement that the total electron number be a constant. On the other hand, we have obtained almost temperature-independent value $n_f = 0.96$ in the $\varepsilon_f = -1200$K case.[5]

The magnetic susceptibility $\chi(\omega)$ can also be calculated in the NCA. Our theory gives

$$\mathrm{Im}\,\chi(\omega) = Z_f^{-1} A(1 - e^{-\beta\omega}) \int_{-\infty}^{\infty} d\varepsilon\, e^{-\beta\varepsilon}\, \eta_J(\varepsilon)\, \eta_J(\varepsilon+\omega), \tag{7}$$

where $A = \pi g_J^2 J(J+1)(2J+1)/3$ with g_J the g-factor. Figure 2 shows the results for $\mathrm{Im}\chi(\omega)/\omega$. It can be seen that the line shape is close

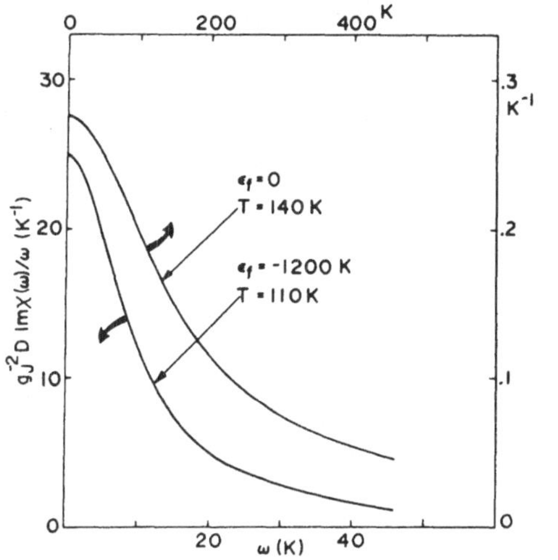

Fig. 2. Dynamical magnetic susceptibilities.

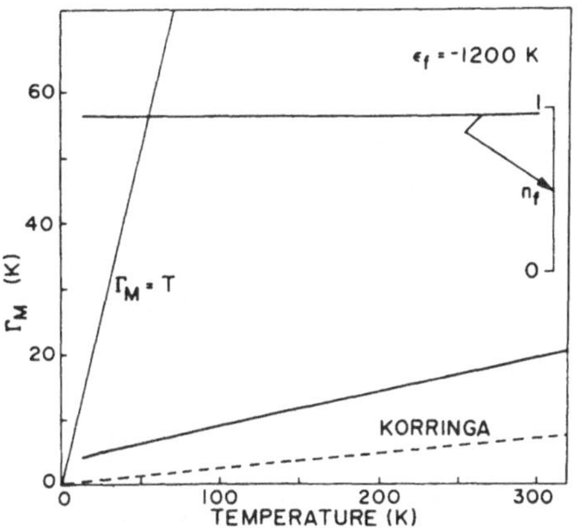

Fig. 3. Temperature dependence of the magnetic relaxation rate Γ_M in the Kondo regime.

to the Lorentzian in both IV and Kondo regimes, but the scales are quite different in the two cases. In Fig. 3 shown is the magnetic relaxation rate Γ_M derived by $\Gamma_M^{-1} = \lim_{\omega\to 0} [\omega\chi(0)]^{-1} \mathrm{Im}\,\chi(\omega)$ in the Kondo case. The Γ_M is almost the same as the half-width of $\mathrm{Im}\chi(\omega)/\omega$ because of the nearly Lorentzian line shape. The relaxation rate is considerably enhanced over the Korringa value $\Gamma_M = 2(2J+1)\pi(W_0/\varepsilon_f)^2 T$ and reproduces the qualitative feature of Γ_M observed in Kondo compounds such as $CeCu_2Si_2$ and CeB_6.[10]

DISCUSSION

Although Ce systems were exclusively discussed in the previous section, it should be stressed that the NCA can give 4f-electron density of states and dynamical susceptibilities also for other rare-earth systems. For all rare-earths at high temperatures the NCA reproduces the lowest-order relaxation rates derived by the Mori method.[11,12] Moreover, the Korringa-type relaxation is obtained in the local-moment regime.[4] The high-temperature approximation of the NCA can thus explain qualitatively the experimental temperature dependence of Γ_M in IV Tm systems, where the typical IV relaxation above $T \gtrsim 100K$ gives way to the Korringa-type one below $T \lesssim 100K$.[1]

The sharp peak of $\rho_{4f}(\varepsilon)$ at the Fermi level for $\varepsilon_f = -1200K$ shown in the previous section should explain the huge coefficient γ of the T-linear specific heat observed in Kondo compounds at low temperatures[13]. In our calculation with W_0=30K and ε_f = -1200K the spike in

$\rho_{4f}(\varepsilon)$ completely vanishes at T = 300K, and there remains a broad maximum. On the other hand in recent photoemission experiment in Ce systems, at room temperature frequently observed are two 4f derived peaks.[14] The Kondo temperature of the order of 300K would retain the double-peaked structure in $\rho_{4f}(\varepsilon)$ up to room temperature, but then difficulty arises to account for the gigantic specific heat. To understand the photoemission results, which involve high-energy excitations, we may have to include other kinds of interactions in addition to the hybridization.

In conclusion, we emphasize the wide scope of the NCA with which one can view the valence fluctuation phenomena in a unified way and decide on the most appropriate model. The extension of this type of self-consistent theory to include intersite interactions will be the next step of the development.

Acknowledgment — Much of this work has been done in collaboration with H. Kojima and M. Tachiki to whom I would express my gratitude.

REFERENCES

1. E. Holland-Moritz and M. Prager, J. Magn. and Magn. Materials 31 - 34, 395 (1983).
2. Y. Kuramoto, Z. Phys. B40, 293 (1981).
3. N. Grewe and H. Keiter, Phys. Rev. B24, 4420 (1981).
4. Y. Kuramoto, Z. Phys. B (in press).
5. H. Kojima, Y. Kuramoto, and M. Tachiki, to be published.
6. P.W. Anderson, in: Valence Fluctuations in Solids, L.M. Falicov, W. Hanke, and M.B. Maple eds., North-Holland, Amsterdam (1981).
7. T.V. Ramakrishnan and K. Sur, Phys. Rev. B26, 1798 (1982).
8. K. Yamada, Prog. Theor. Phys. 53, 970 (1975).
9. F.D.M. Haldane, in: ref.6, p.153.
10. S. Horn, F. Steglich, M. Loewenhaupt, and E. Holland-Moritz, Physica 107B, 103 (1981).
11. Y. Kuramoto, Z. Phys. B37, 299 (1980).
12. E. Müller-Hartmann, in: Electron Correlation and Magnetism in Narrow-Band Systems, T. Moriya ed., Springer, Berlin (1981), p.178.
13. J.M. Lawrence, P.S. Reiseborough, and R.D. Parks, Reports on Prog. in Phys. 44, 1 (1981).
14. R.D. Parks, N. Martensson, and B. Reihl, in: Valence Instabilities, P. Wachter and H. Boppart eds., North-Holland, Amsterdam (1982) p.239.

A NEW APPROACH TO THE MIXED VALENCE PROBLEM

Piers Coleman

Joseph Henry Laboratories
Princeton University
Princeton, N.J. 08544

Trying to solve the mixed valence problem can be likened to climbing a mountain: one needs first and foremost optimism, one also needs good equipment without which one is sure to fail. This short talk presents some "new equipment" for attempting the assault on the mixed valence problem.

A mixed valent crystal can be pictured as a lattice of rare earth (r.e.)ions which can exist in two valence states: one a $2j+1 = N$ fold degenerate state of spin j, the other typically a singlet. There is an extended band of free electrons which hybridises weakly with the r.e. f states causing valence fluctuations

$$f^n \rightleftharpoons f^{n-1} + e^- \, .$$

Despite the weakness of the hybridisation, the mixed valence problem is non-trivial because the valences of the r.e. ions are constrained to two values.

In this talk I shall use the mixed valence impurity problem to illustrate the difficulties and to show how the new approach surmounts them. The standard model Hamiltonian describing the system is the $U = \infty$ Anderson model[1,2] which can be written

$$\hat{H} = \sum_{km} E(k)\hat{c}^{\dagger}_{km}\hat{c}_{km} + \sum_{m} E_f \hat{f}^{\dagger}_m \hat{f}_m + \hat{H}_{mix} \tag{1}$$

where

$$\hat{H}_{mix} = \sum_{km} \left\{ V(k)\, \hat{c}^{\dagger}_{km} (\hat{P}_o \hat{f}_m) + V(k)^* (\hat{f}^{\dagger}_m \hat{P}_o)\, \hat{c}_{km} \right\} \, . \tag{2}$$

The first two terms in the Hamiltonian describe the band electrons and the f state. $\hat{H}_{mix}$ mixes the valence states and the operator P_0 projects out states with no f electrons.

The great difficulty in dealing with this Hamiltonian arises because the composite Hubbard operators[2] $\hat{X}_{om} = P_o \hat{f}_m$ and $\hat{X}_{mo} = \hat{f}_m^+ P_o$ that feature in $\hat{H}_{mix}$ have awkward commutation rules

$$\left[\hat{X}_{m'o}, \hat{X}_{om}\right]_{\pm} = \hat{f}_{m'}^{+} \hat{f}_m \pm \delta_{m'm} P_o \quad . \tag{3}$$

This occurs because the valences are restricted. It means that there is no Wick's theorem for the Hubbard operators so that conventional field theory approaches are inapplicable. The standard approach to the problem is to use time ordered Goldstone perturbation theory first developed by Keiter and Kimball[3], and recently extended by Kuramoto[4] as described in his chapter.

The philosophy of the new approach is to replace the awkward Hubbard operators by a product of a boson and fermion operator, both of which have standard commutation algebras

$$\hat{X}_{om} \longrightarrow \hat{b}^{+} \hat{f}_m \tag{4}$$

$$\hat{X}_{mo} \longrightarrow \hat{f}_m^{+} \hat{b} \ . \tag{5}$$

A few days before this summer school it was learnt that a similar approach was employed by Barnes[5] but was never exploited for a study of the mixed valence problem. The mathematical motivation for this replacement is that the time ordering of these operators can be treated by conventional approaches. The physical interpretation is that the spinless valence state is now represented by a boson that is created when an f electron hops out of the r.e. ion and is destroyed when a band electron hops into the r.e. ion.

The Hamiltonian that this replacement leads to is the same, apart from a new mixing term

$$\tilde{H}_{mix} = \sum_{km} \left\{ V(k) \hat{c}_{km}^{+} \left(\hat{b}^{+} \hat{f}_m\right) + V(k)^{*} \left(\hat{f}_m^{+} \hat{b}\right) \hat{c}_{km} \right\} \tag{6}$$

The boson is assigned zero energy, just like the empty state f^0. The new Hamiltonian commutes with the "charge" operator

$$\hat{Q} = \hat{n}_f + \hat{b}^{+} \hat{b} \tag{7}$$

so this quantity is conserved. $\tilde{H}_{mix}$ converts bosons into f electrons and vice versa, conserving Q. Thus the subspaces F_Q of definite Q are disjoint. In particular, in the subspace F_1, making the identification

$$|f^\circ\rangle \equiv b^+|0\rangle$$
$$|f^1{:}jm\rangle \equiv f_m^+|0\rangle \qquad (8)$$

the new Hamiltonian is exactly equivalent to the old and the resonance $f^1 \rightleftharpoons b^1 + e^-$ corresponds to the valence fluctuation.

Since the subspaces are disjoint, the F_1 subspace is easily projected by associating a real or complex chemical potential with the charge Q, leading to the following expressions for the partition function Z_{MV} of the mixed valence impurity

$$Z_{MV}(\beta) = \underset{\lambda\to\infty}{Lt} \left\{ Tr\left(\hat{Q}\, e^{-\beta[\hat{H}+\lambda(\hat{Q}-1)]}\right)\right\} \qquad (9)$$

$$Z_{MV}(\beta) = \int_0^{2\pi} d\lambda/2\pi \cdot Tr\left(e^{-\beta\hat{H} + i\lambda(\hat{Q}-1)}\right) \qquad (10)$$

The second approach is more aesthetic and probably the most appropriate for the lattice problem where there is one charge per site Q_i. The first approach is used here as it leads to a simple diagrammatic expansion. Expectation values of any operator $\hat{A}$ in the F_1 subspace are easily evaluated as

$$\langle A\rangle_{MV} = \underset{\lambda\to\infty}{Lt} \left\{ \langle \hat{A}\hat{Q}\rangle_\lambda \,/\, \langle \hat{Q}\rangle_\lambda \right\} \qquad (11)$$

where $\langle\ \rangle_\lambda$ denotes the expectation value in the grand canonical ensemble containing all subspaces F_Q, with chemical potential $-\lambda$.

The diagrammatic approach treats $\hat{H}_{mix}$ as an interaction about $\hat{H}_0 = \hat{H}_f + \hat{H}_{band}$. Using the interaction representation

$$\hat{H}_{mix}(\tau) = e^{\tau\hat{H}_0}\, \hat{H}_{mix}\, e^{-\tau\hat{H}_0} \qquad (12)$$

the partition function of the grand canonical ensemble is

$$Z_G(\lambda) = \left\langle T\exp -\int_0^\beta \hat{H}_{mix}(\tau)\, d\tau \right\rangle_\lambda Z(\hat{H}_0;\lambda) \qquad (13)$$

which equals $e^{-\beta\Delta F}$. ΔF is simply the sum of all closed loop connected diagrams generated by $\hat{H}_{mix}$. Fig. 1 shows the propagators and vertices associated with this expansion. Fig. 2 shows the linked cluster expansion of ΔF.

Similarly, the expectation value $\langle\hat{A}\rangle_\lambda$ of any operator $\hat{A}$ can be expanded as a sum of linked clusters generated by contracting with products of H_{mix}

$$\langle\hat{A}\rangle_\lambda = \sum \left(\text{linked clusters generated by contractions of } \hat{A}\right). \qquad (14)$$

When the $\lambda\to\infty$ limit is taken these expansions simplify. In each connected diagram there are a certain number of closed loops around which charge flows. For each of these loops there is a factor of $e^{-\beta\lambda}$, so when the $\lambda \to \infty$ limit is taken, only one loop diagrams survive

$$\underset{\lambda \to \infty}{Lt}\; e^{\beta\lambda} \langle A \rangle_\lambda = \sum e^{\beta\lambda} \left(\text{"single loop" clusters} \right) \tag{15}$$

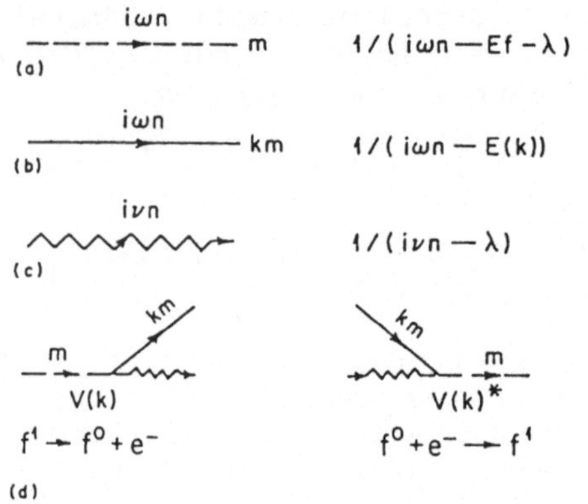

Fig. 1. Notation used in diagrams. (a) pseudo-f propagator, (b) band electron propagator, (c) boson propagator, (d) vertices.

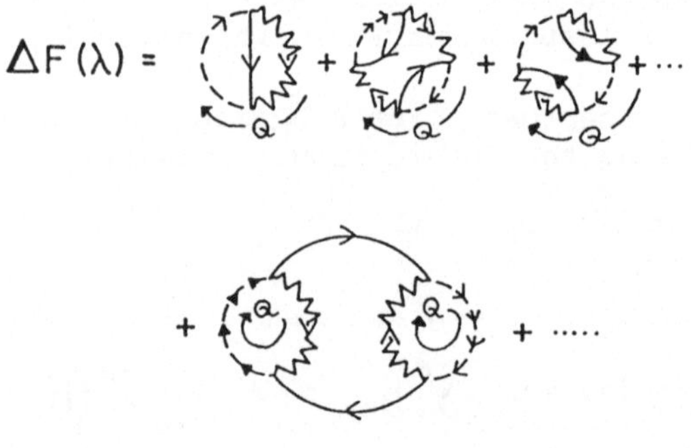

Fig. 2. Expansion of ΔF. The fourth diagram has two loops around which charge Q flows and vanishes as $e^{-2\beta\lambda}$ in the $\lambda \to \infty$ limit.

Two useful results are quickly derived. Firstly, Z_{MV} can be written down in closed form in terms of the clothed boson propagator $\mathcal{D}(i\nu_n)$ and the clothed pseudo-f propagator $G_{fm}(i\omega_n)$. $\mathcal{D}(i\nu_n)$ and $G_{fm}(i\omega_n)$ can be expanded in terms of their self energies

$$\begin{aligned} G_{fm}(i\omega_n) &= \left(i\omega_n - E_{fm} - \lambda - \Sigma_{fm}(i\omega_n) \right)^{-1} \\ \mathcal{D}(i\nu_n) &= \left(i\nu_n - \lambda - \Pi(i\nu_n) \right)^{-1} . \end{aligned} \tag{16}$$

The associated spectral functions are centred around $\omega = \lambda$ and it is convenient to subtract this energy displacement when taking the $\lambda \rightarrow \infty$ limit, defining

$$A_{fm}(\omega) = \frac{1}{\pi} \underset{\lambda \to \infty}{Lt} \ \mathrm{Im}\, G_{fm}(\omega + \lambda - i\delta)$$
$$B(\nu) = \frac{1}{\pi} \underset{\lambda \to \infty}{Lt} \ \mathrm{Im}\, D(\nu + \lambda - i\delta) . \tag{17}$$

Applying equation (9) provides the closed expression for Z_{MV}

$$Z_{MV}(\beta) = \int_{-\infty}^{\infty} d\omega \left\{ B(\omega) + \sum_m A_{fm}(\omega) \right\} e^{-\beta\omega} . \tag{18}$$

Secondly, the spectral function of the true f state can be calculated. In terms of Hubbard operators the full f propagator is

$$\mathcal{G}_m(\tau) = \langle T\, X_{om}(\tau)\, X_{mo}(0) \rangle , \tag{19}$$

which in the boson approach becomes

$$\mathcal{G}_m(\tau) = C_m(\tau) / Z_{MV}(\beta) , \tag{20}$$

where $C_m(\tau)$ is the two particle correlation function shown in Fig 3.

$$C_m(\tau) = \underset{\lambda \to \infty}{Lt} \ e^{\lambda\beta} \langle T\, b^{+}(\tau) f_m(\tau) f_m^{+}(0)\, b(0) \rangle . \tag{21}$$

In general $C_m(\tau)$ incorporates a vertex term $\Lambda(\tau_1, \tau_2)$ which can't be easily calculated. However, in the large degeneracy limit[6,7] the corrections to the vertex are of order $O(1/N^2)$. A well defined large N limit is obtained by rescaling the bare interaction vertex

$$V(k) = \tilde{V}(k) / \sqrt{N} \tag{22}$$

and keeping $\tilde{V}(k)$ fixed as N is made large. Typical r.e. ions have large degeneracies of 6 or 8 justifying this approach. Fig. 3. shows that the lowest order vertex corrections are $O(1/N^2)$, and if the large N limit is well defined, these corrections will vanish uniformly (ultimately, this must be confirmed in final results). By neglecting all vertex corrections as shown in Fig. 4. , we can write down two self-consistent equations for the boson and pseudo-f self energies

$$\Sigma_{fm}(\omega) = (1/N) \sum_k |\tilde{V}(k)|^2 f(E_k)\, D(\omega + E_k)$$
$$\Pi(\omega) = \sum_k |\tilde{V}(k)|^2 f(E_k)\, G_{fm}(\omega + E_k) \tag{23}$$

The equation for $C_m(\tau)$ is now simply

$$C_m(\tau) = \underset{\lambda \to \infty}{Lt} \ e^{\lambda\beta}\, G_{fm}(\tau)\, D(-\tau) \tag{24}$$

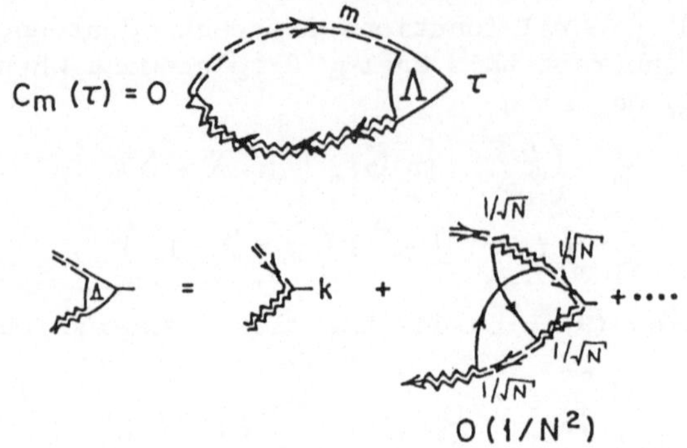

Fig. 3 The real f-propagator showing vertex and lowest order vertex corrections.

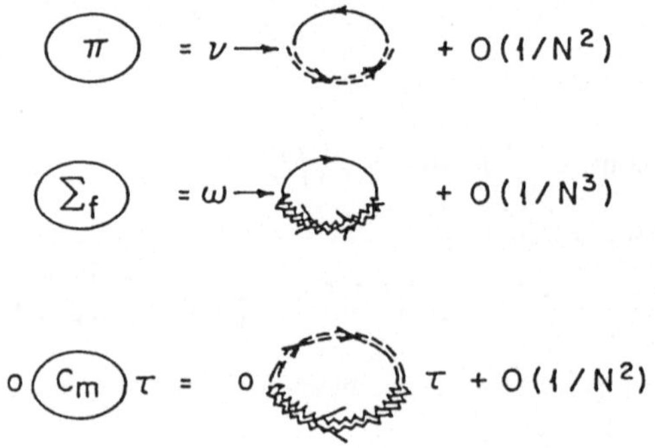

Fig. 4. Diagrams for the self-consistent expansion of propagators and self energies to $O(1/N^2)$.

Equations (23) and (24) hide subtle issues. Clearly they must be solved self-consistently. If this formalism is to extract any non-perturbative features of the mixed-valent system, one cannot simply iterate (23) twice, starting with the bare propagators. Naively this will give a result accurate to $O(1/N^2)$, but this assumes that the new ground state is a simple perturbation of the old.

General spectral considerations show[8] functions A_{fm} and B have the form

$$\begin{aligned} A_{fm}(\omega) &= \theta(\omega - E_o)\, A^+_{fm}(\omega) + \theta(E_o - \omega)\, Z_{MV}(\beta)\, e^{\beta\omega}\, a_{fm}(\omega) \\ B(\nu) &= \theta(\nu - E_o)\, B^+(\nu) + \theta(E_o - \nu)\, Z_{MV}(\beta)\, e^{\beta\nu}\, b(\nu) \end{aligned} \tag{25}$$

where E_o is the ground state energy and A^+, B^+, a and b all have finite low temperature limits. At absolute zero $Z_{MV}\, e^{-\beta\omega} \to 0$ so the last terms in these expressions vanish at T = 0. A^+_{fm} and B^+ are the energy spectra of states formed by adding a single boson or pseudo-f electron to the unperturbed electron sea $|0\rangle$. These spectra contain X-ray singularities at threshold because their introduction changes the charge Q which means that the band electrons experience a new scattering potential, just as in the X-ray problem[9,10]. Whilst no analytic solutions to (23) are available, it is straightforward to show that the self consistent solutions have a threshold with power law singularities[8]

$$\begin{aligned} G_{fm}(\omega) &\sim 1/(\omega - E_0)^{1/(N+1)} \\ D(\nu) &\sim 1/(\nu - E_0)^{N/(N+1)} \end{aligned} \tag{26}$$

These results are a promising sign that the self consistent equations are bootstrapping us into a state that has no overlap with the non-interacting ground-state.

Taking (24) and Fourier transforming it, the spectral function of the real f-state can be evaluated

$$\rho_{fm}(\omega) = (1 + e^{-\beta\omega}) \int_{-\infty}^{\infty} d\nu \, e^{-\beta\nu} A_{fm}(\omega+\nu) B(\nu) / Z_{MV} \tag{27}$$

and this has the zero temperature limit

$$\rho_{fm}^{(0)}(\omega) = \theta(\omega) \int_{E_0}^{E_0+\omega} d\nu \, A_{fm}^{+}(\omega+\nu) b(\nu) + \theta(-\omega) \int_{E_0+\omega}^{E_0} d\nu \, a_{fm}(\omega+\nu) B^{+}(\nu) \tag{28}$$

The analytic forms (26) guarantee a smooth and finite real-f spectral function at the Fermi energy. This is a necessary condition for the formation of a Fermi liquid at low temperatures and it will automatically lead to a linear specific heat and finite susceptibility at T=0 [11,12].

Numerical solutions to equations (23) are obtained by iteration to self-consistency. The results reported here are for a band of constant density of states, centred at the Fermi energy with half-width D. The rescaled resonant width $\tilde{\Delta} = \pi\rho|\tilde{V}|^2$ was taken to be D/5 and the degeneracy was 6. The results shown in Fig. 6. are for $T/\tilde{\Delta} = 0.03$. The sequence of curves show the transition from mixed valence to Kondo behaviour. When $E_f \sim -\tilde{\Delta}$ a rapid transition in the spectral structure occurs, and as the bare f-energy E_f, is reduced, the f-peak moves down below the Fermi surface, leaving a last remnant at the Fermi energy due to spin fluctuations. The shape of this "Kondo resonance" is surprising, as the conventional view pictures the f-resonance as lying above the Fermi energy: this may indicate that the 1/N expansion becomes inadequate in this regime. One check on these results that has yet to be done is a calculation of thermodynamic variables and a comparison with the Bethe Ansatz calculations of Schlottmann and Rajan[13,14].

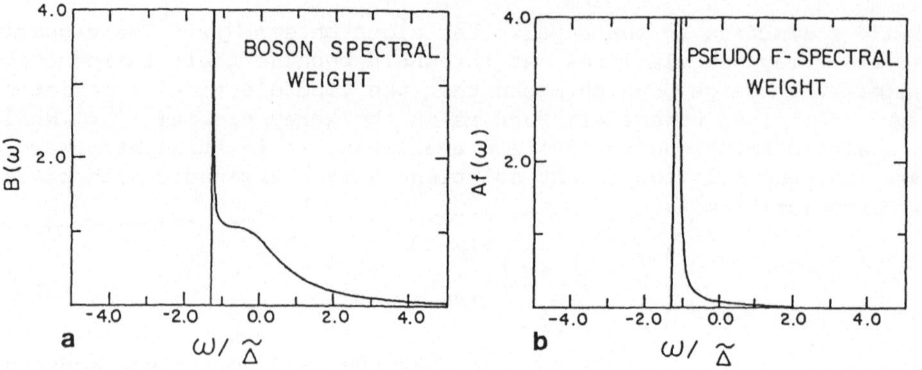

Fig. 5(a) Boson spectral function, (b) pseudo-f spectral function.

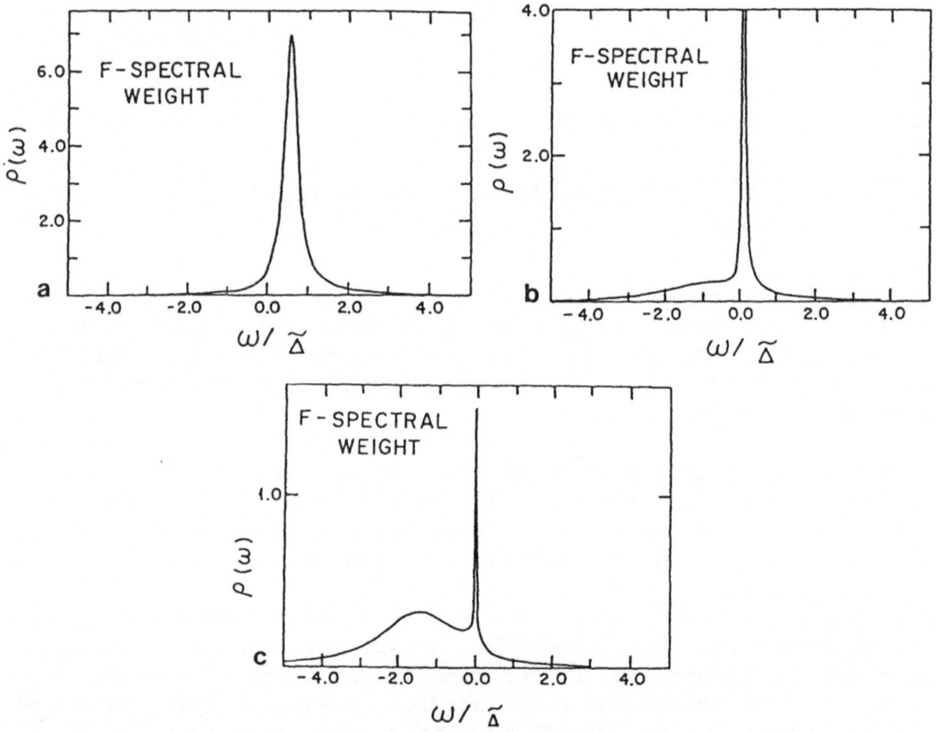

Fig. 6 Real f spectral function. (a) "Empty impurity" $E_f = 0.\tilde{\Delta}$ (b) "Mixed valent" $E_f = -1.1\,\tilde{\Delta}$ (c) "Kondo regime" $E_f = -2.0\,\tilde{\Delta}$.

Many aspects of the boson approach can't be dealt with in this article. If the large N expansion can be shown to work reliably for the Kondo impurity regime, then work on the lattice problem will become justified. Formally the extension to the lattice is straightforward. Naive diagrammatics indicate that two site interactions are O(1/N) effects, whilst many site interactions and the RKKY interaction are $O(1/N^2)$ effects[6,8]. It is too early to make predictions.

ACKNOWLEDGEMENTS

Particular thanks go to P. W. Anderson for his encouragment and advice throughout this project. Many others helped in discussions: T. V. Ramakrishnan who inspired the project and also F. D. M. Haldane, C. M. Varma and Hidenaga Yamagishi. I am grateful to the Science and Engineering Research Council, U. K. for their generous financial support whilst at Princeton.

REFERENCES

1. P. W. Anderson, Phys. Rev. 126 41 (1961).
2. J. Hubbard, Proc. Royal Soc. A277 237 (1964).
3. H. Keiter and J. C. Kimball, Int. J. Magn. 1 233 (1971).
4. See chapter by Y. Kuramoto.
5. S. E. Barnes , J. Phys. F, 6 1375 (1976) and J. Phys. F, 7 2637 (1977).
6. T. V. Ramakrishnan in: "Valence Fluctuations in Solids," eds. L. M. Falikov, W. Hanke, M. B. Maple (North Holland, Amsterdam 1981).
7. F. C. Zhang and T. K. Lee, preprint, Virginia State University (1983).
8. P. Coleman, Phys. Rev. B, 28 5255 (1983).
9. P. W. anderson, Phys. Rev. Lett. 18 1049 (1967).
10. P. Nozieres and C. de Dominicis, Phys. Rev., 178 1097 (1968).
11. J. M. Luttinger, Phys. Rev., 119 1153 (1960).
12. D. Langreth, Phys. Rev. 150 516 (1966).
13. V. T. Rajan, Phys. Rev. Lett., 51 308 (1983).
14. P. Schlottmann, Phys. Rev. Lett. 50 1967 (1983).

MODEL CALCULATIONS OF ELECTRON SPECTRA FOR Ce MIXED VALENCE COMPOUNDS

O. Gunnarsson

Max-Planck Institut für Festkörperforschung
7000 Stuttgart 80
Federal Republic of Germany

Institute for Theoretical Physics
University of California at Santa Barbara
Santa Barbara, CA 93106

K. Schönhammer

I. Institut für Theoretische Physik
Universität Hamburg
2000 Hamburg 36
Federal Republic of Germany

1. INTRODUCTION

The 4f level plays an important role for many properties of Ce compounds, and its occupancy, n_f, and coupling, Δ, to the conduction states are of great interest.[1,2] From thermodynamic data it has been inferred that Δ is small ($\sim$0.01 eV) and that, depending on the compound, n_f can have any value between zero and one.[1,2] Recent spectroscopic measurements have, however, suggested that both n_f and Δ are much larger than believed earlier.[1,2] In the following, we develop methods for calculating electron spectra, and we deduce new values of n_f and Δ from these calculations. The calculations are performed for a generalized Anderson impurity model. Anderson[3] and Ramakrishnan[4] realized that for this model 1/N can be treated as a small parameter, where N is the degeneracy of the 4f level. This idea has been used to develop methods for calculating thermodynamic properties.[4-6] We have presented a method for calculating electron spectra at T = 0, which is particularly suited for a large degeneracy.[7,8] A diagrammatic method was later developed by Coleman[9] for systems where the double occupancy of the f-level is suppressed. Here we briefly describe our method[7,8] and some applications.[10-13]

2. METHOD

We use a generalized Anderson impurity model

$$H = \sum_{\nu=1}^{N} \{\int \varepsilon \psi^{\dagger}_{\varepsilon\nu}\psi_{\varepsilon\nu}d\varepsilon + (\varepsilon_f - U_{fc}(1-n_c))\psi^{\dagger}_{\nu}\psi_{\nu} + \int (V(\varepsilon)\psi^{\dagger}_{\nu}\psi_{\varepsilon\nu} + h.c.)d\varepsilon + U\sum_{\nu<\mu} n_\nu n_\mu\} + \varepsilon_c n_c \quad (1)$$

where ν is an orbital (m) and spin (σ) index. The first term describes the conduction states, ε_f is the bare f-level energy and $V(\varepsilon)$ is a hopping matrix element. The f-f Coulomb interaction is U, ε_c is the core level energy and U_{fc} describes the interaction between the core hole and the f-level. This model is related to the model where the conduction states are labelled by a wave vector index k and σ, instead of ε and ν.[8] The ground-state of (1) is calculated variationally, and we discuss the limit $U = \infty$ below. We introduce a many-electron basis function $|0\rangle$, where the f-level is empty and the conduction states below the Fermi energy, $\varepsilon_F = 0$, are filled. Applying H repeatedly to $|0\rangle$, we introduce

$$|\varepsilon\rangle = N^{-1/2} \sum_{\nu} \psi^{\dagger}_{\nu}\psi_{\varepsilon\nu}|0\rangle \quad (2)$$

and

$$|E\varepsilon\rangle = N^{-1/2} \sum_{\nu} \psi^{\dagger}_{E\nu}\psi_{\varepsilon\nu}|0\rangle$$

$$|E\varepsilon\varepsilon'\rangle = (N(N-1))^{-1/2} \sum_{\nu\neq\nu'} \psi^{\dagger}_{E\nu}\psi_{\varepsilon\nu}\psi^{\dagger}_{\nu'}\psi_{\varepsilon'\nu'}|0\rangle \quad (3)$$

where $\varepsilon \leq \varepsilon_F$ and $E > \varepsilon_F$. A variational calculation using $|0\rangle$ and $|\varepsilon\rangle$ converges in the limit $N \to \infty$, if we introduce a rescaled hopping matrix element $\tilde{V}(\varepsilon)$

$$|\tilde{V}(\varepsilon)|^2 = N|V(\varepsilon)|^2 \quad (4)$$

Inclusion of the states (3) in the variational calculation leads to corrections to the ground-state energy E_0 of the order 1/N, and the error in E_0 is of the order $1/N^2$. If $U < \infty$, additional states must be included.

As an example of the calculation of electron spectra we discuss core level x-ray photoemission spectroscopy (XPS). The spectrum is $\rho_c(\varepsilon) = \mathrm{Im}\, g_c(\varepsilon - i0^+)/\pi$, where

$$g_c(z) = \langle E_o(N)|\psi_c^\dagger(z-E_o(N) + H)^{-1}\psi_c|E_o(N)\rangle \ , \tag{5}$$

$|E_o(N)\rangle$ is the ground-state and ψ_c is a core level annihilation operator. We write (5) as

$$g_c(z) = \sum_{ij} \langle E_o(N)|\psi_c^\dagger|i\rangle\langle i|(z-E_o(N)+H)^{-1}|j\rangle\langle j|\psi_c|E_o(N)\rangle, \tag{6}$$

where $\{|i\rangle\}$ is a basis set. We calculate the matrix elements $\langle i|(z-E_o(N)+H)|j\rangle$ and replace $\langle i|(z-E_o(N)+H)^{-1}|j\rangle$ in (6) by the inverse of this matrix. An important property of this approach becomes obvious if we transform to new states $|n\rangle$ which diagonalize H in the subspace $\{|i\rangle\}$. Our approximate core spectrum can then be written as

$$\rho_c(\varepsilon)=\sum_n |\langle n|\psi_c|E_o(N)\rangle|^2\delta(\varepsilon-E_o(N) + \varepsilon_n) \tag{7}$$

where $\varepsilon_n = \langle n|H|n\rangle$. The spectrum is thus positive and normalized to one if the $\{|i\rangle\}$ span the subspace of valence states used to construct $\psi_c|E_o(N)\rangle$. For a very large but finite system the variational calculation for the case $U = \infty$ including $|0\rangle$ and $\{|\varepsilon\rangle\}$ gives the exact ground state in the limit $N \to \infty$.[7,8] If the states $\{|i\rangle\}$ used in (6) contain the states $\psi_c|0\rangle$ and $\{\psi_c|\varepsilon\rangle\}$ one obtains the exact core spectrum in this limit. For the actual calculation of the core spectra it is important to include states with double occupancy of the f-level.

3. VALENCE SPECTRUM

The method of Sec. 2 can be applied to the f-electron valence spectrum. For $N \to \infty$ and $U = \infty$ we obtain

$$\rho_v(\varepsilon) = (1-n_f)^2\tilde{V}(\varepsilon)^2/(\varepsilon-\delta)^2 \qquad -\delta \le \varepsilon < \delta \ , \tag{8}$$

where $\delta = \varepsilon_f - \langle 0|H|0\rangle - E_o(N) > 0$ is related to the Kondo temperature and $\varepsilon_F = 0$. This result is compared with an exact relation between $\rho_v(\varepsilon_F)$ and n_f. In the spin-fluctuation limit ($n_f \sim 1$), (8) deviates less than 10 per cent from the exact relation for $N \geq 6$, and in the mixed valence regime the agreement is even better.[7,8]

Equation (8) describes a Kondo-peak at an energy δ above ε_F. In valence photo emission, which measures the spectrum for $\varepsilon \leq \varepsilon_F = 0$, this peak shows up as a sharp rise at ε_F, with an energy scale δ (see Fig. 1). If $\tilde{V}(\varepsilon)^2$ is reduced while $\varepsilon_f < 0$ is kept fixed, so that we approach the spin-fluctuation limit, $\rho_V(\varepsilon_F)$ grows. Simultaneously δ is reduced, and the structure seen in photoemission becomes narrower and loses weight.[7,8] For $\varepsilon_f << -\pi|\tilde{V}(\varepsilon_f)|^2$, these results are very different from the $N = 1$ case, for which the only contribution to $\rho_V(\varepsilon_F)$ is the tail of the peak at ε_f.

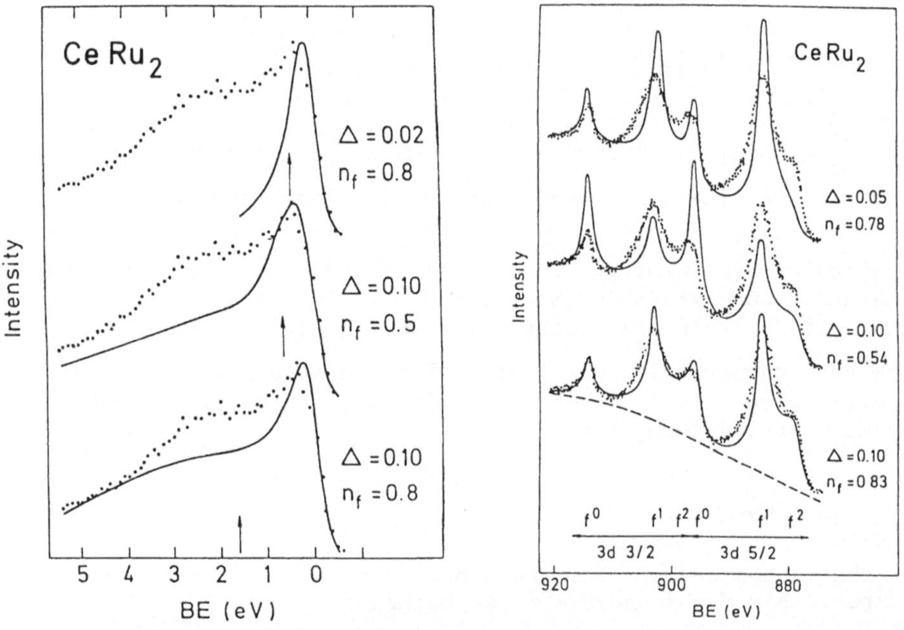

Fig. 1. Theoretical (full line) and experimental (dots) results for the 3d core (left) and valence (right) photoemission spectra of $CeRu_2$. For the core spectrum we have added an inelastic background and a 1.8 eV (FWHM) Lorentzian broadening and for the valence spectrum we used a Gaussian broadening of 0.47 eV (FWHM). The experimental results are due to Fuggle et al[12] and Allen et al.[16]

We now discuss[8] the peak weights and widths for $\varepsilon_f \lesssim -\pi|\tilde{V}(\varepsilon_f)|^2$, neglecting double occupancy in the initial state. The spectrum has an "ionization" peak at $\sim\varepsilon_f$, the peak in Eq. (8) corresponding to final states of mainly f^1 character and a peak several eV above ε_F of f^2 character. The weight of the spectrum below ε_F is n_f, and the f^1 and f^2 peaks have the weights $N(1 - n_f)$ and $(N - 1)n_f$, respectively, if we neglect the transfer of weight from the f^2 to the f^1 peak due to hybridization. The f^1 peak is very asymmetric with respect to ε_F, with only a small fraction ($\sim 1/N$) of the weight below ε_F. The "ionization" peak is usually quite non-Lorentzian, with its shape determined by $\pi\tilde{V}(\varepsilon)^2$. The half-width of the f^1 peak, $(1 - n_f)\pi V(\varepsilon)^2$, is very small and proportional to $(1 - n_f)/N$. Finally, the half-width of the f^2 peak has a hopping contribution $2\pi V(\varepsilon)^2 \sim 2/N$.

4. ESTIMATES OF n_f AND Δ

We have also studied $3d \rightarrow 4f$ x-ray absorption spectroscopy (XAS),[11] BIS[13] and the static, T = 0 susceptibility, χ.[8] The results are compared with experiment, to deduce the values of U_{fc}, U, $V(\varepsilon)$ and ε_f, in (1). The energy-dependence of $V(\varepsilon)$ is estimated from valence XPS, and the coupling strength, Δ, is defined as the average of $\pi V(\varepsilon)^2$ over the occupied band. If the other parameters are fixed, ε_f determines n_f. Thus n_f, rather than ε_f, is used as the basic parameter. Since the values of U_{fc} and U, needed to describe experiment, are in fair agreement with ab initio calculations,[15] our principal parameters are n_f and Δ.

To the left in Fig. 1 we show the 3d core level spectrum of $CeRu_2$. The peaks are classified as f^0, f^1 and f^2, according to the main character of the corresponding final states. Two theoretical spectra are superimposed to describe the 3d spin-orbit splitting. The lower part shows the spectrum for the values of n_f and Δ, giving the best agreement with experiment. To test the sensitivity of the spectrum, we reduce n_f by 0.3 and Δ by a factor of two, which leads to large changes for the f^0 peak (middle) or the f^2 shoulder (top). Owing to this sensitivity, values for both n_f and Δ can be deduced.

Fig. 1 also shows the valence spectrum for the "best" XPS parameters and two other sets. The experimental spectrum was taken for the f-resonance photon energy, and the conduction band emission was not subtracted, although it may still be of importance. The nature of the peak close to ε_F has been controversial. Our results suggest that it has a substantial f-character. Although n_f and Δ cannot easily be determined from the valence spectrum, a small Δ is inconsistent with experiment.

In BIS and 3d → 4f XAS, an f-electron is added, and the weight of the f^1 peak is related to the weight, $W(f^0)$, of the f^0 component in the initial state. This relation is complicated by final state effects. For instance, $W(f^0) = 0.2$ corresponds to the f^1 peak weights 0.37 (BIS) and 0.08 (XAS) for $\Delta = 0.1$ eV. Since it is harder to obtain Δ from these experiments than from XPS, we keep the XPS value of Δ and vary n_f to optimize the agreement with experiment. Table I shows that the values of n_f, deduced[10] from the different experiments, vary typically by 0.1. The BIS values tend to be lower than the others. Usually, n_f is the order 0.7 or larger. Our estimate of Δ for $CePd_3$ (0.11 eV) lies between two recent, very different estimates, which gave 0.007 eV[17] and ∿1 eV,[18] respectively.

5. CONCLUDING REMARKS

We have presented a method for calculating electron spectra for a large degeneracy, and we have given evidence (Ref. 8) that the method is usually rather well converged for $N \geq 6$. For $U = \infty$, the calculations can be performed analytically in the limit $N \to \infty$. We have compared theoretical and experimental spectra to determine n_f and Δ. In this comparison final state effects are essential. The parameters in Table 1 may have an appreciable uncertainty, which is partly due to problems in the comparison with experiments, such

Table 1. Estimates of n_f and Δ

	Δ_{av}	n_f			
	XPS	XPS	3d XAS	BIS	χ
$CeRu_2$	0.10	0.83	----	0.59	0.78
$CePd_3$	0.11	0.91	0.86	0.97	0.82
$CeNi_2$	0.10	0.84	0.79	0.78	0.74
$CeNi_5$	0.09	0.79	0.81	0.61	----

as background subtraction. However, the parameters could also be influenced by effects not included in the model. Since this influence may depend on the experiment, it is interesting to see if the model (1) can describe different experiments, using one set of parameters. The fairly small spread in the estimates in n_f in Table 1 is encouraging in this respect. It would be interesting to incorporate additional effects, such as multiplet couplings, to see how this would influence the parameters.

ACKNOWLEDGEMENT

We thank R.O. Jones for a critical reading of the manuscript.

REFERENCES

1. "Valence Fluctuations in Solids" (Eds.: L.M. Falicov, W. Hanke and M.B. Maple, N. Holland, Amsterdam, 1981).
2. "Valence Instabilities" (Eds.: P. Wachter and H. Boppart, N. Holland, 1982).
3. P.W. Anderson, Ref. 1, p. 451.
4. T.V. Ramakrishnan, Ref. 1, p. 13.
5. T.V. Ramakrishnan and K. Sur, Phys. Rev. B26, 1798 (1982).
6. F.C. Zhang and T.K. Lee (to be published).
7. O. Gunnarsson and K. Schönhammer, Phys. Rev. Lett. 50, 604 (1983).
8. O. Gunnarsson and K. Schönhammer, Phys. Rev. B28, 4315 (1983).
9. P. Coleman (to be published and this volume).
10. O. Gunnarsson, K. Schönhammer, J.C. Fuggle, F.U. Hillebrecht, J.-M. Esteva, R.C. Karnatak and B. Hillebrand, Phys. Rev. B28, 7330 (1983).
11. J.C. Fuggle, F.U. Hillebrecht, J.-M. Esteva, R.C. Karnatak, O. Gunnarsson and K. Schönhammer, Phys. Rev. B27, 4637 (1983).
12. J.C. Fuggle, F.U. Hillebrecht, Z. Zolnierek, R. Lässer, Ch. Freiburg, O. Gunnarsson and K. Schönhammer, Phys. Rev. B27, 7330 (1983).
13. F.U. Hillebrecht, J.C. Fuggle, G.A. Sawatzky, O. Gunnarsson and K. Schönhammer (to be published).
14. F.D.M. Haldane, Ref. 1, p. 153; R.M. Martin, Phys. Rev. Lett. 48, 362 (1982); A. Yoshimori and A. Zawadowski, J. Phys. C15, 5241 (1982); D.C. Langreth, Phys. Rev. 150, 516 (1966).
15. J.F. Herbst and J.W. Wilkins, Phys. Rev. Lett. 43, 1760 (1979).
16. J.W. Allen, S.-J. Oh, I. Lindau, M.B. Maple, J.F. Suassuna and S.B. Hagstrom, Phys. Rev. B26, 445 (1982).
17. D.M. Newns and A.C. Hewson, J. Phys. F10, 2429 (1980).
18. S.-J. Oh and S. Doniach, Phys. Rev. B26, 2085 (1982).

DYNAMIC SUSCEPTIBILITY OF A MAGNETIC IMPURITY

Pedro Schlottmann[+]

Fachbereich Physik, Freie Universität Berlin
Arnimallee 14, 1000 Berlin 33, W.-Germany

I. INTRODUCTION

In this paper we briefly review some calculations of the dynamic magnetic susceptibility of impurities in metals. We limit ourselves to the best known limits, i.e. the case of a spin-1/2 Kondo impurity (section II) and the mixed-valence impurity with orbital degeneracy (section III). In the mixed-valence regime we summarize the main results for Ce and Tm impurities and discuss the validity of the approximation by comparing with the exact Bethe-ansatz solution for the ground-state.

II. SPIN-1/2 KONDO LIMIT

A magnetic impurity in a metal shows a crossover from the asymptotically free spin at high T to a singlet state at low T. The susceptibility and the impurity lifetime are finite at zero temperature since the magnetic moment of the impurity is compensated by the conduction electrons. The qualitative change from an infinite to a finite lifetime or susceptibility at T = 0 represents a symmetry breaking which cannot be achieved by simple perturbation theory. In a series of papers /1-3/ we transformed the s-d Hamiltonian such that the renormalization can be started with a finite relaxation time. The transformed Hamiltonian consists of a resonant level (Toulouse limit) and a large perturbation. The resonance width acts like an infrared cutoff such that the perturbation ex-

[+]Heisenberg-fellow of the Deutsche Forschungsgemeinschaft

pansion at T = 0 converges term by term. For an anisotropic Kondo coupling the results are in good agreement with previous ones by Schotte and Schotte /4/ and Götze and Schlottmann /5/. With a careful diagrammatic analysis the spin-rotational invariance has been restored. We obtain two coupled nonlinear first-order differential equations for the invariant coupling and the resonance width. At low temperatures a Fermi-liquid type solution is obtained. The characteristic energy scale is the Kondo temperature. The static susceptibility and the impurity relaxation rate have been calculated, reproducing perturbation theory in leading and next-leading logarithmic hierarchy. At low T both χ_0 and T_1 are finite, indicating a singlet ground state and a decrease with temperature as T^2. There is a monotonic and smooth crossover from a singlet state to the asymptotically free spin, giving evidence for the gradual breaking up of the spin compensation. The susceptibility is in fair agreement with Wilson's numerical diagonalization /6/. The Kondo resonance, i.e. the f-density of states can be extracted. One finds logarithmic deviations from a lorentzian shape at low T and large ω. The width grows with T, being T_K at low T.

III. THE MIXED-VALENCE IMPURITY

We consider a simple model consisting of two multiplets with total angular momenta J_1 and J_2, that correspond to the Hund's rule groundstates of the $4f^{n+1}$ and $4f^n$ configurations, respectively, which are hybridized through the conduction electron states. The model and the notation is the same as in /7-9/. The Hamiltonian is given by $H = H_o + H_v$, where

$$H_o = \sum_{km} \varepsilon_k c^+_{km} c_{km} + \sum_{M_1} E_{J_1M_1} B_{J_1M_1} + \sum_{M_2} E_{J_2M_2} B_{J_2M_2}$$
$$H_V = V(2J_2+1)^{1/2} \sum_{kmM_1M_2} (M_2 m|M_1)[A^+_{M_1M_2} c_{km} + c^+_{km} A_{M_1M_2}]. \quad (\text{III}.1)$$

Here B_{JM} are projectors onto the ionic states and E_{JM} the corresponding energies. The operator c_{km} denotes the annihilation of a conduction electron with momentum k, total angular momentum j and z-component m. Only one of the two possible partial waves (j = 5/2 and j = 7/2) is considered, namely j = 5/2 for the light and j = 7/2 for the heavy rare-earth. This corresponds to a particular jj-coupling instead of the usual Russell-Saunders scheme. V is the hybridization matrix element. We will use the following short-hand notation for the Clebsch-Gordan coefficients

$$(J_2 M_2 j m | J_2 j J_1 M_1) \equiv (M_2 m | M_1) \quad (\text{III}.2)$$

and $|J_1M_1\rangle = |M_1\rangle$, $|J_2M_2\rangle = |M_2\rangle$ for the ionic states. We have then $B_{JM} = |M\rangle\langle M|$ and $A^+_{M_1M_2} = |M_1\rangle\langle M_2|$ is a f-charge creation operator (not obeying fermion anticommutation rules).

The two multiplets J_1 and J_2 are mixed via the hybridization matrix element, the valence of the impurity being noninteger in this way. For the sake of simplicity we assume $J_1 > J_2$. If the ground-state energies of the two configurations then satisfy $E_{J_1} > E_{J_2} + \varepsilon_F$, where ε_F is the Fermi energy, the mixing builds up a "bonding" multiplet of degeneracy $2J_2+1$ and an "antibonding" multiplet of angular momentum J_1. The energies of the mixed states $\tilde{E}_{J_2}$ and $\tilde{E}_{J_1}$ are lower than the respective energies E_{J_2} and E_{J_1}. The energy shifts, if calculated consistently are related by

$$(2J_2+1)(\tilde{E}_{J_1}-E_{J_1}) = (2J_1+1)(\tilde{E}_{J_2}-E_{J_2}) , \qquad \text{(III.3)}$$

such that the manifold with degeneracy J_2 has a larger energy reduction and will be the groundstate of the system. A charge excitation at the impurity at low T then requires the energy $\Delta E = \tilde{E}_{J_1} - \tilde{E}_{J_2}$. Note that the renormalized energies are temperature dependent, since they involve the Fermi function. The Boltzmann statistics of the energy levels straightforwardly yields the thermodynamics of the impurity. This is essentially the result of the Brillouin-Wigner theory /10,11/ and the renormalization group /12/ in the mixed valence regime.

The relevant operators for the magnetic response are the total angular momentum operators of the two ground-multiplets

$$S_{J_1} = \sum_{M_1} M_1 B_{J_1 M_1} \quad , \quad S_{J_2} = \sum_{M_2} M_2 B_{J_2 M_2} . \qquad \text{(III.4)}$$

We consider a 2x2 matrix susceptibility and obtain the magnetic response by multiplying the vector of Landé factors (g_{J_1}, g_{J_2}) from the right and from the left. Using Mori's formalism we express the dynamic matrix susceptibility in terms of the static matrix susceptibility, $\hat{\chi}^o$, and a relaxation kernel matrix, $\hat{m}(z)$

$$\hat{\chi}(z) = [z\hat{1} + \hat{m}(z)\hat{\chi}^{o-1}]^{-1}\, \hat{m}(z) . \qquad \text{(III.5)}$$

Note that for $z \rightarrow 0$ Eq. (III.5) yields the static matrix susceptibility. Mori's method is the subject of Müller-Hartmann's /13/ chapter and will not be repeated here. Details of its application to the present example can be found in /7-9/. The components of the static susceptibility are obtained through the Brillouin-Wigner partition function discussed above and the components of the relaxation kernel matrix are approximately expressed by one scalar relaxation function N(z). The relaxation function N(z) contains the information on the dynamics of the f-electrons. The charge susceptibility and the electrical resistivity are also determined by N(z). At high temperatures, i.e. $T \gg \Gamma = \pi \rho V^2$, perturbation theory is valid and the imaginary part of N(z) is given by

$$N''(\omega) = -(\Gamma/2\omega)(\langle B_{J_1}\rangle + \langle B_{J_2}\rangle)\left[\mathrm{Tanh}\left(\frac{\omega-\Delta E}{2T}\right) + \mathrm{Tanh}\left(\frac{\omega+\Delta E}{2T}\right)\right]. \qquad \text{(III.6)}$$

The relaxation rate is then proportional to $N''(\omega=0)$ divided by the susceptibility of one of the multiplets. This result was first derived by Kuramoto and Müller-Hartmann /14-16/.

The expression (III.6) shows already some of the interesting features of the dynamic susceptibility. Beside the quasi-elastic peak at zero frequency governed by $N''(0)/\chi^o$ there are inelastic structures in the susceptibility described by the Tanh. They correspond to transitions from the hybridized bonding states with angular momentum J_2 to the hybridized antibonding multiplet J_1. This transition requires the energy ΔE. At high temperatures the energy is provided by the thermal bath. When T is lowered a threshold for excitations /17/ builds up at $|\omega| \gtrsim \Delta E$. At low T the energy for the transition must be provided by an external source, i.e. neutrons or photons.

Part of this threshold behavior remains valid if (III.6) is calculated in higher order in V than just the second. The evaluation of (III.6) involved a free f-propagator, i.e. we assumed the f-state to be infinitely lived. An improved expression can be obtained by replacing this propagator by a dressed one. The conduction electron states can be considered as not affected by a single impurity and the hybridization is not renormalized according to Haldane's /12/ arguments. In this way we may interpret N(z) as the decay of a f-number operator into a f-particle mode and a conduction electron mode, which we evaluate in the spirit of a mode-mode coupling approach. Details of this approximation scheme are found in /18/.

An alternative way to evaluate the relaxation function N(z) in the interacting system is the Brillouin-Wigner method. Here N(z) is approximated by a sum of four continued fractions /7,8/. The result is very similar to the mode-mode coupling approach.

The dynamic susceptibility of a Ce-impurity is shown in fig. 1 for a high and a low temperature. At high T the response function has a lorentzian shape, while at low T we have the quasielastic peak at $\omega = 0$ and the inelastic bump due to the transitions from the bonding to antibonding states. The relaxation rate, i.e. the width of the quasielastic peak, is only weakly temperature dependent. The inelastic peak is a very broad structure, which however covers 60 % (shaded area) of the susceptibility sum-rule

$$\int \frac{d\omega}{\pi}\, \chi''(\omega)/(\omega \chi^0) = 1 \quad . \qquad \text{(III.7)}$$

The deviations from a lorentzian shape could in principle be used as a criterion to distinguish among a mixed-valent Ce-impurity (with valence $\nu > 3.6$) and a Kondo impurity for which the deviations are only logarithmic and hence negligible.

The inelastic transition also appears at low T in the dynamic susceptibility of a Tm-ion. It is even broader than for a Ce-impurity since there are more hybridizing channels for $J_2 = 7/2$ than for $J_2 = 0$. Its width and height cannot possibly account for the experimentally observed inelastic peak in TmSe and in dilute Tm-systems /19,20/. Since the ion is magnetic at low T the relaxation rate follows a Korringa law and is linear in T. For $T > \Delta E$ perturbation theory is valid and the relaxation rate saturates. This behavior is qualitatively in agreement with the experimental facts and shown in fig. 2.

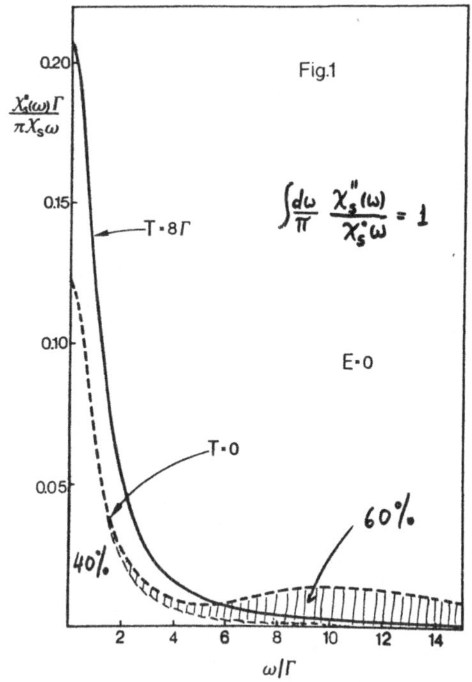

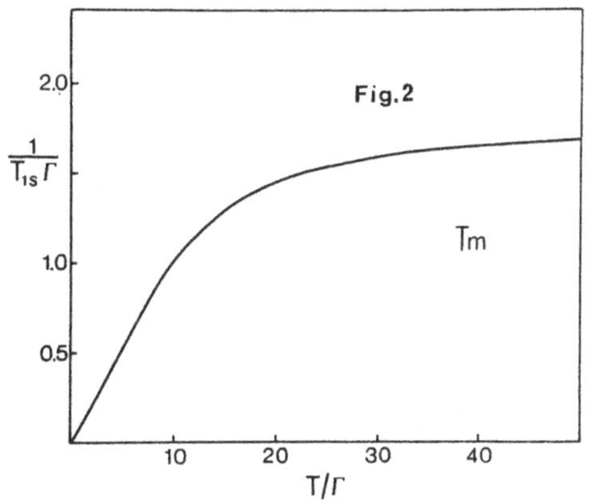

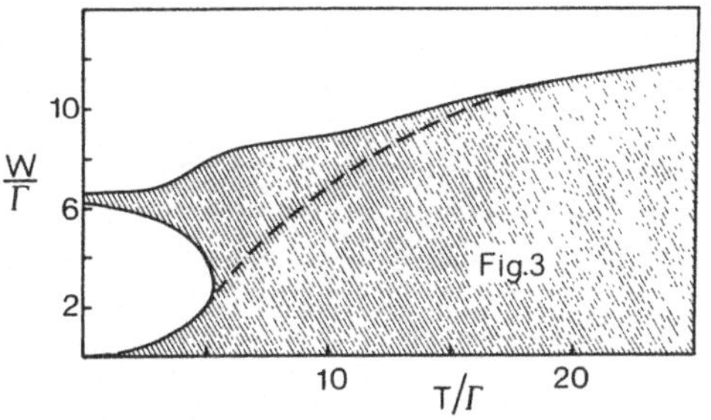

Fig. 1: Dynamic susceptiblity of a Ce-impurity as a function of frequency for high and low T. At low T an inelastic peak develops at $|\omega| \gtrsim \Delta E$. Note that the area under the curves is normalized to 1. The parameters are $J_1=j=5/2$, $J_2=0$, $g_{J_1}=6/7$, $D=100\pi\Gamma$ and $E_{J_1}=E_{J_2}$.

Fig. 2: Spin-relaxation rate for a Tm-impurity as a function of T. Parameters: $J_1=6$, $J_2=7/2$, $g_{J_1}=7/6$, $g_{J_2}=8/7$, $D=100\pi\Gamma$ and $E_{J_1}-E_{J_2}=10\Gamma$.

Fig. 3: Schematic representation of elastic and inelastic crystal field peaks of a mixed-valent Tm-impurity. Parameters as in fig. 2 and $10b_4=4\Gamma$.

An alternative explanation for the inelastic peak in TmSe (which in some samples is observed, in others not) is a crystal field transition. Since the Zeeman-Hamiltonian does not commute with the crystal field Hamiltonian a large number of inelastic scattering transitions between the crystal field levels take place. The crystal field splitting is only important at low T, where most of the levels have only a weak occupancy within the regime of validity of our approximation. The intensity of most of the transitions is then small at low temperatures. Since ΔE is large the most important one occurs within the J_2-multiplet. We make the following approximations (not all of them have a clear justification): (i) we consider only the susceptiblity of the ionic J_2-multiplet, (ii) neglect the crystal field splitting of the ionic J_1-multiplet and (iii) consider only fourth order cubic crystal field splitting of the ionic J_2-multiplet. The neglect of the susceptiblity of the J_1-multiplet strongly affects quantitatively the results /9/ at low T and ω. In fig. 3 we show the main peaks and their half-widths for $10b_4=4\Gamma$ and otherwise the same parameters as in fig. 2.

We should finally discuss the validity of the approximations made above. The exact groundstate properties of the model for $J_2=0$ and arbitrary $j=J_1$ have been obtained by means of Bethe's ansatz /21/. The spin susceptiblity consists of two terms, a Kondo contribution and a mixed-valence contribution. The Kondo effect is not included in the approximations discussed in the present paper. The approximations are then valid if the Kondo part is much smaller than the mixed-valence contribution. This is the case for $n_f < 0.3$ at T=0 or $E=E_{J_1}-E_{J_2} > 0$ for $J_1=5/2$. The range of validity is therefore considerably smaller than postulated by Ramakrishnan /11/. For the Tm the approximation is restricted to $E > 5\Gamma$, since the convergence of perturbation theory gets worse with increasing J_2.

REFERENCES

1. P. Schlottmann, J. Magn. Magn. Mater. 7:72 (1978); J. Phys. (Paris) 39:C6-1486
2. P. Schlottmann, Solid State Commun. 38:1087 (1981)
3. P. Schlottmann, Phys. Rev. B 25:4815 (1982); Phys. Rev. B. 25:4828 (1982); Phys. Rev. B 25:4838 (1982)
4. K. D. Schotte and U. Schotte, Phys. Rev. B 4:2228 (1971)
5. W. Götze and P. Schlottmann, Solid State Commun. 13:17 (1973); J. Low Temp. Phys. 16:87 (1974)
6. K. G. Wilson, Rev. Mod. Phys. 47:773 (1975)
7. P. Schlottmann, in Valence Instabilities, edited by P. Wachter and H. Boppart (North-Holland, Amsterdam, 1982), p. 471
8. P. Schlottmann, Phys. Rev. B, in press
9. P. Schlottmann, preprint
10. A. Bringer and H. Lustfeld, Z. Physik B 28:213 (1977)
11. T. V. Ramakrishnan and K. Sur, Phys. Rev. B 26:1798 (1982)
12. F. D. M. Haldane, Phys. Rev. Lett 40:416 (1978)
13. E. Müller-Hartmann, chapter in this book
14. Y. Kuramoto, Z. Phys. B 37:299 (1980)
15. Y. Kuramoto and E. Müller-Hartmann, in Valence Instabilities in Solids, edited by L. M. Falicov, W. Hanke and M. B. Maple (North-Holland, Amsterdam, 1981), p. 139
16. E. Müller-Hartmann, in Electron Correlation and Magnetism in Narrow-Band Systems, Proceedings of the Third Tanaguchi International Symposium, Mount Fuji, Japan (Springer-Verlag Berlin, 1981), p. 178

17. J. Mazzaferro, C. A. Balseiro and B. Alascio, Phys. Rev. Lett. 47:274 (1981)
18. P. Schlottmann, Phys. Rev. B 25:2371 (1982); ibid 26:3474 (1982)
19. M. Löwenhaupt and E. Holland-Moritz, J. Appl. Phys. 50:7456 (1979)
20. E. Holland-Moritz, private communication
21. P. Schlottmann, Z. Phys. B 51:49 (1983)

SUPERCONDUCTING GROUND STATE OF A STRONGLY INTERACTING ELECTRON SYSTEM : UBe_{13}

H.R. Ott* and H. Rudigier*

Laboratorium für Festkörperphysik
ETH-Hönggerberg, 8093 Zürich
Switzerland

Z. Fisk+ and J.L. Smith+

Los Alamos National Laboratory
Los Alamos, New Mexico 87545

INTRODUCTION

Very recently it has been demonstrated that UBe_{13}, in spite of its seemingly unfavourable normal-state properties, is a bulk superconductor with a critical temperature T_c of about 0.9 K[1]. The occurrence of superconductivity in UBe_{13} is particularly surprising since the non-magnetic compounds $LaBe_{13}$, $LuBe_{13}$ and $ThBe_{13}$ are not superconducting above 0.45 K[2]. Both the superconducting and normal state properties of UBe_{13}, respectively, clearly indicate that they must be dominated by the presence of 5f electrons on the U ions. The low-temperature properties of UBe_{13} are indeed very similar to those of $CeCu_2Si_2$, a compound whose anomalous behaviour is thought to arise from strong interactions of conduction electrons with the 4f electrons of the Ce ions[3]. Here we should like to discuss some of the experimentally determined properties of UBe_{13}, mainly intending to initiate further theoretical work considering the possible ground states of strongly interacting electronic systems in solids.

---

*Work supported by the Schweizerische Nationalfonds zur Förderung der wissenschaftlichen Forschung

+Work performed under the auspices of the U.S. Department of Energy

EXPERIMENTAL RESULTS AND DISCUSSION

From x-ray measurements we deduce a room-temperature lattice constant of our UBe_{13} samples of about 10.26 Å. According to the crystallographic work of Bänziger and Rundle[4] this results in a nearest U-U distance of 5.13 Å, prohibitive for a sizeable direct f-f overlap between U ions. It is therefore not surprising but expected that, upon cooling from room temperature, the magnetic susceptibility χ follows a Curie-Weiss law, as shown in Fig. 1.

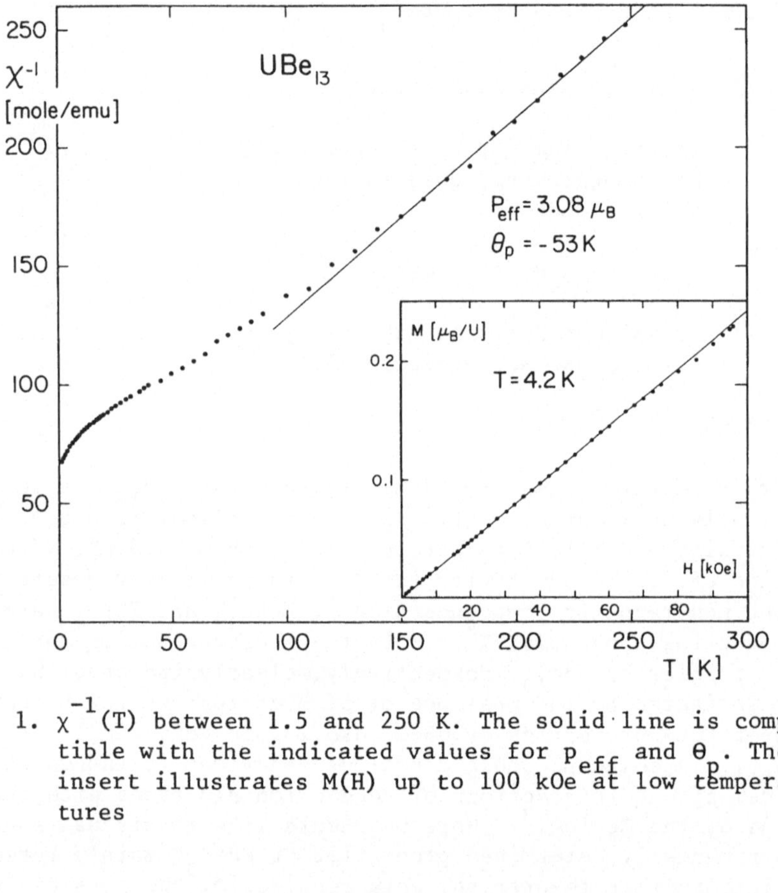

Fig. 1. $\chi^{-1}(T)$ between 1.5 and 250 K. The solid line is compatible with the indicated values for p_{eff} and Θ_p. The insert illustrates M(H) up to 100 kOe at low temperatures

The slope of $\chi^{-1}(T)$ above 150 K corresponds to an effective moment $p_{eff} = 3.08\ \mu_B$, and the data in this temperature range are consistent with a paramagnetic Curie temperature $\Theta_p = -53$ K. The quoted value for p_{eff} is larger than the effective moment of a free U ion with one localized 5f electron of 2.54 μ_B, but also considerably lower than 3.58 μ_B and 3.62 μ_B, the free-ion moments for an ionic configuration of the U ions with 2 and 3 f electrons, respectively.

Below 150 K, $\chi^{-1}(T)$ is reminiscent of that of various rare-earth compounds where such a behaviour is successfully attributed to crystal electric field (CEF) effects on a well defined Hund's rule ground state of localized f electrons[5]. At low temperatures, χ rises steadily to about 1.5 x 10^{-2} emu/mole before entering the superconducting state.

Low-temperature magnetization M(H) data (an example is shown in the insert of Fig.1) raise first doubts that the properties of UBe_{13} can be reasonably well described by assuming a definite 5f configuration of the U ions and using conventional calculations taking into account possible CEF effects[6]. As may be seen, M(H) varies linearly up to 100 kOe, a feature which is definitely not expected if possible dispersions of f electron states with increasing magnetic field are considered. Moreover it should be noted that the apparent moment at 100 kOe has only about half the value of that of UAl_2 at the same temperature[7]. Both the large negative value of Θ_p and the insensitivity of χ with respect to external magnetic fields indicate that strong antiferromagnetic interactions play a significant role.

Quite anomalous for an intermetallic compound is the temperature dependence of the electrical resistivity ρ as shown in Fig.2. $\rho(T)$ steadily increases with decreasing temperature below room temperature, saturating at about 230 $\mu\Omega$cm at 10 K. At still lower temperatures, a distinct anomalous increase of ρ with a maximum at 2.4 K is observed. The subsequent sharp decrease is then intercepted by the superconducting transition at 0.9 K.

It has been realized for some time that many superconducting compounds containing d- or f transition metals reveal anomalous temperature dependences of ρ below 300 K[8]. In addition, similar resistive behaviour to that shown in Fig. 2 has also been observed in various Ce compounds, including as examples, $CeA\ell_3$[9] and $CeCu_2Si_2$[10,11]. In those cases, Kondo-type interactions[9,12], virtual-bound-state formation[13] and scattering of conduction electrons at CEF levels of the 4f electrons[11] or combinations of these scattering mechanisms[9,11-13] were alternatively made responsible for the observed features. $\rho(T)$ curves with general features like those shown in Fig. 2 were, in context with $Ce_{1-x}La_xA\ell_3$ compounds[9] reproduced by calculations applying the concept of Kondo side-bands[14], which takes into account the possible CEF splitting of the f electron Hund's rule ground state. Since, however, preliminary inelastic neutron-scattering experiments[15] show absolutely no sign of any CEF excitations in the required temperature and energy ranges, we doubt that an explanation based on Kondo side-bands is appropriate. Of course, it is also rather unlikely that an ordinary electron-phonon interaction can give rise to an electrical resistivity as observed and any solution of this problem will have to consider the temperature dependence of the interactions of a highly correlated electron system.

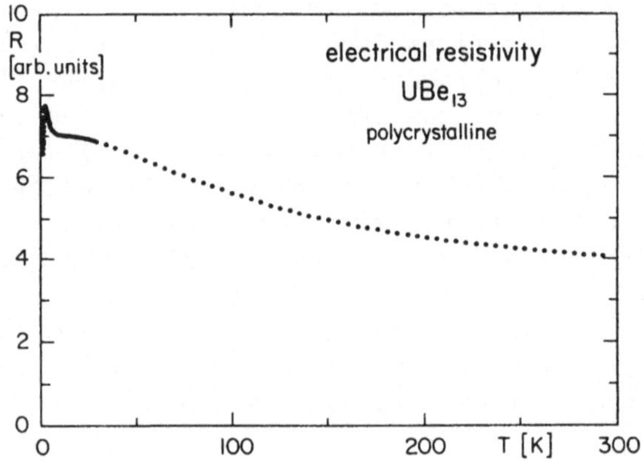

Fig. 2. Temperature dependence of the electrical resistivity of UBe_{13} between 1 and 300 K.

This conjecture is mainly based on results of low-temperature specific-heat measurements, which are shown in Fig.3. If we compare these data with the specific heat of $ThBe_{13}$, consisting of an electronic and a lattice contribution and reaching about 250 mJ/mole K at 13 K^2, we realize that UBe_{13} has an extremely enhanced low-temperature specific heat which, in principle, might be of conventional magnetic origin. As we have indicated above and shall discuss below, various experimental facts are in contrast with such an interpretation

First we consider $c_p(T)$ above 1 K. A plot of c_p/T vs T^2 is shown in Fig.4, in comparison with experimental data for $CeAl_3$ [16]. Above 7.5 K, c_p/T = 185 mJ/mole K^2 and is constant with temperature. Below 7.5 K, c_p/T rises with increasing slope, leveling off at approximately 0.9 J/mole K^2 around 1 K. This indicated that at elevated temperatures UBe_{13} has a high renormalized density of electronic states leading to an electronic specific-heat parameter γ = 185 mJ/mole K^2.

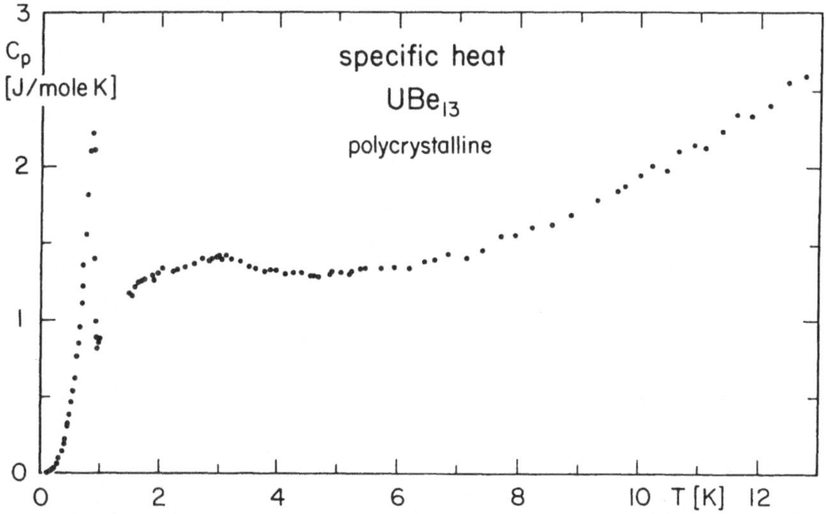

Fig. 3. Low-temperature specific heat of UBe_{13}.

The rather strong increase of c_p/T below 7.5 K is probably due to an extremely narrow resonance in the density of electronic states curve, situated directly at the Fermi energy E_F, as was postulated previous-

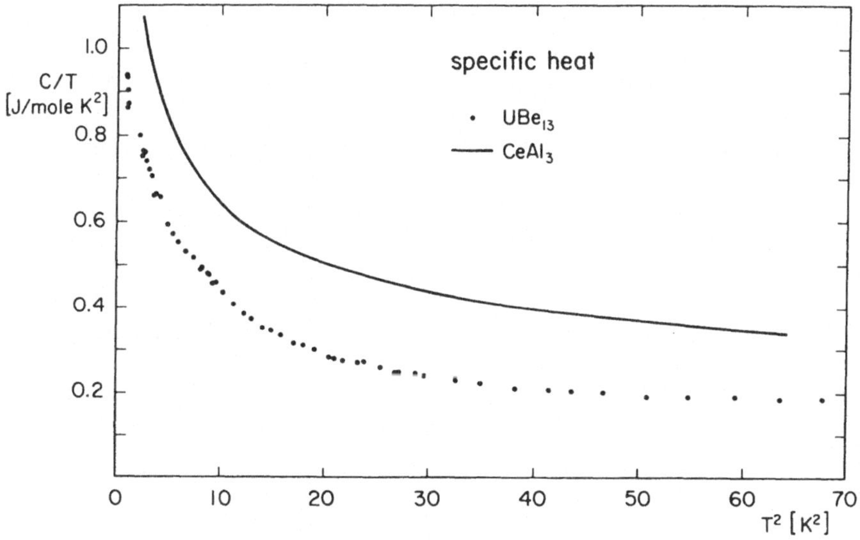

Fig. 4. c_p/T vs T^2 for UBe_{13} and $CeAl_3$[16]. Lattice contributions have been subtracted.

ly for $CeAl_3$[13]. The data for this latter compound (see Fig.4) underline the similar behaviour of these two materials above 1 K. They appear to provide a physical realization of the many-body resonance at E_F forming in the energy spectrum of strongly interacting electronic systems. While $CeAl_3$ appears to adopt a highly correlated state below 1 K it remains normal conducting to below 10 mK[13]. UBe_{13}, however, is superconducting below 0.9 K[1] and the bulk nature of this superconducting state is most convincingly demonstrated by the observation of the specific-heat anomaly at the transition as shown in Fig. 3. As happens in $CeCu_2Si_2$[3], the superconducting state of UBe_{13}, how-formed among particles with exceptionally large effective masses $m^* = 200\ m_e$[1], as indicated by the enormous step of c_p at the transition and the subsequent reduction of the entire entropy still available at T_c. Accordingly the slope $\partial H_{c2}/\partial T$ at T_c is gigantic and amounts to -260 kOe/K[1].

The facts presented here suggest that the superconducting state of UBe_{13} may not primarily be due to conventional electron-phonon coupling but rather results from interactions of magnetic origin, probably of antiferromagnetic nature. More experiments, intended to verify this conjecture are currently under way.

REFERENCES

1. H.R. Ott, H. Rudigier, Z. Fisk, and J.L. Smith, Phys.Rev.Lett. 50, 1595 (1983)
2. E. Bucher, J.P. Maita, G.W. Hull, R.C. Fulton, and A.S. Cooper, Phys.Rev.B 11, 440 (1973)
3. F. Steglich, J. Aarts, C.D. Bredl, W. Lieke, D. Meschede, W. Franz, and H. Schäfer, Phys.Rev.Lett. 43, 1892 (1979)
4. N.C. Bänziger, and R.E. Rundle, Acta Crystallogr. 2, 258 (1949)
5. see e.g. F. Hulliger and H.R. Ott, Z.Phys.B 29, 47 (1978)
6. K.R. Lea, M.J.M. Leask, and W.P. Wolf, J.Phys.Chem.Solids 23, 258 (1962)
7. J.J.M. Franse, P.H. Frings, F.R. de Boer, A. Menovsky, C.J. Beers, A.P.J. van Deursen, H.W. Myron, and A.J. Arko, Phys.Rev.Lett. 48, 1749 (1982)
8. Z. Fisk, and A.C. Lawson, Solid State Commun. 13, 277 (1973)
9. K.H.J. Buschow, H.J. van Daal, F.E. Maranzana, and P.B. van Aken Phys.Rev.B 3, 1662 (1971)
10. B.C. Sales and R. Viswanathan, J.Low Temp. Phys. 23, 449 (1976)
11. W. Franz, A. Griessel, F. Steglich, and D. Wohlleben, Z.Phys.B 31, 7 (1978)
12. B. Cornut, and B. Coqblin, Phys.Rev.B 5, 4541 (1972)
13. K. Andres, J.E. Graebner, and H.R. Ott, Phys.Rev.Lett. 35, 1779 (1975)
14. F.E. Maranzana, Phys.Rev.Lett. 25, 239 (1970)

15. J.K. Kjems, H.R. Ott, and S.K. Sinha, unpublished results
16. P. Delsing, H.R. Ott, and H. Rudigier, unpublished results
17. see e.g. K.D. Schotte, and U. Schotte, Phys.Lett. 55A, 38 (1975)

PRESENT STATUS OF THEORY: 1/N APPROACH

Philip W. Anderson

Bell Laboratories
Murray Hill, N. J. 07974
and
Joseph Henry Laboratories
Princeton University
Princeton, N. J. 08544

When I began to prepare these lectures, and to wonder what I could possibly say that hadn't already been said by all the brilliant lecturers you have already had, I began to realize how many years I have actually been associated with the problem of "Moment Formation". As I have often explained, I don't actually quite go back all the way to the first exploration of this problem, since the virtual state concept was originated in fact by Blandin and Friedel in early 1959[1] or so, and I heard a discussion of it by Blandin in 1959 at Oxford, which stimulated me to develop the model of which so much is made in the mixed valence field.

In all of these years I have learned a number of theorems and quasi-theorems about the moment problem, and developed a rough physical understanding of what is going on, which is usually proof against the great complexity of what we see in real systems. In fact, until recently I had come to think that the mixed valence problem was almost a trivial exercise in physical understanding, but more recently I have come to believe that there are real and very interesting complexities which are much worth discussing, and which do indeed go beyond the primitive, fundamental understanding which one can achieve with simple physical concepts.

Let me list some of these concepts in some kind of order, mostly historical, so that you will get some idea of where I am going and whether or not to go off and play tennis instead of listening.

(1) The Compensation Theorem. Concepts of admixture and of polarization and that admixture ≈ polarization: so net Δm is in localized state.

(2) Friedel sum rule and general idea that chemical energies >> magnetic energies: net occupation is fixed, only dynamics changes. (For "impurity" systems and Coulomb interactions this leads to the Friedel overall sum rule, but in fact mostly each type of electron is conserved: "rule of fixed valence".)

(3) Friedel and Luttinger theorems. Friedel: number of electrons of a given type = (sum of phase shifts/π); Luttinger: Fermi surface volume = number of electrons. General, relating statement: charge of quasiparticles is not renormalized: no e* (in spite of Falicov).

(4) Scaling ideas.

(A) Ground State Fixed-Point Theorem: Ground states are fixed points of scaling. Three types:

(a) Fermi Liquid
(b) Self-trapped (Ferromagnetic Kondo)
(c) Broken Symmetry. (Magnetic or Superconducting)

(B) Kondo Scaling: From magnetic impurity to Fermi liquid.

(C) Haldane Scaling: Sinking of the singlet and the role of degeneracy.

(5) Finally, the $1/N_f$ idea: weakness of spin-spin interactions and validity of perturbation theory ensured by large degeneracy of the f level.

To complete my preview of what I'm going to say, I should list my more recent problems and thoughts about why it is not all so simple as that:

(1) The Haldane-Kondo crossover and emergence of the "Kondo resonance" in experiment and in 1/N theories.

(2) Absence of phonon effects: is this wavefunction renormalization and the weakness of the Kondo resonance?

(3) Superconductivity: what is the role of finite U and the N = 2 singlet?

----

(1) Let me start out, then, with the compensation theorem and

the birth of the Anderson model. The compensation theorem is contained in two abstracts by Al Clogston and myself written in late 1960,[2] trying to understand magnetic resonance data on such magnetic impurity systems as Mn in Cu. Magnetic resonance, both nuclear and electron, was what we had in those days to study magnetic systems, and there was a great puzzlement as to what extent the free electron spins were part of the local moments of the Mn, to what extent they relaxed them (the Korringa relaxation and the Kondo model of free-electron Mn-spin interaction, actually introduced in those days by Yosida), and to what extent the Mn spins polarized the free electrons. These abstracts were the origin of the Anderson model, which was there first studied by simple perturbation theory. But what they emphasized is that a local (say "f") electronic state has, in simple perturbation theory, two effects on the free electron band, not one: first, its wavefunction is admixed into the free electrons' wavefunctions and vice versa, which leads to no net change in moment but a change in the wavefunction or form factor, interchanging spin in the two states; and second, the effective energies of the free electron states are modified, shifting them relative to the Fermi level and causing a net polarization which - since it effectively takes place at $r = \infty$ - is surely only in the free electrons and is antiferromagnetic due to the fact that the filled spin state is below the Fermi energy and repels electrons. It is a common tendency to neglect one or the other of these two effects - in the case of mixed valence, most often admixture, which in many cases gives the whole of the effect - but between the two of them, in the case of a reasonably broad band they approximately cancel and the appearance is of a net decrease in local moment with no effect on the surroundings - a result which usually remains true right through the Kondo and mixed valence regimes, and has confused and disappointed successive generations of NMR specialists. This is why the recent experimental results on $CeSn_3$ are exciting.

(2) The second basic physical concept which is quite important and has not been adequately treated is the question of sum rules and chemical stability. The simplest and most rigorous expression of this idea is the Friedel sum rule, which simply expresses the idea that a metal must always be locally neutral. (Even an insulator normally sustains only slight deviations from local neutrality.) The main "Anderson" and "Kondo" models do not contain charge neutrality automatically and must be carefully handled to do so, which is the first reason which makes "rigorous" solutions of these models somewhat less relevant than is apparent on the surface. The phase shift sum rule of Friedel which I will shortly discuss is absolute. More generally, it is very rare that the valence state of an atom really changes substantively as a function of ordinary temperatures or pressures. The energy differences among valence states, while they can be relatively small, are still expressed in electron volts rather than degrees, and except in very exceptional circumstances, a true "valence transition" by a whole unit will

almost never really take place. This is true even in much simpler and less obvious cases: for instance, the valence state at Na in Na metal and in Na^+Cl^- is essentially indistinguishable within a sphere about the Na ion of size equal to the metallic radius, and similarly for Cl.

The distinction between real and nominal valence is like that between "bare electrons" and "quasiparticles" in Fermi liquid theory. The number of "bare" d or f electrons around an atom is very firmly fixed to within at most a couple of tenths of an electron; but the nominal a valence can change quite radically. The most spectacular example of this is the case of transition metal impurities in semiconductors, where magnetic valence can change by several units, while, as Haldane and I suggested[3] and George Watkins proved,[4] the actual number of d electrons changes not at all.

The nominal valence is describable in a number of ways. One way is to ask what the appropriate starting point would be for a perturbation theory which would continuously connect with the actual electronic state of the system. For NaCl, for instance, the appropriate starting point is a set of Wannier functions resembling the Cl^- atomic functions; these will spill over into empty s functions on the Na^+, but one could essentially start putting the solid together from Na^+ and Cl^- ions and never encounter a discontinuity. Na metal, on the other hand, would be best described by starting from Na^+ ions and a free electron gas.

It is in this sense that a substance like SmB_6 or SmS in the insulating state has a nominal valence Sm^{++}, even though when one studies the actual number of f electrons on Sm it is more nearly equal to Sm^{+++}, and the metallic state of SmS has a nominal valence intermediate between the two. In the pressure transition the number of bare f electrons does not actually change very much, as is also the case in Ce metal, because in spite of the large volume change there is not much electronic energy involved.

The final discussion on Friday left the subject of valence in considerably clearer state than my first thoughts above. I think my definitions above remain clear and valid, but it was pointed out that the real valence can change rather radically under some circumstances, as, e.g., in the pressure transition of SmS.

The conclusion we came to is that in normal chemical bonding the matrix elements connecting the various atomic states are much more sensitive to volume or crystal structure than are the atomic states, and hence what will usually happen is a change in the bonding character - in the sense of ionic to covalent, for instance, as in one of the borderline II-VI compounds. A perfect example of this kind of thing was given by Wohlleben, where he shows that the large volume change in Ce metal is accompanied by a very large change in the mixing matrix element and hence in T_F, but little or no net change in valence. That is, the f shell is changing from essentially metallic to ionic bonding character.

On the other hand, where the matrix elements are and stay small, either because of strong intra-f-shell correlations as in Sm or because of a smaller f shell as in Yb, the real valence can change fairly radically. However, it is interesting that in few or none of these compounds does real valence change with T. For example, in SmB_6 there is a big change in nominal valence as defined above, from Sm^{++} at low temperature to 2.6 at high, as determined from the magnetic symmetry character, but no change at all in real valence.

At this point it is perhaps necessary to discuss a side development which has not yet taken its proper place in the mythology or, to be quite honest, in my own mind, the "first order" valence transitions which seem to appear in Hartree-Fock theory when one takes screening properly into account. These were first emphasized by Duncan Haldane[5] and a more realistic, but physically less satisfactory, model has been studied by Schluter and Varma[6]. I think Haldane had the physics correct in his oversimplified model and I will follow him.

Haldane's way of dealing with them was to observe that screening of the "f" charge by the s-d channels could be quite accurately modeled by including a corresponding Falicov-Kimball type of term of form $\lambda n_f n(o)_{free}$ in the Anderson Hamiltonian and then writing $n(o)_{free}$ as the sum of the amplitudes of a set of Tomonaga bosons. In a static Hartree-Fock approximation that just gives you an effective "negative U" contribution $-c(\overline{n}_f)^2$ to the f-electron energy, and this, when balanced against the magnetic energies, makes the valence transition of the Anderson model somewhat first order where, in the simple case, it is second order (I show you, in Figure 1, typical phase boundaries). This is not surprising since

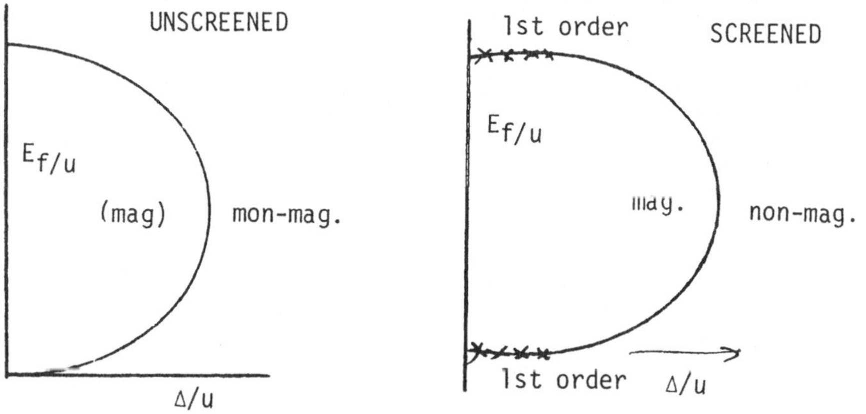

Fig. 1

at the edge of the magnetic region n_f is changing vertically with the parameter E_f and one might expect an effective negative U to recurve it.

I need hardly remind you that much physics is missing here, and specifically that the original "second order" transitions were spurious in that quantum fluctuations spread them out into an entire Kondo-like crossover phenomenon. This will also be true of the first-order jumps. Nonetheless, I believe with Haldane that the screening phenomenon plays a role of considerable but unknown magnitude in at least two ways: in sharpening up the valence transition, so that more physical cases occur near to it, and in reducing the effective f-level width Δ. One may in fact question whether these two roles cannot be combined by simply remarking that all calculations - such as those of Piers Coleman - seem to give f-level mixing integrals V_{ff} and Δ's which are too large, from both points of view, and the renormalization due to screening is most welcome.

When we ask ourselves - as I often did in the early days - why, if the parameter E_f needs to be so very close - within a few hundred ^0K - to the central value, so many valence transitions take place, this kind of scale compression is perhaps useful. But actually I now feel this is a non-problem, since the real physics should be much more directly based on the parameter $\bar{n}_f$, which leaves one looking at the Anderson model graph from quite a different perspective (see Figure 2). Screening makes it even flatter.

The confusing f wavefunction transition of Varma and Schluter appears in their model because they do not adequately allow for an f-character free electron channel, which is necessary to accommodate the change in f amplitude in the Haldane first-order transition.

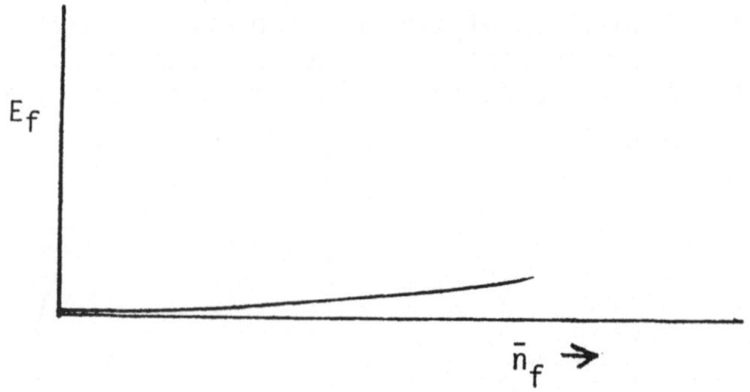

E_f varies slowly with $\bar{n}_f$

Fig. 2

(3) The third basic physical principle is the rigorous theorems which follow from the fact that interactions cannot modify the charge of a quasiparticle, i. e., they conserve numbers of electrons. This means that the "real" state and the "nominal" state must refer to exactly the same number of electrons. For the nominal, or "reference" state, the number of electrons may be obtained essentially by counting nodes of the wavefunctions of the last occupied electronic states, by an ancient theorem of differential equation theory. This is the real source of the Friedel and Luttinger theorems. In the case of an individual atom in a metal, the form such a "node-counting" theorem takes is the Friedel identity for the sum of scattering phase shifts:

$$n = \left(\sum_{\ell,m} \delta_{\ell,m} \right) / \pi$$

Thus, if, as in the Kondo system, there is a transition between a magnetic state at high temperature and a non-magnetic one at low temperature, the sum of the scattering phase shifts for the two signs of spin must remain constant and as a result the two values $\delta_{occ} = \pi - \varepsilon$ and $\delta_{empty} = 0 + \varepsilon$ (giving n = 1) must change to $\delta_{\uparrow} = \delta_{\downarrow} = \pi/2$. It follows that the resistance due to magnetic scattering

$$\rho \propto \sum_{\ell,m} \sin^2\delta_{\ell m} \ ,$$

must rise spectacularly and reach the unitarity limit for the appropriate channels. This theorem is especially effective in the mixed valence case, in that, once we recognize that the number of f channels is large (n_f = 6, 8 or 14 depending on spin-orbit coupling) and that we are always interested in accommodating only one electron at most, we recognize that in the non-magnetic state $\delta_f \simeq \pi/n_f$, or that if there is a single "f" scattering resonance, the Fermi level must lie on the edge of that resonance. This is the primary source of the advantage of "large n_f" methods: that the resonance never actually sits right <u>at</u> the Fermi level. (Although, as we will see, in the "Kondo limit" the resonance may be very close by.)

The corresponding theorem for the mixed valence or Kondo lattice problem is the Luttinger theorem: that the volume of the Fermi surface must remain unchanged - the volume of the Fermi surface in k space again being a sophisticated way of counting nodal surfaces of wavefunctions. The most spectacular instance of the operation of this theorem occurs in the mixed valence insulators SmS and SmB_6, where Sm, in the mixed valent ground state, assumes the "nominal" valence 2, even though its real f occupancy is closer to Sm^{+++}. Hence it assumes a non-magnetic, J = 0 ground state, and the electron count becomes exactly the equivalent of a simple ionic compound $(X)^{++}$ S^{--} or $(X)^{++}$ $(B_6)^{--}$. Since B_6^{--} and S^{--} are closed shell ions, there can be - and, in fact, should be - a gap at the Fermi surface, which, though extremely small, is observed in both cases. The equivalent, though less spectacularly, will in general hold for all non-magnetic mixed valence ground states: the Fermi surface volume and symmetry will be the equivalent of that for an ordinary non-magnetic atom of the same nominal valence, even though the effective masses can be very much different.

One may very heuristically relate the two phenomena by observing that in the various forms of Green's functions or muffin tin-based methods, and even more in pseudopotential methods, the atom is generally replaced by its effective nonlocal potential, which is simply the phase shifts or quantum defects for the different angular momentum channels. For states precisely at the Fermi level, these must add up to give the appropriate number of f electrons by having the appropriate phase shifts in the f channel.

On the other hand, you will note a slight difference between the single atom version of these theorems, and the coherent one which is going to hold for the ground state of a regular crystal in the Fermi liquid regime. In the incoherent, atomic case, the f resonance actually intersects the Fermi surface and there are necessarily a small number of "nominal" f electrons, i.e., f quasiparticles, although they exist only as the edge of a scattering resonance, admixed into the free electron band. In the Fermi liquid case, recognizable bands of tight-binding, f-like symmetry need not, and in general will not, cross the Fermi surface; instead there will be a strong f-like pseudopotential which affects the ordinary s-d bands of the background metal.

(4) Scaling ideas.

About 1970 we began to understand that the secret to the conceptual understanding of the various Hamiltonians which had been proposed for moment formation problems is the basic idea of scaling or renormalization. This actually was the first entry of the renormalization group into condensed matter physics, quite independently of, and slightly before, the Wilson work on phase transitions. Although this was by no means the way it came about, the great liberating concept which scaling gives us is the idea of states as representing fixed points of a scale transformation, and in particular of ground states as the various kinds of possible stable, low-energy and low-frequency fixed points of a quantum system of the appropriate kind. We know that at high enough temperatures, the magnetic state will try to arrange itself so as to have maximum entropy, by decoupling whatever coupling constant (J in the Kondo case, or V_{mix} in the Anderson model) is involved: but the high-entropy decoupled state usually turns out to be an unstable fixed point and the system goes through a crossover to a stable fixed point.

The appropriate stable fixed point for the single impurity problems, both Kondo and mixed valence, is the so-called Fermi liquid limit. A local spin interacting with a Fermi sea cannot be a low-energy fixed point because of the quantum process of spin-flip scattering. Thus by far the most common and important $T = 0$ fixed point is the Fermi liquid: the replacement of the magnetic spin by an effective non-magnetic center. For visualization purposes, we imagine the local spin binding a free electron into a bound singlet state, which then provides a rather strong scattering center for the remainder of the free electrons, but has no internal dynamics at least at $\omega \cong 0$. The strength of the scattering center at the Fermi surface is controlled by Friedel sum rules and symmetries, as was beautifully explained by Nozieres in what may be the single most important paper ever written on the Kondo effect.[7] For practical purposes, MV centers will have π/N_f scattering.

For the lattice of Kondo or MV centers a possible fixed point is also the unmagnetized Fermi liquid or, in the case where the valence is satisfactory, simply a non-magnetic gapped insulator. We emphasize that in this case the vital sum rule is the Luttinger one - the Friedel sum rules control the pseudopotentials but not the actual Fermi surface - a rather tricky point. Bands do not have pure f or d character, and a valence band which is perfectly normal and effectively non-magnetic can compensate for the extra f-electron phase shift which must lie around say a Ce atom.

Of course, in the lattice case magnetic or superconducting Fermi liquids are also allowed. As we shall see - and as Rama will emphasize for SmB_6 - either external magnetic fields or exchange fields can seriously disrupt the scaling process, and the band version of this is to point out that in the presence of an exchange gap, the magnetic state can of course become an infrared stable fixed point. In fact, any broken symmetry SDW, CDW, or other state can intervene once the Fermi liquid becomes coherent between sites.

To anticipate the 1/N physics which you have been hearing about, it is important to note that the establishment of coherence, and in fact all interatomic interactions, are basically a 1/N effect: the electron goes out in 1 of N channels and has to come back in the <u>same</u> one to be coherent. Thus we hope there is a temperature regime of incoherence where the single impurity is a valid point of view, above a temperature regime where coherence and order sets in.

Finally, there is indeed the possibility of a <u>decoupled</u> ground state - the ferromagnetic Kondo is the canonical case, but we can also imagine a self-trapped small polaron type of state;

such states are of importance in other problems and the chemists' mixed valence compounds are usually of that sort - but for the nature of our conference we ignore them.

(B) Kondo Scaling.

The "Kondo" scaling is appropriate when the mean field theory of the Anderson Hamiltonian predicts a stable moment - i.e., well inside the original Hartree-Fock stability region. In the f-series the appropriate spin interaction was derived by Coqblin and Schrieffer. The physics of Kondo is that in the normal case of antiferromagnetic interaction, the spin attracts the electrons with which it can undergo spin-flip scattering, hence increasing the effective interaction indefinitely for the lowest-energy electrons. The simplest way to do the first stages of this scaling is by the "poor man's method" of continually reducing the cutoff D, leading to the famous scaling equations for the two (dimensionless) coupling constants $J_\pm$ and J_z, which are of course equal for isotropic exchange:

$$\frac{dJ_\pm}{d\,\ln D} = -J_\pm J_z \qquad \text{(+ higher order)}$$

$$\frac{dJ_z}{d\,\ln D} = -J_\pm^{\,2} \qquad \text{(+ higher order)}$$

These lead to the famous logarithmic Kondo terms in resistivity, etc., when the scaling is carried out using kT as a lower limit on D. It was first recognized by Anderson, Yuval and Hamann that large J is essentially a weak coupling limit in the equivalent space-time scaling problem where one follows the history of the local spin in imaginary time rather than the coupling constant in energy space, and is hence trivial. The crossover from small to large J limits was calculated approximately by Anderson and Yuval and reasonably accurately by Armytage and then, in a famous paper, by Wilson. The space-time approach is an enormous aid to visualization and also, in itself, leads to interesting statistical problems which are of some recent interest (see J. Cardy,[8] and B. G. Kotliar's thesis[9]).

Nozieres outlined the general nature of the ground state of the Kondo system as equivalent to a moderately interactive local non-magnetic center, satisfying a number of symmetry rules, and the important sum rules mentioned above, which often are enough to determine the state nearly exactly. This is a kind of localized "Fermi liquid", and in the corresponding lattice case we of course expect the ground state also to be a highly modified and moderately strongly interacting Fermi liquid.

(C) Haldane Scaling.

F. D. M. Haldane, in his thesis and in some papers based on it,[10] pointed out the key element in the mixed valence case, after identifying the asymmetric Anderson model as the correct model for this problem, and, as we showed earlier, making it plausible that the borderline regime of that model would be fairly common. He observed that in the "poor man's scaling" regime of the asymmetric Anderson model, one encounters a relatively rapid scaling behavior when the electronic transitions carry one between atomic levels with different degrees of degeneracy N_f, and especially so when one of the levels - as in Ce, Sm and Yb - is non-degenerate and hence non-magnetic. As observed by Varma, the system always scales toward the non-magnetic valence. Haldane showed that the reason was that there are N_f possible electronic transitions out of the non-degenerate state, which lower its energy by perturbation theory, while each of the degenerate states can only hop to the non-degenerate one. Scaling the cutoff D at high energies, we obtain equations such as those given by Coleman:

$$\frac{dE_f^*}{dD} = -\frac{\Delta}{N_f \pi} \; \frac{1}{D-(E_f-E_b^*)}$$

$$\frac{\delta E_b^*}{\delta D} = \frac{\Delta}{\pi} \; \frac{1}{D+(E_f-E_b^*)}$$

Note there is no N_f factor in the second, so E_b (the effective singlet energy in this theory) scales N_f times as fast. These scaling equations lead to trajectories as shown in Figure 3 (from Haldane's thesis). Even where the non-degenerate state starts higher, we find that it scales either to a purely non-magnetic case, or to the crossover region into a Kondo limit - which itself then scales to the non-magnetic case, by Kondo scaling.

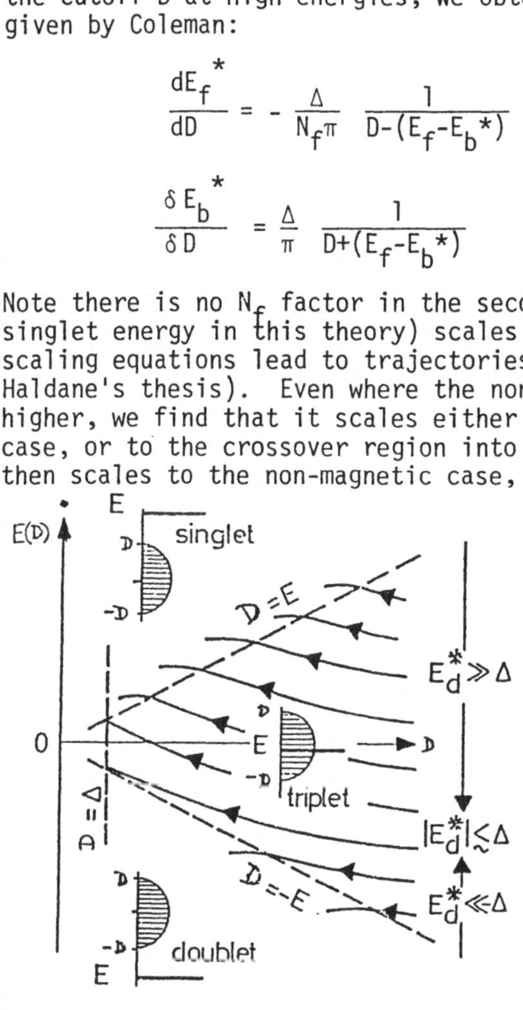

Fig. 3. Scaling trajectories E(D), ending at crossovers (broken lines) to a $\langle n_d \rangle \simeq 0$ singlet regime for $E_d^* >> \Delta$, to a $\langle n_d \rangle \simeq 1$ doublet local moment regime for $E_d^* << -\Delta$, and to a mixed valence Fermi liquid regime for $|E_d| \lesssim \Delta$.

In poor-man's scaling the crossover region is a difficult one, since here the effective "f" energy level passes out of the band through one of the cutoff energies $\pm D$. One of the energy denominators diverges and the scaling becomes more rapid as we approach the crossover. This is, however, not conceptually difficult in the space-time theory originated by Yuval, Anderson and Hamann. In fact, in his thesis Haldane shows how to carry the two theories in precisely parallel fashion and to control the crossover region. When the scaling crosses over into the Kondo region he avers that T_K is given by

$$T_K \sim \text{const.} \times \sqrt{W\Delta} \exp \frac{-\pi|E_f|}{2\Delta}$$

where W is the overall cutoff. One expects that "2" here is N_f, so the answer in mixed valence terms comes out

$$T_K \sim \sqrt{W\Delta} \exp \frac{-\pi|E_f|}{N_f\Delta}$$

which can often be quite large and is compatible with the observed situation.

The observed behavior of most mixed valence systems is compatible with the idea that the separate atoms undergo this scaling procedure more or less independently. One ends up at T_K or T_f with a set of effective non-magnetic centers, which only then establish coherence and form a band-type Fermi liquid at a lower temperature one might call T_b and might estimate to be $\sim 1/N_f$ smaller.

(5) Usefulness of $1/N_f$.

This brings us to the point of $1/N_f$ theory. In field theory, it has become customary to take advantage of the "large" degeneracy of the quarks and gluons in such theories as color gauge theory to establish a relatively simple limit where various kinds of corrections become small. Since much of the physics is still contained in the limit of large degeneracy number N_d, it is hoped (piously) that qualitative results will be correct for the observed $N_d = 3$. This was not in fact the inspiration for my suggestion of large N_f as a simplification, but it indicates additional value in it. In fact, what I initially suggested was based on the misapprehension that most mixed valence systems were in the top part of the Haldane diagram, with the scaled $E_f > 0$. In that case, many simplifications follow from large E_f. In particular, the sum rules require that the phase shift at E_{fermi} in any one channel be only π/N_f at low temperatures, which is small enough that the resonances can be treated in the original Anderson-Clogston perturbation theory, i.e., the effective scattering potential in any one channel is $\sim 1/N_f$, and one need not worry about the strong resonances which are away from the Fermi energy

A second obvious improvement is that the interactions between atoms are considerably reduced. To influence the next atom a scattered electron must travel to it and return in the channel from which it left. (This becomes obvious in any of the possible perturbation theory representations of RKKY etc. type interaction.) This means that there is an effective "dilution factor" of $1/N_f$, and this also represents a factor making incoherent random scattering more effective than coherent Bragg scattering, so that band formation takes place only after the atoms have settled down (the factor of N_f between T_b and T_F already referred to). This appears to possibly have been seen in $CeSn_3$, where the form factor change and χ peak occurs at a temperature a factor of 5 below T_F.

Now for my last thoughts. (1) As I have already remarked, I think some combination of space-time theory and $1/N_f$ techniques may get us through the Haldane-Kondo crossover without too much difficulty. The Kondo lattice problem which then results, however, is undoubtedly a hard one.

(2) The f-like quasiparticles in the Kondo lattice must be expected to have a large wavefunction renormalization factor Z (or actually, it is Z^{-1} which is large). To my knowledge, this has not been calculated but the spectral densities given by the $1/N_f$ calculations show that the density associated with the Kondo peak is very small, probably of order T_K/Δ. The real state of the f site is very close to $n_f = 1$, with the electron typically hopping on and off for brief periods of order $1/E_d$. The phonons cannot follow this actual motion, only the very slight changes in mean occupation which occur on a time scale of order $\hbar/T_K$. Thus the electron-phonon interaction is renormalized away in the Kondo lattice cases. In the more straightforward mixed valence cases of, for instance, Ce metal itself, T_f is higher than the Debye temperature and, again, the occupancy fluctuates too rapidly for phonons to follow. Thus we never expect the local displacements to follow the "valence" fluctuations. To a great extent, here is again a manifestation of the fact that real valence does not actually change much as the nominal valence fluctuates.

Finally, I have promised to say something about superconductivity in the heavy electron systems. The Haldane and Wilson-Nozieres theories leave a repulsive pseudopotential for electron-electron interaction at the effective center, reflected in a (χ/γ) ratio greater than one. It is noteworthy that in UBe_{13} this ratio decreases as we approach T_c, reflecting perhaps some kind of local singlet pairing.

I could guess that what might be occurring is an effective attraction between pairs of f electrons on the same ion in singlet states. This cannot be primarily a phonon interaction, I think, because of the above arguments and because phonons are weaker than the intrinsic Coulomb U once the time-scales of both have been lengthened by Kondo renormalization. A far-fetched idea is the following.

From nuclear shell theory we know that two particles in an open shell (or, in fact, any even number) have a paired-up singlet state which Bohr, Mottelson and Pines showed to be a close analogy of the BCS state. It may be written (for $J = L \pm 1/2$)

$$\Psi_{\text{sing.}} = \sum_{M>0} c_M^{J^+} c_{-M}^{J^+} \Psi_0$$

In the nuclear case with attractive interactions, this is the lowest state and is normally seen. (Hence the I = 0 value for even-even nuclei.) In atoms, under Hund's rule Coulomb interactions, this is a high-energy state due to repulsive exchange interactions. But this state, like the empty singlet, can be subject to Haldane renormalization by a quantity of order $\Delta \ln W/\Delta$, since it connects with more magnetic substates than vice versa. In the case of UBe_{13}, particularly, we suspect the valence of U to be mixing between 1 and 2, not 0 and 1, and also we expect Δ to be quite large. I have made no calculations but I would expect, under these conditions, the effective Wilson interaction to be attractive. Ce is a more difficult case in $CeCu_2Si_2$, but the photoemission people have found evidence for some $(f)^2$ in several compounds and this one may be the most biased towards $(f)^{1+\varepsilon}$ of any. That would be a testable prediction of this theory.

REFERENCES

1. A. Blandin and J. Friedel, J. Phys. Radium 20, 160 (1959).
2. A. M. Clogston and P. W. Anderson, Bull. Am. Phys. Soc. 6, 124 (1960).
3. F. D. M. Haldane and P. W. Anderson, Phys. Rev. B 13, 2553 (1975).
4. G. C. de Leo, G. D. Watkins, W. B. Fowler, Phys. Rev. 23, 1851 (1981).
5. F. D. M. Haldane, Phys. Rev. B 15, 281, 2484 (1977).
6. M. A. Schluter and C. M. Varma, Helv. Phys. Acta 56, 147 (1983).
7. P. Nozieres, J. Low Temp. Phys. 17, 31 (1974).
8. J. Cardy, J. Phys. A 14, 1407 (1981).
9. B. G. Kotliar, Thesis, Princeton University (1983).
10. F. D. M. Haldane, Thesis, Cambridge University, 1977, J. Phys. C 11, 5015 (1978).

PARTICIPANTS AT "MOMENT FORMATION IN SOLIDS"
HELD AT LESTER B. PEARSON COLLEGE OF THE PACIFIC
August 21 - September 2, 1983

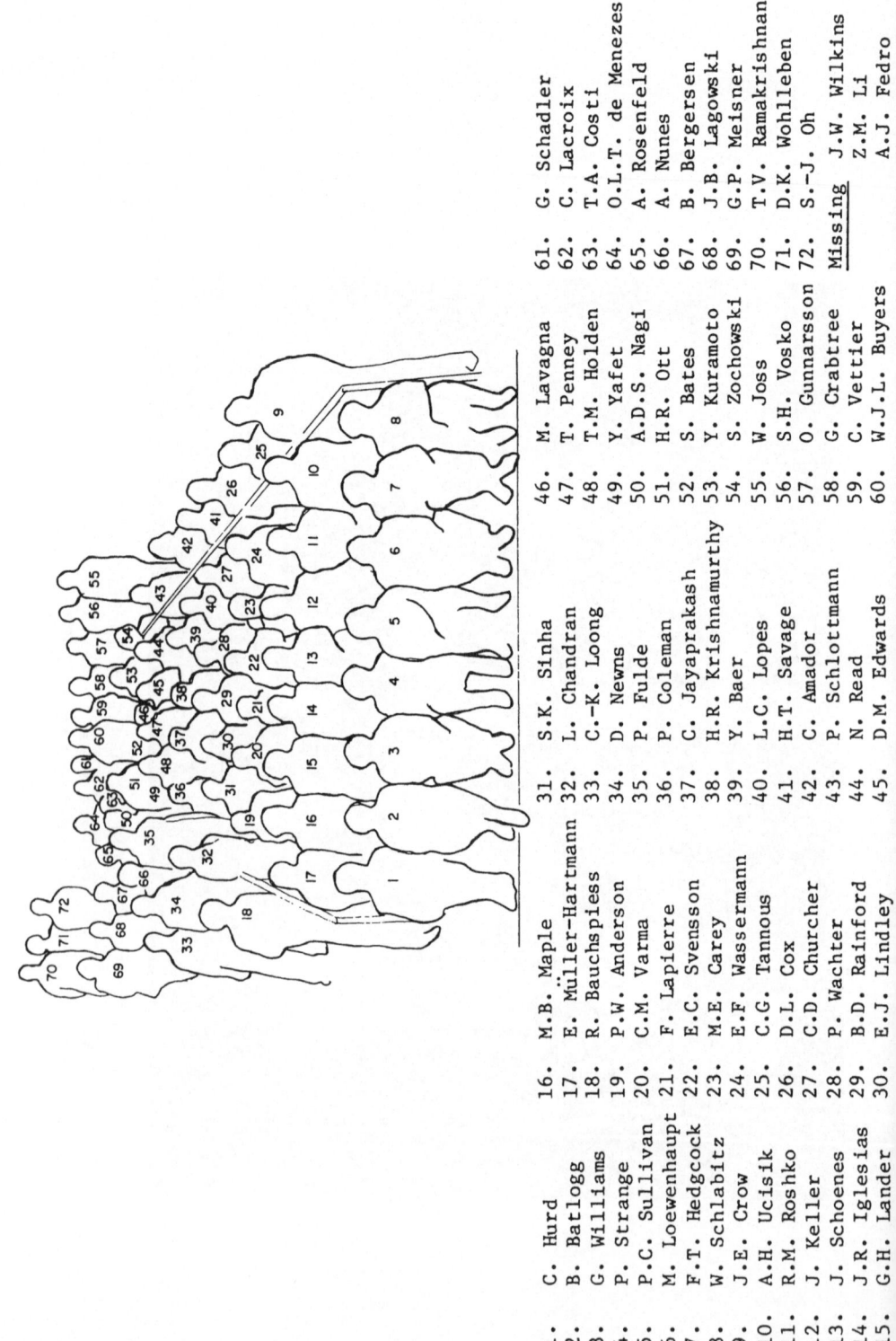

1. C. Hurd
2. B. Batlogg
3. G. Williams
4. P. Strange
5. P.C. Sullivan
6. M. Loewenhaupt
7. F.T. Hedgcock
8. W. Schlabitz
9. J.E. Crow
10. A.H. Ucisik
11. R.M. Roshko
12. J. Keller
13. J. Schoenes
14. J.R. Iglesias
15. G.H. Lander
16. M.B. Maple
17. E. Müller-Hartmann
18. R. Bauchspiess
19. P.W. Anderson
20. C.M. Varma
21. F. Lapierre
22. E.C. Svensson
23. M.E. Carey
24. E.F. Wassermann
25. C.G. Tannous
26. D.L. Cox
27. C.D. Churcher
28. P. Wachter
29. B.D. Rainford
30. E.J. Lindley
31. S.K. Sinha
32. L. Chandran
33. C.-K. Loong
34. D. Newns
35. P. Fulde
36. P. Coleman
37. C. Jayaprakash
38. H.R. Krishnamurthy
39. Y. Baer
40. L.C. Lopes
41. H.T. Savage
42. C. Amador
43. P. Schlottmann
44. N. Read
45. D.M. Edwards
46. M. Lavagna
47. T. Penney
48. T.M. Holden
49. Y. Yafet
50. A.D.S. Nagi
51. H.R. Ott
52. S. Bates
53. Y. Kuramoto
54. S. Zochowski
55. W. Joss
56. S.H. Vosko
57. O. Gunnarsson
58. G. Crabtree
59. C. Vettier
60. W.J.L. Buyers
61. G. Schadler
62. C. Lacroix
63. T.A. Costi
64. O.L.T. de Menezes
65. A. Rosenfeld
66. A. Nunes
67. B. Bergersen
68. J.B. Lagowski
69. G.P. Meisner
70. T.V. Ramakrishnan
71. D.K. Wohlleben
72. S.-J. Oh

<u>Missing</u> J.W. Wilkins
Z.M. Li
A.J. Fedro

INDEX